$60

Australian Battalion Commanders in the Second World War

In *Australian Battalion Commanders in the Second World War*, Garth Pratten explores for the first time the background, role and conduct of the commanding officers of Australian infantry battalions in the Second World War. Despite their vital role as the lynchpins of the battlefield, uniting the senior officers with the soldiers who fought, battalion commanders have so far received scant attention in contemporary military history. This book redresses the balance, providing a gripping, meticulously researched and insightful account that charts the development of Australia's infantry commanding officers from part-time, ill-prepared amateurs to seasoned veterans who, although still not professional soldiers, deserved the title of professional men of war.

Drawing on extensive archival material, Pratten recreates battle scenes and brings to light personalities as diverse as the fiery Heathcote 'Tack' Hammer who, during a solo patrol at Alamein, sustained a serious gunshot wound to the face yet went on to capture three German prisoners of war, and the softly spoken Theo 'Myrtle' Walker, who jumped from the last boat leaving Crete rather than leave his battalion behind.

The story of the Australian battalion commanders is one of improvisation, adaptation and evolution; of an army learning from hard-won experience to integrate men and technology and overcome both human enemy and hostile terrain. Most of all, it is a story of men confronting the timeless challenges of military leadership – mastering their own fear and discomfort in order to motivate and inspire their troops to endure the maelstrom of war.

Dr Garth Pratten is senior lecturer in the Department of War Studies at the Royal Military Academy, Sandhurst. He has worked for both the Australian Army and the Australian War Memorial, and taught history at Deakin University. He has been a member of the Australian Army reserve for seventeen years. The grandson of a veteran of the siege of Tobruk, Garth's principal historical interest is the Australian experience of the Second World War, with a particular focus on the conduct of military operations. In April 2006 he was awarded the Australian Army's C. E. W. Bean Prize for Military History for his PhD thesis.

OTHER TITLES IN THE AUSTRALIAN ARMY HISTORY SERIES

(Series editor: David Horner)

Phillip Bradley *The Battle for Wau: New Guinea's Frontline 1942–1943*
Mark Johnston *The Proud 6th: An Illustrated History of the 6th Australian Division 1939–1946*

Australian Battalion Commanders in the Second World War

GARTH PRATTEN

CAMBRIDGE UNIVERSITY PRESS
Cambridge, New York, Melbourne, Madrid, Cape Town, Singapore, São Paulo, Delhi

Cambridge University Press
477 Williamstown Road, Port Melbourne, VIC 3207, Australia

Published in the United States of America by Cambridge University Press, New York

www.cambridge.org
Information on this title: www.cambridge.org/9780521763455

© Garth Pratten 2009

First published 2009

Designed by Rob Cowpe Design
Typeset by Aptara Corp.
Printed in China by Printplus

A catalogue record for this publication is available from the British Library

National Library of Australia Cataloguing in Publication entry
Pratten, Garth, 1973–
 Australian battalion commanders in the Second World War / Garth M. Pratten.
 9780521763455 (hbk.)
 Australian Army history series
 Includes index.
 Bibliography.
 Australia. Army. Infantry – History – World War, 1939–1945.
 Command of troops.
 World War, 1939–1945 – Australia.
355.33041

ISBN 978-0-521-76345-5 hardback

Reproduction and communication for educational purposes
The Australian *Copyright Act 1968* (the Act) allows a maximum of
one chapter or 10% of the pages of this work, whichever is the greater,
to be reproduced and/or communicated by any educational institution
for its educational purposes provided that the educational institution
(or the body that administers it) has given a remuneration notice to
Copyright Agency Limited (CAL) under the Act.

For details of the CAL licence for educational institutions contact:

Copyright Agency Limited
Level 15, 233 Castlereagh Street
Sydney NSW 2000
Telephone: (02) 9394 7600
Facsimile: (02) 9394 7601
E-mail: info@copyright.com.au

Reproduction and communication for other purposes
Except as permitted under the Act (for example a fair dealing for the
purposes of study, research, criticism or review) no part of this publication
may be reproduced, stored in a retrieval system, communicated or
transmitted in any form or by any means without prior written permission.
All inquiries should be made to the publisher at the address above.

Cambridge University Press has no responsibility for the persistence or accuracy of URLs
for external or third-party internet websites referred to in this publication and does not
guarantee that any content on such websites is, or will remain, accurate or appropriate.
Information regarding prices, travel timetables and other factual information given in this
work are correct at the time of first printing but Cambridge University Press does not
guarantee the accuracy of such information thereafter.

For my grandfather,
VX 10581 Staff Sergeant Richard Edward Pratten.
11 December 1911–15 December 2000
A hard, practical and decent man. Not an infantryman,
but still a soldier who embodied their ethos.
A member of a remarkable generation.

I was told that a particular battalion had not done well and so I went along to see why. I found this battalion just behind the battle line, where they had been brought out. The men were sitting about, they were very tired, very dirty, a lot were wounded. They were hungry and miserable. I looked around, walking amongst those men, and I could not see an officer anywhere, and I thought, as sometimes happened, all the officers had been killed. I went around a corner and I found a little bunch of officers. They were sitting there having a meal, and they were having a meal before their men had fed. Then I knew why that was a bad battalion.

Field Marshal Sir William Slim

Contents

List of maps and figures	ix
Military symbols on maps	xi
Acknowledgements	xii
Notes on the text	xv
List of abbreviations	xvii
Introduction	1
1 'Completely untrained for war': Battalion command in the pre-war army	30
2 The foundations of battalion command: Forming the 2nd AIF, 1939–40	48
3 'We were learning then': The AIF's Mediterranean campaigns, 1941	77
4 Desert epilogue: El Alamein, 1942	112
5 Victims of circumstance: Battalion command in the 8th Division	130
6 'No place for half-hearted measures': Australia and Papua, 1940–42	165
7 'There is no mystery in jungle fighting': The New Guinea offensives	197
8 'Experienced, toughened, competent': 1945	235
Conclusion	270
Appendix 1 The demographics of Australian battalion commanders	277

Appendix 2 Periods of command	313
Notes	329
Bibliography	390
Index	408

Maps and figures

Maps

3.1:	The Mediterranean theatre	78
3.2:	2/6 Battalion positions on the Wadi Muatered at nightfall on 3 January 1941	97
3.3:	The 2/8th Battalion's defence of Vevi, 12 April 1941	108
4.1:	Australian battalion operations at El Alamein, 23 October – 4 November 1942	121
5.1:	Southern Malaya and Singapore	148
5.2:	The 2/30th Battalion's dispositions forward of Gemas	155
5.3:	Australian dispositions on Singapore, 8 February 1942	158
6.1:	Papua and the Kokoda Track	180
6.2:	The battle of Eora Creek, 23–27 October 1942	187
6.3:	Australian attacks at Gona, 29 November – 8 December 1942	189
7.1:	Australian area of operations in New Guinea, 1943–44	199
7.2:	2/12th Battalion attack on Prothero, 20–21 January 1944	208
7.3:	2/6th Battalion positions, Mubo area, morning of 21 June 1943	211
8.1:	2/14th Battalion operations at Manggar airfield, 4–9 July 1945	252
8.2:	2/1st Battalion operations around Nambut Ridge, January–February 1945	261
8.3:	9th Battalion operations towards Mosigetta, 23 January – 1 February 1945	264

FIGURES

3.1:	Outline organisation of the AIF in the Middle East	79
3.2:	The Australian infantry battalion, 1941	81
7.1:	The Australian tropical scales infantry battalion, 1943	204
A1.1:	Decline in average age of COs	310

Military symbols on maps

Function symbols

⊓ headquarters

⊠ infantry

▭ armour

△ anti-tank artillery

▪ artillery

⬭ defensive position

Strength indicators

•	section
•••	platoon/troop
I	company/squadron
II	battalion
III	regiment
×	brigade
××	division
×××	corps
(+)	reinforced
(−)	sub-unit(s) detached

Abbreviations

AS	Australian
GK	Greek
IN	Indian
MY	Malayan
RTR	Royal Tank Regiment
SSVF	Straits Settlements Volunteer Force
UK	United Kingdom

Acknowledgements

This book was only ever going to be dedicated to one person, and I wish that he was still alive to see its completion. The experience of the Second World War was the frame of reference through which my grandfather viewed the rest of his life. I have forged a career seeking to understand that experience, and I hope this work contributes to our knowledge of it.

The process of writing a book is a little like the experience of war. In the midst of battle an individual soldier may feel alone, but he cannot survive – let alone be victorious – without the comrades who fight beside him, the colleagues who support him in the field, and the family and friends who long for his return home. Hence, there are many people I need to thank for sustaining me throughout my long campaign that began with an idea for a PhD thesis and ended here.

If not for the influence of two men I might not even have pursued history as a career, let alone completed the doctorate that led to this book, so my first thanks belong to Philip Jackson and Peter Stanley: for their enthusiasm and commitment to the teaching and study of history, and for being both mentors and friends. Thanks also go to 'Auntie Joan' Beaumont for her guidance, patience, wise counsel, staunch support, and the odd tipple from her wine cabinet in moments of need during my PhD candidature.

David Horner has been another driving force in this book's publication, from his energising support in the last days of my PhD to his sagely advice on preparing it for publication. I owe him many thanks.

The research on which this book is based was completed under a co-operative scholarship offered by Deakin University and the Australian War Memorial. Hence I must also acknowledge the support provided by the Memorial as an institution and, in particular, that of the Director, Steve Gower, and the Assistant Director, Public Programs, Helen Withnell.

The completion of a book can also be seen as a journey that forms just a small part of the longer journey of life. My two companions through part of the larger journey both played their part in a successful end to the shorter one. My thanks to Rosalind for providing a sounding board when this book was just a jumble of loosely connected ideas and for continuing to share the benefits of her experience as she trod the PhD path ahead of me; and to Michelle for her love, interest and support at a time when I was not really quite sure in what direction either journey was heading. Michelle's forbearance also needs to be recognised for ignoring the mess I tramped across her kitchen floor after scaling Singapore's largest compost heap for a better view.

Any good historical work must be founded on solid archival research. Thus, this book would not have been possible without the assistance of Jim Allen, Kevin Canny and Andrew Park at the Australian Army's Central Records Office; and the staffs of the following libraries and archives: the Imperial War Museum; the National Archives of the United Kingdom; the Rylands Library of the University of Manchester; the Liddell Hart Centre for Military Archives – particularly Sarah Presscott; the National Library of Australia; the National Archives of Australia reading rooms in Canberra, Melbourne and Sydney – particularly David Bell and Yvonne Wise at the former; and the Research Centre at the Australian War Memorial – everybody there helped me at one time or another.

In no particular order, I'd also like to thank the following people:

My former colleagues in the Military History Section at the Australian War Memorial (with assorted ANU hangers-on): Steve Bullard, Eric Carpenter, Anne-Marie Condé, John Connor, Miesje de Vogel, Ashley Ekins, Dan Flitton, Matthew Glozier, Ian Hodges, Peter Londey, Brad Manera, Robert Nichols, Libby Stewart and Craig Wilcox – one would have to go a long way to find a more collegial bunch of colleagues.

My new colleagues at the Royal Military Academy Sandhurst, whose warm welcome and commitment to operational history has provided me with new and refreshing academic stimulous.

My esteemed friend Mark Johnston – a scholar and a gentleman if ever there was one – for his enthusiasm, encouragement and advice.

Learned colleagues throughout the Commonwealth for their fellowship and assistance: Paul Addison, Martin Alexander, Patrick Brennan, Carl Bridge, Jeremy Crang, David Dilks, Brian Farrell, Kent Fedorowich, David French, Jeff Grey, Glyn Harper, Geoff Hayes.

Roger Lee and the staff of the Army History Unit for their generous support, which allowed me to conduct research in the United Kingdom and visit Second World War battlefields in Libya.

Anthony Bright from the Australian National University's cartography unit for the maps.

Ian Campbell, whom I've never managed to meet, but whose work on Frederick Galleghan was an inspiration for me.

The many fine soldiers I have served with during my seventeen years in the Australian Army Reserve, and while working with both the Australian and British armies. I have learned something from you all, and I hope that what I say in the pages that follow is the better for it. Particular thanks go to my own COs in recent years – Lieutenant Colonels Angus Bell, Laurie Lucas, Dominic Teakle and Schon Condon – and to Majors Steve Brumby and Dave Wilson. Above all, I would like to acknowledge the comradeship of Major Lyle Dahms and the officers and other ranks of Charlie Company, 4/3rd Battalion, the Royal New South Wales Regiment, during 2002–03. They gave me an inkling as to just what a military unit based on trust and mutual respect can achieve.

The individuals who have done me little favours along the way: Carla Marchant from the Art Section at the Australian War Memorial; Jane Peek, the Memorial's Senior Curator of Heraldry; and Jacqui Hyne and Martine Brieger at the Ku-ring-gai Council.

Pauline de Laveaux, Jodie Howell and Thuong Du at Cambridge University Press, and Cathryn Game, my editor, for guiding me patiently through the publishing process.

My friends Alice, Ben, Caroline, Craig, the two Emmas, Gordon, Jayne, Jo, Jonathon, Melinda, Michelle C., Nicole, Scott and Lisa, and the two Tracies, for doing what friends do, despite the erratic mood swings and tardy communication my thesis, then this book, produced.

My greatest thanks belong to two groups of people. The first is my family – Mum, Dad, Neely, Renee, Kelly, Leia, Marina, Richard and Callum – for putting up with me throughout my long campaign, for their unflagging support, for their pride in the award of my PhD. I am especially grateful to Mum and Dad. My interests have been honed and directed by several people whom I have already mentioned, but it was Mum and Dad who were there at the beginning, and who I know will be in the front row when this book is launched.

The second is the veterans of the Second World War who welcomed me into their homes and shared their experiences. Many of them are no longer with us, and Australia is all the poorer for their passing. This is my book, but it is their story.

Notes on the text

Measurements

Apart from where they were part of the official designation of a weapon or piece of equipment, metric units of measure have been used throughout this work. While each generation deserves to be understood on its own terms, this must be done in a language that the current generation understands.

Terminology

At the outbreak of the Second World War the formal title of Australia's army was the Australian Military Forces (AMF). The AMF was composed of a small regular component – the Permanent Military Forces (PMF) – and a large part-time volunteer force known simply as the militia. AMF was often corrupted to Australian Militia Force(s). In some quarters the militia was also known as the Citizen Military Forces (CMF), which was a vestige of the part-time conscripted force that existed between 1921 and 1929.

The militia was restricted by legislation to service in Australian territory and so, with the coming of war, a special volunteer force was raised for overseas service, which became known as the Second Australian Imperial Force (AIF). In the wider community AMF was increasingly used to refer solely to the militia, although this was incorrect because the AMF embraced not just the militia but the AIF and PMF as well. In February 1943 an order was promulgated to clarify the army's organisation nomenclature. AMF remained the title for the entire army, and the use of AIF and PMF remained unchanged, but the militia's title was formally changed to CMF. Complicating matters, however, was another policy change that allowed militia troops to volunteer for unrestricted overseas service. Where 75 per cent of a unit's personnel were such volunteers it was transferred to the AIF and entitled to place '(AIF)' after its name.

To avoid confusion, the term 'militia' has been used throughout this book to refer to the units that began the war as part of the extant part-time army. AIF refers only to the units originally raised for unrestricted overseas service in 1939–40 that would come to form the 6th, 7th, 8th and 9th Divisions.

Abbreviations

AAF	Australian Army form
ADFA	Australian Defence Force Academy
Adjt	Adjutant
AIF	Australian Imperial Force
AMF	Australian Military Forces
AMR	Australian Military Regulation
AMR&O	Australian Military Regulations and Orders
AS	Australian
AT	anti-tank
AWM	Australian War Memorial
Bde	brigade
BEF	British Expeditionary Force
BGS	brigadier, general staff
BM	brigade major
Bn	battalion
Capt	captain
CARO	Central Army Records Office
CB	Commander of the Order of the British Empire
CGS	Chief of the General Staff
CMF	Citizen Military Forces
CMG	Companion of the Order of St Michael and St George
CO	commanding officer
Comd	commander
Coy	company
CRA	Commander Royal Artillery
C&S School	Command and Staff School
CSM	Company Sergeant Major
CTC	Middle East Combined Operations Training Centre
cwt	hundredweight
DAAG	Deputy Assistant Adjutant General

xvii

Div	division
DSO	Distinguished Service Order
ED	Efficiency Decoration
fn	footnote
Gen	general
GOC	General Officer Commanding
GRO	general routine order
GS	general staff
HE	high explosive
HQ	headquarters
IN	Indian
KBE	Knight Commander of the Order of the British Empire
KMSA	Keith Murdoch Sound Archive
KMSA(t)	Keith Murdoch Sound Archive – transcript
LHCMA	Liddell Hart Centre for Military Archives
LHQ	Land Headquarters
LMG	light machine gun
LNS	League of National Security ('White Army')
LT	line telegraphy
Lt	lieutenant
Lt Col	lieutenant colonel
Lt Gen	lieutenant general
MA	Master of Arts
Maj	major
Maj Gen	major general
MBA	Military Board agendum
MBE	Member of the Order of the British Empire
MBI	Military Board instruction
MC	Military Cross; motor cycle
MESS	Middle East Staff School
METS	Middle East Tactical School
MMG	medium machine gun
MO	medical officer
MS	manuscript
MT	motor transport
MUR	Melbourne University Rifles/Regiment
MY	Malay
NAA	National Archives of Australia
NAUK	National Archives of the United Kingdom (formerly the Public Record Office)

NCO	non-commissioned officer
n.d.	no date
NLA	National Library of Australia
NZEF	New Zealand Expeditionary Force
OC	Officer Commanding
Offr	officer(s)
O Group	orders group
OR	other rank(s)
para(s)	paragraph(s)
PhD	Doctor of Philosophy
Pl	platoon
QM	Quartermaster
RAP	Regimental Aid Post
reg	regulation
RP	Regimental Police
RQ	Regimental Quartermaster Sergeant
RSM	Regimental Sergeant Major
RTD	Recruit Training Depot
SMG	sub-machine gun
SNCO	senior non-commissioned officer
SSVF	Straits Settlements Volunteer Force
SUR	Sydney University Regiment
SWPA	South-West Pacific Area
Tac HQ	tactical headquarters
TEWT	tactical exercise without troops
Tpt	transport
UK	United Kingdom
UNSW	University of New South Wales
VD	Volunteer Decoration
VDC	Volunteer Defence Corps
WD	war diary
WE	war establishment
WT	wireless telegraphy

Introduction

> Just as the education and chance in life of the civil family depends very much on the head of that family, so is it with this military family represented by the Battalion and whose head is the Battalion commander. What manner of man he is determines to a great extent the character of the officers he has under him. On the officers of the Battalion depend the efficiency of the unit at work, and its happiness and contentment at rest.
>
> William Devine, *The Story of a Battalion*, p. 80

Tom Daly was described by Gavin Long, Australia's Official Historian of the Second World War, as 'the outstanding CO' of the Balikpapan campaign of July 1945.[1] Just before 9am on 1 July, Daly and the 2/10th Battalion landed at Balikpapan with orders to capture a steep hill, nicknamed Parramatta, that dominated the landing beaches. Almost immediately Daly was confronted with a dilemma. The first knoll on the way to Parramatta – Hill 87 – loomed before the 2/10th but the battalion's supporting tanks were bogged on the beach, and artillery and naval gunfire support was likewise unavailable. Should he wait for the fire support but give the Japanese time to recover from the pre-invasion bombardment? Or should he push on without it, with all the risks that entailed? Each option had the potential for heavy casualties.

Knowing the danger an organised Japanese force atop Parramatta could pose to the landings, Daly ordered an assault. Relying almost solely on the tactics of fire and movement to maintain momentum, the attacking

'Thank God for the 2/10th Battalion.' Troops of the 2/10th Battalion atop Parramatta. Balikpapan, Borneo, 1 July 1945. (AWM Neg. 018831)

troops surged up Hill 87; it was in Australian hands by 12.40pm. By this time the tanks had been freed and moved forward, and the artillery was also in action. With this support in train, the 2/10th was firm on the summit of Parramatta by 2.12pm.[2] Visiting Parramatta the next day, and viewing the fields of fire it provided over the landing beaches, Lieutenant General Leslie Morshead, commander of I Corps, commented, 'Thank God for the 2/10th Battalion.'[3]

Daly walked Long through the attack on Parramatta a few days after the end of the war. He noted that Daly was 'a quiet, shy chap' but that concealed beneath this persona was 'great drive and sound military sense'.[4] These qualities were still evident when I spoke to Daly 55 years later. His answers to my questions were considered, honest and self-effacing. Reflecting on the assault on Parramatta, Daly recalled the 'warm glow' he felt when he issued orders and they were carried out without hesitation: 'I had never been so proud of the chaps . . . all their battle drill was just exactly as we'd trained. They were spread out; the commanders were just behind their chaps.' Despite his success that day, he still pondered over his actions, admitting he often lay awake at night thinking about how he could have done better: 'I wonder if I had been more efficient, if I'd really been more on top of my job, been more professional, some chaps wouldn't have died.'[5]

Daly's actions, and his reflections on them, illustrate some of the challenges faced by Australian infantry battalion commanders during the Second World War: to train soldiers well; to employ supporting arms to the best effect; to improvise when plans went awry; to motivate soldiers to do things that self-preservation argued against; and to value the life of the individual, yet have the courage to risk it in pursuit of larger objectives. This book examines how the commanding officers (COs) of Australian infantry battalions dealt with these and other challenges. It explores their origins, conduct and experience and assesses their effectiveness on the battlefield and their influence on the Australian Military Forces (AMF) as a whole. Ultimately, it seeks to determine what the elements of successful battalion command were and whether these embodied a distinctly Australian style. As COs were usually lieutenant colonels it can be assumed that all individuals mentioned in this book held that rank unless otherwise stated.

The infantry

The AMF of the Second World War was the largest army ever assembled by Australia, and it was primarily an infantry force. Soldiers, armed with rifle, bayonet, grenade and machine gun, dependent on their feet as their primary means of tactical mobility, provided its striking power. On the battlefield, their role was to seek out and close with the enemy, kill or capture their personnel, and occupy and hold ground.[6] The role of all other troops in the AMF, either directly or indirectly, was the support of infantry operations. Although the First Australian Armoured Division was formed in 1941, it never saw operational service; the few Australian armoured units that were eventually deployed in the South West Pacific theatre were distributed among infantry formations to provide mobile fire support. The destructive power of artillery is unquestioned, but Australian artillery units rarely operated in an offensive role in their own right. When artillery was employed it was at the behest of the infantry, and in most cases infantry was still required to close with and occupy a position at the end of a bombardment or, at the very least, confirm its destruction.

As in all the British Commonwealth armies of the war, the infantry battalion, the strength of which hovered around 800 men, was the principal fighting unit; its structure and its place within the wider organisation of the AMF will be explored in chapter 2. As David French has argued,

Infantry. Men of the 2/48th Infantry Battalion, the most highly decorated Australian unit of the Second World War, stand at ease during an awards parade. Ravenshoe, Qld, 23 October 1944. (AWM Neg. 081729)

the operational significance of the infantry battalions was 'out of all proportion to their numbers'.[7] In 1942 an Australian infantry division had an approximate strength of 15,500 men, but the frontline infantry in its nine battalions numbered merely 6750.[8] If the battalions could not advance, neither could the rest of the division. Reaching beyond the battlefield, a single Australian division required a further 33,000 personnel to maintain it on active operations. Roland Griffiths-Marsh has estimated that fewer than 100,000 Australians saw service as infantrymen during the war from a total of 726,543 enlistees in the AMF.[9] Casualty statistics highlight the critical role of the infantry. Of the 10,674 fatal casualties sustained by the AMF as a result of enemy action, 8086 (or 76 per cent) were members of infantry battalions.[10]

Beside the battalions there were other units in the AMF that could be labelled infantry: motor regiments, pioneer battalions, machine gun battalions, independent companies and commando squadrons. This book is concerned only with the standard infantry battalions. The AMF's motor regiments never saw active service. The machine gun and pioneer battalions were essentially support troops, the former providing medium-range direct fire support and the latter a light military engineering capacity. Although, at times, the troops of both the machine gun and pioneer battalions were required to fight as conventional infantry, they were neither trained nor organised for this role. The soldiers of independent companies and commando squadrons were highly trained infantry soldiers often recruited from the battalions. Their units, however, were roughly a quarter of the strength of a battalion and were usually employed as raiding forces, to conduct reconnaissance or to protect the flanks of the formations they supported.

Given the operational significance of the infantry battalions, their COs were among the AMF's most critical command appointments. The battalion CO was the highest level of command in the AMF in which the incumbent had regular personal contact with the men he commanded. Fred Chilton, CO of the 2/2nd Battalion and, later, commander of the 18th Brigade, reflected that as a brigade commander one was 'elevated' and out of touch with the men, whereas a 'battalion commander, he's everything'.[11] Accompanying his men into battle, the CO was the highest appointment that could routinely and directly influence the actions of combatant soldiers.[12] Ultimately, the responsibility for the battalion's conduct, both in and out of battle, resided with him. He was responsible for establishing and maintaining standards through the application of discipline and the coordination of training, and for ensuring success on the battlefield through the judicious and timely employment of the battalion's subunits and weapons and any support with which it was provided. The CO was often seen as the father of the battalion and was commonly nicknamed the 'old man'.

In this book, an infantry CO is defined as any officer officially appointed to command an infantry battalion in the AMF between the formation of the 2nd AIF in October 1939 and the end of the war in August 1945. An official appointment was one that was promulgated in the various promotions lists published throughout the war.[13] The experience of men who held acting battalion commands will also be considered, although not in the demographic analysis that appears at appendix 1, due to deficiencies in records. Periods of acting command were rarely noted in an officer's personal service record, and they are inconsistently recorded in other primary and secondary sources.

WRITING ABOUT BATTALION COMMAND

Despite their key role, infantry COs have received scant attention in histories of the AMF in the Second World War. Studies of command have largely been confined to the operational and strategic spheres. The field is dominated by David Horner's *Crisis of Command* and *High Command*, the thoroughness of which seems to have largely stifled the development of any fresh approaches. Horner's edited volume *The Commanders: Australian Military Leadership in the Twentieth Century*, examines 16 Australian commanders of all three services, but focuses on their achievements at the highest levels of military command. There has only been one consolidated study of Australian battalion commanders in any conflict, which is David Butler, Alf Argent and Jim Shelton's detailed examination of 3 RAR's COs in Korea: *The Fight Leaders*. One of Butler's chapters deals briefly with Charles Green's command of the 2/11th Battalion in 1945,[14] but Ian Campbell's MA thesis, 'A model for battalion command: Training and leadership in the 2nd AIF: A case study of Lieutenant-Colonel F.G. Galleghan', is the only sustained examination of Australian battalion command in the Second World War yet produced. Campbell's work is thorough and insightful, but does not compare Galleghan's performance to other COs in Malaya, nor examine it in the light of the experience of Australian COs in other theatres.

Biographies of the AMF's most senior commanders, numbering at least 13, overwhelm those of the COs. Only seven of the 276 Second World War infantry COs are the subjects of published biographies. These, however, are principally anecdotal character studies. Although some attempt to assess the performance of their subjects as commanders, these biographies largely fail to analyse their tactical ability or their leadership in any detail. Most of them seem to have resulted from the great affection the authors felt for their subjects, and their analysis is often coloured by it.[15]

Australia's experience of the Second World War has produced a rich and diverse collection of unit histories, but COs are seldom discussed at any length. Battalion histories vary widely in form and substance. Some are little more than scrapbooks of photographs and anecdotes, while others, generally those produced in the twenty years after the war, are detailed and reflective accounts of the battalion's wartime experience. In such works it could be expected that the COs would be prominent characters, but they make surprisingly few appearances. A battalion's first CO often receives the most attention, which perhaps indicates that his influence, in setting the battalion's original standards, was the most marked. For example, the larger-than-life figure of Vivian England, the 2/3rd Battalion's first CO, is accorded several admiring passages in the 2/3rd's unit history, *War Dance*.[16] Yet England was replaced in 1941.

His three successors barely rate a mention in the rest of the book and are not even included in its nominal roll.

This emphasis on a battalion's first CO, however, might also be a result of generic form. Most of the battalion histories begin with a 'birth narrative' that constitutes their first chapter. The battalion is born with the appointment of the first CO, who then oversees its recruitment and training. This narrative concludes as the battalion reaches maturity, an event signified by either its deployment overseas or its preparations for its first battle. From this point on the battalion takes on a life of its own and the significance of the CO as a father figure seems to decline. The arrival of a new CO might be acknowledged with a short biography, and his departure is similarly accorded a brief summary of his influence on the battalion. From time to time a CO might make a cameo appearance in other sections of the narrative. Humorous anecdotes featuring the CO are recounted with relish, and individual achievements, such as honours or awards, or remarkable bravery or endurance, are generally also acknowledged. These short passages, however, provide no more than snapshots of the character of the COs. They offer a brief insight into the style of leadership these men employed but say little of their tactical abilities or of their ultimate effectiveness as a commander.

Veiled criticism of COs is infrequent in the battalion histories and direct criticism rare. To criticise a former CO in a public forum runs contrary to the almost chivalrous code of honour shared by many Second World War veterans.[17] Further, to criticise a CO would be akin to criticising the battalion, thus bringing it into disrepute. The influence of former COs in the post-war battalion associations also needs to be considered. Particularly in the 20 years immediately following the war, many were stalwarts of battalion associations and gave generously of their time, money and influence. Battalion histories were generally the product of committees of which former COs were often members.

The manner in which these relationships could cloud judgement is demonstrated by the example of Alex Falconer of the 24th Battalion. Falconer was critical to the establishment of the battalion association and served on its committee until 1959. An obituary in the battalion newsletter in 1987 lauds Falconer's role in moulding the 24th into a 'competent fighting machine'.[18] Yet there is no mention here, nor in the battalion history, that Falconer's superiors found him to be a temperamental CO, that he was removed from command for placing unrealistic demands on his battalion,[19] and that he was dismissed from the Army following a prolonged period of absence without leave in early 1944.[20]

This is not to say that all COs have been universally well treated by the histories of their battalions. Silences can be instructive. Rather than taint the story of the battalion, authors have often simply denied unpopular COs a place in it. It is instructive to note that the most sustained criticism of a CO in a battalion history – that of Peter Webster in the 57/60th Battalion's history *Hold Hard, Cobbers* – appears in a work not authored by a former member of the battalion.[21]

The battalion histories largely follow in the 'democratic' tradition of Australian military history established by Charles Bean with his epic series *The Official History of Australia in the War of 1914–1918*. The exploits and experience of soldiers and junior leaders who formed the bulk of the battalion are accorded the lion's share of the narrative in both the work of Bean and the Second World War battalion historians. Bean did not overlook COs in his work, but was more interested in their character and leadership qualities, and addressed them in pithy vignettes. Of the legendary CO of the 48th Battalion, Ray Leane, he wrote: 'Like many other strong men, he possessed a somewhat "difficult" temperament; but stern, clean, intensely virile, with a jaw as square as his great shoulders and a cheek muscle that seemed to be always clenched, with a sense of duty which constantly involved his battalion in the most dangerous tasks, and a sense of honour which always ensured that he or one of his family bore the brunt.'[22]

The Second World War official histories are more bland in their treatment of COs. Like Bean, their authors generally devote scant prose to analysing the decisions or actions of battalion commanders, but their work is also largely devoid of the pen portraits that Bean presents. In the light of the papers of the Second World War official historians, which inform much of this book, these are puzzling omissions. The official historians consulted extensively and widely circulated their drafts for comment. Their papers are replete with thoughtful and candid reflections by COs, their subordinates and superiors, and insightful observations by the historians themselves, particularly Gavin Long, who spent much of the latter years of the war visiting troops in forward training areas and at the front. There are vivid character studies and heated debates over tactical decisions. Incompetence, exhaustion and personal animosity are all discussed, yet little of this material made it into the published volumes. For example, there is no trace in Dexter's *New Guinea Offensives* of the schism in the 2/43rd battalion that resulted in the dismissal of Robert Joshua in December 1943, nor is there reference in Long's *Final Campaigns* to the mutiny in the 61st Battalion on Bougainville and the physical and mental breakdown of its commander, Walter Dexter.

There is little evidence of an explicit editorial policy regarding such matters, although it can be surmised that several factors were at work, including Australian notions of equality, a traditional reticence to apportion blame to individuals, the continuing influence of those officers who remained active in the military, government or commerce, and the threat of libel charges.[23] Prevailing attitudes to what should be included and excluded in an official history can be gauged from the views of the official historians' correspondents. Roland Oakes, writing to Lionel Wigmore, author of *The Japanese Thrust*, supported the notion that an official history was not the place to apportion blame. He made reference to the 'unseemly bickering and criticism' that had ensued following the surrender at Singapore and enjoined everyone involved to recognise that they made mistakes: 'Who doesn't in the fog of war? Providing they did what they thought was right at the time, to hell with other people's criticism.'[24] Tom Louch, commenting on Long's draft of the action at Derna for *To Benghazi*, displayed a rather cynical attitude to the official histories but highlighted the general expectation that they should avoid controversy. He observed that if 'some controversial matters' were omitted it was 'probably better so'. Louch had 'no intention of raking up the past' and for that reason had already destroyed his own papers. He concluded: 'Official Histories are not in my line:– one has to be so respectful. If, however, when you have finished this job, you care to embark on an Unofficial History I will give you all the assistance I can.'[25]

For Norman Jeanes, respect was central to what the official histories were about. He criticised Long for highlighting the hardships faced by Australian troops on the passage from Morotai to Brunei Bay in 1945. Jeanes did not disagree with what Long wrote, but nevertheless considered its publication in the official history 'undesirable' and that it would do 'great harm' to the reputation of Australia's troops. Jeanes believed that an admission that troops resented 'hardship, discomfort, and sometimes exasperating inefficiency' would weaken the esteem in which his men were held and asked for the section to be 'considerably modified'.[26]

In the face of Jeanes's criticism, Long stood his ground,[27] but there is evidence that the wide circulation of the drafts did serve to dilute or silence dissenting voices and lop tall poppies. After criticism from Major General George Wootten, a reference to orders for Australian not to take prisoners during the fight for Giarabub was dropped from *To Benghazi*.[28] Brigadier Victor Windeyer took David Dexter to task for momentarily focusing on Alexander MacDonald, CO of the Jungle Warfare Training Centre, whom he described as 'ruthless but efficient'. He noted that few lieutenant colonels were accorded individual treatment and that there were many who deserved the distinction of a description as much as or more than MacDonald.[29] MacDonald's three words of fame did not appear in the published version of *The New Guinea Offensives*.[30]

Windeyer's remarks, however, do indicate that he believed COs deserved more attention in the official histories than they were accorded. Brigadier Frederick Galleghan was similarly displeased with the innocuous approach of the official historians. Galleghan had a particularly testy relationship with both Long and Wigmore and criticised the latter's account of the Gemas ambush for its 'timidity and undue modesty'. The nub of his criticism was that he personally, as the CO of the battalion involved, had not been accorded the laurels he thought he deserved: 'Incidentally, the DSO to the CO was the only immediate [award][31] in the AIF and I think one of three immediates in the campaign. I think you should read the citation for the DSO and then consider if this Chapter does really portray the Bn leader.'[32] The democratic style of both Bean and Long's teams is echoed by much of the writing that explores the AMF's operational experience during the Second World War. The focus is the platoon and the company, and on the experience of junior officers and other ranks. As Michael Evans has observed, Australian war literature has focused on 'individual over institution'.[33] The one exception is Eric Lambert's *Twenty Thousand Thieves* in which Lieutenant Colonel Ormond 'Groggy Orme' Fitzroy is a prominent character. Although *Thieves* is distinctly anti-establishment in its tone, when read in conjunction with historical sources there is an air of authenticity about many of the command challenges that confront Fitzroy and the expectations his troops have of him; Lambert served in both the 2/2nd Machine Gun and 2/15th Infantry battalions.

The two major historical works dealing specifically with the experience of Australian soldiers in the Second World War, John Barrett's *We were There* and Mark Johnston's *At the Front Line*, say little about the influence of COs on the battlefield, but they do feature fleetingly in both works. Barrett lists among his respondents 'Infantry Colonel', who is quoted on only two occasions, in reference to attitudes towards smoking and sex.[34] Johnston devotes a short, informative discussion to the 'Influence of commanding officers',[35] but this is limited to the CO's role as a disciplinarian when the battalion was out of the line, which, given that Johnston's main source was the letters and diaries of soldiers, is in itself telling.

Margaret Barter, in *Far Above Battle*, is one of few authors who has sought to understand the internal dynamics of an Australian Second World War infantry battalion: the 2/2nd. Like Barrett and Johnston, she devotes considerable attention to battlefield action as recalled by individual soldiers and draws some insightful conclusions, but there is no discussion of the role of the CO. For instance, Fred Chilton receives only two scant mentions in her discussion of the fighting for Bardia and Tobruk,[36] despite having been awarded a DSO for his role in the former battle during which his 'coolness, courage and initiative' was said to have 'set an excellent example' and been 'an inspiration' to the battalion.[37] Joan Beaumont also explores the group dynamics of a single battalion – the 2/21st – in *Gull Force* but, unlike Barter, the influence of the CO is a key component of her analysis. Beaumont's examination of the command of Len Roach is diverse, ranging from an exploration of his character and leadership style through to the factors influencing the battalion's dispositions on Ambon.[38] The emphasis of *Gull Force*, however, is the experience of Japanese captivity, and the performance of Roach's successor, John Scott, during the fighting on Ambon is not examined in detail; given the overwhelming superiority of the Japanese invasion forces, Beaumont considers Scott's operational performance to be largely irrelevant to her wider narrative.[39]

In the studies of particular campaigns and operations the role of the battalion commander is acknowledged but rarely explored. Peter Stanley's *Tarakan: An Australian Tragedy* focuses principally on what he has termed 'the corporals' war'. When considering command decisions, he accords most attention to those made at the strategic level. In his treatment of COs, Stanley follows the example of Bean. The COs appear in snapshots – the frank pre-landing address given to his men by Bob Ainslie of the 2/48th Battalion; George Warfe standing in full view of the Japanese on Anzac Highway to inspire the tired men of his 2/24th Battalion; the distribution of captured Japanese swords by George Tucker of the 2/23rd Battalion to the men he felt most deserved them[40] – but only a faint trace of their influence on the course of operations is visible. Stanley, with his co-author Mark Johnston, does delve more deeply into the experience of the battalion CO in *Alamein: The Australian story*. The COs still flit in and out of the story, but many of the tactical issues they faced are discussed[41] and hence their influence is a little more quantifiable.

Peter Brune's four works on the Papua campaign,[42] which collectively represent one of the most sustained bodies of recent Australian operational scholarship, demonstrate an awareness of the role of the CO. He interviewed several in the course of his research, quotes them liberally and makes an effort to understand their tactical decisions – with varying degrees of success.[43] Brune's works, however, are still largely soldiers' histories, and it is their reflections that predominate. Further, he does not use these as a means to explore small group dynamics or battlefield motivation in the way Barter attempts to do, and as a result there is no significant analysis of the leadership of COs. Brune's greatest weakness is his desire to turn the Papuan campaign into a great national myth, replete with heroes and villains. The men of the battalions, including the COs, are Brune's heroes, and his empathy for their experience seems to have prevented him from criticising their actions. Brune insightfully notes the dilemmas confronted by COs in action, but his analysis rarely looks behind it. For instance, he states that at Gona three COs were sacked for protesting against 'the slaughter of their troops', and dismisses the possibility that their actions might have been partly responsible with a glib reference to their 'prior impressive experience and performance'.[44]

The most extensive exploration of the role, experience and influence of Australian COs on operations is found in John Coates's examination of the Huon Peninsula campaign, *Bravery Above Blunder*. Coates adopts a vignette approach similarly to Stanley, but ranges beyond character and personality into issues of command, tactics and responsibility.[45] COs are still not major players in his narrative and some barely rate a mention, but Coates provides an insight into the demands of battalion command at a level of detail seldom found in other works. The attention that Coates accords battalion-level command is no doubt informed by his own long career in the Australian Army, culminating in serving as CGS, which points to one possible explanation as to why it has largely been ignored by other Australian historians. Many authors seem more comfortable examining the battalions as social rather than military entities. Possibly historians with military experience not only feel more confident grappling with the intricacies of unit level command but also recognise its critical influence on the battlefield. Although Frank Sublet's *Kokoda to the Sea* is principally a summary of Dudley McCarthy's official history *South-West Pacific First Year*, what stands out is his shrewd commentary on battalion tactics, most likely a product of his own experience as a CO.

The neglect of command at the unit level is a significant weakness not only in histories of the AMF in the Second World War but also of the experience of all the major Western allies. The majority of published literature on the operations of both the British Commonwealth and United States armies focuses on soldiers and junior leaders, or on the machinations of generals and politicians and the translation of strategy into operational concepts. This polarisation of Western military history was ushered in with the 'New Military History' movement in the 1960s, which emphasised the experience of the rank and file over the perspective of senior commanders. Battalion commanders would seem to have been left in a strange netherworld. As senior officers, made distinct from their men by their authority, status and relatively small numbers, they are not considered to be fighting soldiers. Yet the battalion commanders routinely shared all the danger and privations of their soldiers. Infantry COs were actually the most senior 'fighting' soldiers on the battlefield.

Ironically, battalion commanders seem to have been excluded from traditional command-based military history because they were relatively junior in the command hierarchy. Works dealing with command seldom venture below the rank of general, although the theories they develop can, with extrapolation, be applied to more junior officers. Battalion commanders have not been included among the ranks of military history's 'great captains'. It could be argued that historians of command consider a battalion too small a command to be of any significance when compared with divisions, corps and armies. It was, however, on the shoulders of COs that the reputations of the Second World War's great captains were built. Formation commanders provided the vision; their staffs turned the vision into operational plans; and, as a CO of a United States infantry battalion has pointed out, it was the battalion and brigade commanders who were the 'principal executors of any plan of action adopted'.[46]

The nature of the archival record might also have contributed to the neglect of the role of Australian battalion commanders. They have not left a large volume of personal papers. This is possibly because many were too busy to keep diaries, but it does not account for the relatively small number of post-war memoirs. The significance of COs does not seem to have been highly regarded by the Australian Army. Attempting to explore their selection, assessment and promotion through the archival record is laborious and frustrating. Shortsighted archival policies have meant that the vast majority of the original paperwork relating to these matters has been either lost or destroyed. Highlighting two examples, the Office of the Military Secretary, ultimately responsible for all officer promotions and appointments, has left little archival trace of its wartime activities; and the Central Army Records Office (CARO) culled confidential reports from officers' service dossiers in the mid-1970s. British military historians are not served any better. The remaining files of both the British Army's Adjutant General's and Military Secretary's departments, held by the National Archives, throw little light on how this army was selecting and training its commanding officers during the war. My investigations in other British War Office file series, with the aim of providing a means of comparative analysis for the AMF's practices, proved largely fruitless.

Although, to borrow a phrase from the hapless General Sir Ian Hamilton, naked truths are not there for the asking, the background and experience of the Australian COs can still be reconstructed from the surviving archival record. This book is founded on detailed statistical evidence derived from the personal service records that were originally held by CARO and now reside with the National Archives of Australia. The analysis of this data, which appears in full at appendix 1, revealed distinct trends and patterns in the appointment, training and career progression of Australia's infantry COs. In order to determine why these trends were manifest, and the relation between them and events on the battlefield, other records needed to be explored. For an insight into the world of a Second World War CO the researcher must turn to the operational records held by the Australian War Memorial. The influence of COs, however, is not readily apparent and has to be discerned from briefings, instructions, orders and reports. It is a process that requires an intimate knowledge of the AMF's operating procedures in the Second World War combined with imagination, patience and just a little luck. These requirements, combined with the sheer bulk of the records, make them a daunting prospect, one that might have deterred researchers in the past from attempting to explore the nature of battalion command.

The most basic operational record of Australian battalions of the Second World War are battalion war diaries: daily narratives usually maintained by the intelligence officer or adjutant.[47] War diaries have two inherent weaknesses. The first is that the demands of the tactical situation often meant they were compiled from rough notes, signal logs or simply memory days, weeks or even months after the events they detailed had occurred.[48] The war diaries' other weakness – and the one which has the most bearing on this book – is that COs were required to approve each month's war diary before it was dispatched to the AMF's Military History Section. COs had authority to make changes, and they did.[49] War diaries were often a medium in which senior officers could justify their decisions, and they were well aware that what was recorded therein would influence how they and their battalions would later be treated in the official history.[50]

In order to 'fill the gaps in the official records',[51] Long and his historians conducted many interviews – Long's notes from the interviews he personally conducted form the principal content of 116 notebooks.[52] These interviews have been revisited for this book, and I also conducted a new series with much the same aim: to address deficiencies in the archival record and gain an insight into personal attitudes towards command. Oral history, however, as admirably discussed by Alistair Thomson in *Anzac Memories*, is not without its own inherent challenges for the historian; it is not simply a window on 'how it really was'.[53] As has already been discussed, officers were particularly conscious of how the experience they represented would affect the way they and their units would be treated in the official history. Some of my own interviewees were reluctant to discuss certain topics so as not to malign the reputations of fellow officers, and others confused dates, events and locations. As Thompson has pointed out, memories, and the way in which they are recounted, are influenced by the expectations of society, the expectations of the interviewer and the interviewee's interpretation of both.[54] In this book, oral history has been considered like any other historical source, and its assertions have been weighed and balanced against those of secondary literature and archival sources. The influence of personal agendas and experience have also been taken into account. There is not the scope in this book to analyse the interviews in the manner of a 'popular memory' approach, although the factors influencing the way experiences have been remembered, then recounted, have informed the use of them.

COMMAND

On the basis of the operational records at the Australian War Memorial and interviews with ten of the twelve Second World War COs still living in 2000, this book examines the experience and effectiveness of Australian battalion commanders during the Second World War. The beginning of this process is to understand what military command, at all levels, entails. The subject has long provided fodder for historians, military theorists and serving officers alike. Some works seek to explain command with a list of attributes, actions and character traits of successful commanders. Others discuss it as a complex relationship between troops, commanders and the battlefield environment. There is no doubt that command is vital to the operation of military forces. Martin van Creveld's very first words attempting to define the nature of command in *Command in War* are that it 'is a function that has to be exercised, more or less continuously, if the army is to exist and operate'.[55] Gary Sheffield has described command a little more specifically as the direction, coordination and effective employment of military force.[56] The current definition of command employed by the Australian Army, derived from a glossary of standardised terminology developed by NATO, states that command is 'the authority vested in an individual for the direction, co-ordination and control of military forces'.[57] Although a little abstract, this definition highlights the cornerstone of military command: authority. At the most basic level, command is the exercise of authority.[58]

The authority of a military commander has a legal and constitutional status. At the simplest level, authority within a military organisation grants the holder the right to enforce the compliance of soldiers of subordinate rank. Armies can be seen as possessing a pool of such authority, granted by the state, which is then distributed in decreasing portions throughout the organisation, from the highest rank to the lowest – the higher the rank a soldier holds, the greater the authority accorded and the larger the number of subordinates who are subject to it. Additional authority can be delegated to subordinates for the purpose of carrying out specific tasks or functions. Although every soldier is notionally subject to the authority of another with higher rank, in practical terms this authority is limited by the organisation. A commander is able to directly enforce the compliance of only a defined group of subordinates, known, depending on size, as a subunit, unit or formation. The exercise of authority over this group, and the equipment, supplies and resources allocated to them, constitutes the act of command.

The corollary of authority within a military organisation is responsibility. It is a fundamental concept of command.[59] Provided with the means, in the form of their authority over a given set of human and materiel resources, commanders are responsible for achieving the ends determined by their superiors. In addition they are also responsible for both the actions and the well-being of those over whom they exercise authority. In an ideal military command structure, authority and responsibility are inextricably linked. The complexity and extent of the tasks allocated to a commander should be proportionate to the extent of the authority that he holds and vice versa.

Van Creveld divides the responsibilities of a commander into two categories. The first includes the provision of everything that a military unit needs to exist: from food and water to the maintenance of sanitary conditions and the operation of an effective system of military justice.

Van Creveld's second set of responsibilities relate to what he brands the 'proper mission' of a military force: to inflict the maximum amount of death and destruction on the enemy in the shortest possible time while incurring the minimum casualties to itself. These responsibilities include the gathering of intelligence and the planning and monitoring of operations. The first group of responsibilities are labelled function related and the second outcome related.[60]

Van Creveld's model has its weaknesses. The distinction between the two groups of responsibilities is somewhat artificial. The function-related administrative responsibilities could also be termed outcome related. The purpose of these administrative actions is to produce fit, well-trained and equipped soldiers, ready for war – a very definite outcome. Van Creveld does, however, point out that these two groups of responsibilities are mutually dependent. A military force cannot defeat the enemy without well-disciplined and trained troops, and troops cannot be trained well without the provision of sound intelligence relating to the enemy and his methods.

It is more instructive to see command as a group of interdependent responsibilities or functions, leading to a single outcome: the successful conduct of operations on the battlefield. Many theorists define command simply on the basis of its operational aspects and, in particular, the process of tactical decision-making. In this fashion, current British Army doctrine espouses a triangular model of command consisting of three overlapping functions: decision-making, control and leadership. Decision-making is self-explanatory. Leadership is the process by which subordinates are motivated to achieve what has been decided on, and control is the process by which their actions are directed and coordinated.[61] Decision-making is founded on training and experience, which provide the mental patterns by which situations are assessed.[62] As the functions of control and leadership indicate, command cannot occur without the subordinates required to transform orders into actions. To develop a well-rounded model of command we need to add a function that embraces a commander's administrative responsibilities.

Therefore the model of command that underpins this book consists of four basic functions: decision-making, control, leadership and resource management. As will be discussed later, 'motivation' might perhaps be a more apt descriptor than 'leadership' as subordinates can be motivated to act without any real leadership being exercised by a commander. Resource management embraces all of a commander's responsibilities to maintain his force in a battleworthy condition, including both training and equipment maintenance; ultimately, soldiers, weapons and other kit combine to form a single resource on the battlefield. This is a unitary model of command, sympathetic to the way it was conceived during the Second World War. It is devoid of the tautologous fragmentation that has taken hold in Western military doctrine since the 1980s. The currently voguish term is C^4ISR: command, control, communications, computers, intelligence, surveillance and reconnaissance. All of its constituent elements, however, contribute to the act of command. Control has already been discussed; intelligence, surveillance and reconnaissance all form the basis of decision-making; and communications and computers are simply aids to both these functions.

The manner in which the functions of command are integrated combined with the means by which they are carried out constitute a command system. As Sheffield has pointed out, in order to ensure that decisions are made that result in effective actions, a commander needs an efficient organisation to provide him with information, process it, then turn his orders into reality.[63] Van Creveld has described command systems in the terms of three means of command: 'organizations, such as staffs or councils of war; procedures, such as the way in which reports are distributed inside a headquarters; and technical means, ranging from the standard to the radio'.[64] One of his main arguments is that while the basic functions of command have not changed throughout history, the means have. The effectiveness of command systems has been determined by their ability to evolve in a manner consistent with the nature of the warfare they have sought to direct.

Assessing the effectiveness of military command, however, can be difficult, which is perhaps another factor contributing to the relative neglect of command studies by military historians. Victory on the battlefield, itself often notoriously difficult to quantify, is not a reliable measure of the success of command. Using a hypothetical example, a CO may have attacked following a sound plan, with motivated, well-equipped and well-supported troops, yet still been defeated. His superior headquarters might have given him a task beyond the means at his disposal, the enemy might have had access to superior intelligence of which he was not aware, or the weather might have simply defied the best efforts of meteorologists to predict its fickle moods. Similarly, casualties, so often used as a simple means of quantifying battlefield action, cannot be used to measure the effectiveness of command. For instance, a successful attack might have resulted in large numbers of casualties, due not to any failure of command but simply to the nature of the opposition encountered. Command could be said to be effective as long as a unit continues to function in

A command system. A battalion commander (second from left) and elements of his headquarters on exercise, Darwin, Northern Territory, September 1942. The three basic elements of a command system – organisation, procedures and means – are illustrated. The CO consults his map while conferring with one of his officers, an NCO prepares a report or return, while radio operators look on, ready to transmit or receive information. (AWM Neg. 097421)

accordance with the directions of its commander and not fragment into smaller unresponsive groups. Thus, command can continue to be effective in the face of defeat, yet it could also be seen to break down when troops have apparently secured victory.

To assess command in a qualitative fashion, we must progress beyond the effective/non-effective baseline. Some command studies measure their subjects against a set of prerequisite qualities for successful command, usually culled from the memoirs, musings and philosophies of other successful commanders and military theorists.[65] COs, however, are but one part of a command system, and their personal effectiveness is inextricably linked to the appropriateness of the system as a whole. In *Command in War*, Van Creveld sets up a situational matrix to analyse command across history by posing a series of generic questions. These questions will inform later analysis of command in AMF battalions:

What demands were made on a command system by the existing state of the art of war? In what ways were these demands met? What organization, if any, was provided for the purpose? What technical means... and what procedures were employed? How was intelligence procured and processed, and in what manner were plans arrived at? What means of communication existed, and how did their characteristics affect the transmission of information? How was the execution of orders monitored, and what control if any, did the commander exercise over the course of events?[66]

By embracing such situational factors, we are able to judge military competence, which has been defined by Richard Gabriel as the ability to plan for reasonable contingencies, thereby minimising the inherent risks of the battlefield and increasing the probabilities of success.[67]

The most basic question that needs to be resolved is how appropriate was a command system to the prevailing operational situation. This book charts the evolution of the command system within Australian infantry battalions during the Second World War in response to changes in technology, tactics and the operational environment. The manner in which Australian COs exercised their command varied widely throughout the war and was most effective in 1945 partly due to the refinement of the command system and its appropriateness to the type of war then being fought.

LEADERSHIP

The most prominent companion of command in extant military theory is the concept of leadership. Like control, it is often portrayed as an equal partner of command, and the terms 'command' and 'leadership' are often confused or used interchangeably. As Peter Watson has pointed out in *War on the Mind,* however, leadership, like control, is just one aspect of command and is 'seriously deficient as a description of all the psychological tasks which face an officer'. An officer requires not only the 'personal skills to make the most of his men', which is leadership, but also skills that are the prerogative of his position such as decision-making or the direction of complex group activities. These tasks are 'more accurately problems of command rather than leadership: a man can be good at one without necessarily being good at the other'.[68] Ideally, an officer should exercise leadership in carrying out all the functions of command. The successful and thoughtful execution of the functions of command often constitute acts of leadership. Hence it is easy to see why the division between command and leadership is often blurred.

There is a vast body of literature examining leadership in the military context, in addition to the masses of work that have been produced exploring its application in the political and business spheres. One noted theorist has commented that leadership is one of the 'most observed and least understood phenomena on Earth'.[69] There is no universally accepted definition of what it is, nor is there a dominant theory as to how it works. At the most basic level, leadership is a process by which an individual is able to influence others to act in a certain way of their own free will.[70] The critical importance of leadership to the exercise of command has been highlighted by the eminent British military historian Sir Michael Howard, himself an infantry officer during the Second World War: 'Leadership is the capacity to inspire and motivate; to persuade people willingly to endure hardships, usually prolonged, and incur dangers, usually acute, that if left to themselves they would do their utmost to avoid.'[71]

Early attempts to develop theories of leadership concentrated on the personality traits and behaviour of successful leaders. Although often based on thorough observation, these were theories in name only as they were simply lists of skills or attributes believed to constitute successful leadership. They shared many common features, but no two were the same. These trait theories essentially represent the state of understanding of leadership in British Commonwealth armies before the Second World War. The theories of leadership that will be discussed presently are largely the product of post-war scholarship. They are introduced as a means by which to analyse leadership as a function of command in Australian battalions during the Second World War. It would be anachronistic and misleading to create the impression that Australian infantry officers in that war understood the concept of leadership in these terms.

Studies have found that the correlation between personality traits and leadership is too small to serve any predictive purpose.[72] Trait-based theories are further undermined by their portrayal of leadership as a one-way process, something that a leader enacts on passive followers. Many theorists have subsequently recognised that leadership is an interactive relationship, and some have even suggested it would be more aptly titled 'followership' to better reflect the power held by followers to refuse to be influenced.[73] While a series of landmark studies conducted around the beginning of the Second World War found that the same group of people, given the same task, would behave differently under different leaders,[74] another set conducted in the 1970s revealed that the way a leader conducted himself was as much a product of the expectations of followers as it was of the situation or the leader's own attitudes.[75] Most modern writing on leadership describes it as a relationship between a set of variables represented by a group of people, the task at hand and the environment in which they are to conduct it. The resulting theories are known as contingency, or situational, theories.[76] They generally posit that there is no single best way for a leader to act, although some expound complex matrixes to guide a leader's behaviour when confronted with a particular situation.[77]

Group dynamics are integral to many modern leadership theories, which emphasise the imperative for a leader to integrate the personal motivations with the goals of the larger group.[78] A key concept is that of the formal and informal group. A formal group is one that has both membership and leadership imposed by an outside authority, a military unit being a prime example. Informal groups are the building blocks of human socialisation and can exist either outside or inside a formal group. Their membership is confirmed from within, and they have their own standards and accepted modes of behaviour. Recognising the ancient roots of human society, a large body of literature suggests that leaders will spontaneously emerge within informal groups. The much-vaunted ideal of Australian mateship is a manifestation of the informal group. A need to be accepted by such groups is critical to the sense of personal well-being of most people.[79]

The challenge faced by the military leader is to integrate informal groups in the formal structure, influence their standards and modes of behaviour, and harness their motivations so that they respond to the demands of the formal group and hence achieve its objectives. The ideal is to make the informal group and the formal group, including their leaders, one and the same. At a basic level, this process is achieved through military training. A soldier's individuality is subordinated before the needs of his unit, and his individual well-being is presented as being inextricably linked to the unit's fortunes. Hard work, long days, spartan conditions and collective punishments for individual transgressions serve to build a group identity in the face of a common ordeal.[80]

Despite the military training process, however, the nature of people is such that individuality or the bonds of small informal groups are rarely completely destroyed. Away from the intensity of the training camp, and influenced by the innumerable pressures of a soldier's life – from homesickness to fear – individuality is prone to reassert itself. M. J. Nicholson has argued that a military subordinate is subject to a wide range of pressures and that the key to leadership is understanding these and working to

Reward power. Lieutenant Colonel Roy Gordon, CO of the 2/3rd Machine Gun Battalion, awards the Military Medal to Corporal Glen Lambert (right) and Sergeant Edgar Halyett, Wewak, New Guinea, 2 October 1945. (AWM Neg. 097421)

eliminate or at least mitigate their effects.[81] Military leaders must continually work to reinforce the formal structure and engage with individual motivations. In 1959 John French and Bertram Raven advanced a model of leadership founded on the exchange of 'social power' – power being defined as the potential to influence others – which provides a sound foundation for conceptualising the development of leadership in the military context. French and Raven described five separate bases of social power on which a leader could draw:
- *legitimate power*: the potential for influence derived from being the incumbent of a position or office
- *reward power*: the potential for influence from having the capacity to bestow or withhold valued objects
- *coercive power*: the potential for influence from being able to impose or withhold punitive actions
- *expert power*: the potential for influence derived from possessing information or skills, and
- *referent power*: the potential for influence derived from being personally liked and respected.[82]

It is instructive to compare this model with the three styles of leadership developed by Frederic Bartlett in his 1927 work *Psychology and the Soldier*. Bartlett wrote of the 'institutional' leader who relied on his hierarchical status; the 'dominant' leader who relied on force of personality; and the 'persuasive' leader who trusted in his ability to influence others through rational argument.[83] If we recognise that legitimate, coercive and reward power are all essentially derived from an officer's legal authority, there is a significant degree of congruence between the two models. They differ, however, in that Bartlett describes three self-contained leadership styles, while French and Raven describe the constituent elements of a multitude of styles, reflecting the true dynamism of leadership. Echoing French and Raven, the Australian Army has defined leadership as a 'mixture of personal example, persuasion and compulsion'.[84] The power bases can also be seen as a continuum that parallels the training process. An officer begins with a store of legitimate, reward and coercive power; as training proceeds he is able to develop and demonstrate his own administrative and tactical skills, thereby accumulating a store of expert power; ultimately, the manner in which he draws on the first four power bases will contribute to the strength of the last: referent power.

Dependent on emotion, referent power is perhaps the most intangible component of leadership. It is, however, arguably the one with the most potential influence. Referent power is essentially synonymous with the more mainstream concept of trust, which has been described as being one of the 'mainstay virtues in the commerce of mankind' without which no significant human relationship would exist.[85] Mark Rocke has defined trust in the specific context of military leadership. He wrote: 'Trust is confidence that expectations of integrity, competence and predictability, held by both leaders and led, will be met.'[86]

Integrity has long been considered one of an officer's most essential qualities. Officers are expected to be honest in their dealings with their subordinates and to conduct themselves in accordance with the standards they espouse. A key determinant of an officer's perceived integrity is the sincerity of his attitude towards his subordinates. Sherman Baldwin has argued that whatever style of leadership an individual adopts it must complement their personality, otherwise a leader will appear 'awkward', 'unnatural' and insincere.[87] Anecdotal evidence suggests that only genuine concern for soldiers' welfare will suffice as they will see through artifice and contrivance.[88]

The link between an officer's behaviour and the performance of his subordinates is clear. Most social relations theorists acknowledge that recognition of the individual and their efforts is critical to their sense of self-worth and thus performance. Ultimately, on the battlefield, a soldier needs to feel that his life – and maybe his death – is of consequence. Brigadier Henry Wilson, Assistant CGS of the British Expeditionary Force (BEF) in France in 1914, noted that the exercise of command turned on 'the recognition that those who are asked to die must not be left to feel that they die alone'.[89] Wilson's comments highlight the significance of personal example on the battlefield; as John Keegan states in *The Mask of Command*, 'those that impose risk must be seen to share it'.[90] Malham Wakin has argued that neither honesty nor courage can be instilled 'by contract'.[91] By being prepared to stand beside their subordinates under fire a military leader reinforces the congruence between his principles and actions. In doing so, he also caters for a more primitive human need of his subordinates to turn to others for solace and direction when confronted by fear and confusion.

This 'imperative of example',[92] however, raises a contradiction in the role of the military leader. Although subordinates expect sincerity, sometimes a leader may have to hide his own fear and confusion in order to sustain his followers. He may have to maintain a false persona of courage, calmness, decision or aggression – a 'mask of command'. Further emphasising the situational nature of leadership, Keegan has written that this mask must 'mark him to men of his time and place as the leader they want and need'.[93]

If a leader is to maintain trust his actions must, above all, remain predictable. Every time that an officer acts in accordance with his stated philosophies or prior actions, thereby meeting the expectations of his subordinates, he builds trust. Trust can even be created around the exercise of coercive power. As long as an officer is fair and consistent in his application of disciplinary sanctions, even when harsh, the predictability of his actions builds a sense of trust. The interrelation between leadership and the other functions of command is clearly visible. For example, a succession of competent tactical decisions builds a base of expert power, develops trust in the commander's abilities and thus contributes to successful leadership.

A similar interrelation between the elements of leadership and the functions of command is inherent in an officer's delegation of authority to his subordinates. Giving subordinates the scope to act on their own initiative is widely considered essential for the management of complex tasks and the maintenance of flexibility on an unpredictable battlefield; in modern military terminology, the practice is known as 'directive control' or 'mission command'. As an act of command, delegation also partly satisfies an officer's responsibility to train his subordinates for higher appointments. In the context of leadership, delegation contributes to a subordinate's sense of self-worth by demonstrating an officer's interest in their development and trust in their capabilities.[94]

Some models have used the degree of delegation as the determinant of leadership styles,[95] yet leadership is much more complex. Leadership can be defined in terms of the five social power bases, but there are a wide variety of actions and behaviour that constitute these bases. A leadership style, then, can be defined partly by the extent to which a leader exercises and delegates his authority but also by the example he sets his followers; by the consideration he exhibits to their needs; and by the manner in which he balances his loyalty to his followers, with his loyalty to his leaders – for most leaders are also followers. Finally, a leadership style is characterised by the degree of congruence between the leader's personality, ideals and actions. Leadership styles will change to reflect different situations, and their ultimate measure of success is simply whether the led continue to act in the manner their leader requires.

A few words as to what leadership is not are also appropriate. Although compulsion may play a part in the dynamics of leadership, there is a general acceptance that a leader who needs to rely too heavily on the utilisation of legitimate or coercive power to achieve his will is not a leader at all.[96] Military units have been commanded with an iron fist and propelled into action by fear of punishment. It has been argued that repressive coercion can instil a habit of obedience that will function amidst the worst the battlefield has to offer, and it also has the ability to develop strong group cohesion among the repressed. Fear, however, works as a motivator only when it is stronger than the fear that it opposes; coercive authority is likely to lose its effect on the battlefield when individual soldiers are exposed to imminent personal danger and removed from the immediacy of the threatened punishment.[97] Additionally, the blind obedience instilled by the employment of coercive power limits the effectiveness of a military unit because subordinates have neither the will nor the ability to react to a changing situation.[98] Making a simple semantic point, troops who are continually coerced into action are pushed and not led. James McGregor Burns makes this point eloquently: 'Much of what commonly passes as leadership – conspicuous position-taking without followers or follow-through, posturing . . . manipulation without general purpose, authoritarianism – is no more leadership than the behaviour of small boys marching in front of a parade, who continue to strut along Main Street after the procession has turned down a side street toward the fairgrounds.'[99] The 'genius' of leadership, he explains, is the way in which leaders see and act on their own and their followers' values and motivations. Leadership, unlike naked power-wielding, is thus inseparable from followers' needs and goals.[100]

An Australian style of command?

Contemporary notions of Australian command in the world wars, heavily influenced by the established Anzac legend, emphasise leadership to the virtual exclusion of all other command functions. According to John Laffin, a prolific proponent of the traditional Anzac virtues, Australian officers did not shout, drive or emphasise their hierarchical status and seldom resorted to punishment. 'Above all else' the Australian officer was 'a leader'.[101] Summarising Australian command 'folklore', Carl Bridge, Glyn Harper and Iain Spence noted that Australians would follow only an officer they respected, always had to know what their commanders had in mind and approve of it, and would be 'led but not driven'.[102] As Mark Johnston has stated, it was obedience 'based on respect and consent, rather than coercion'.[103] Although its origins are rarely discussed, tactical skill is said to be a major source of such respect, highlighting the relationship between expert and referent power. Thus, the command hierarchy in Australian units was based on merit, with little recourse to harsh discipline or the formal distinction of rank.

Bridge, Harper and Spence have warned, however, that an attempt to identify 'typically Australian characteristics' is a problematic pursuit. Unless wide-ranging research into the practices of other armies is conducted there is a 'serious risk of identifying as distinctively Australian features which are in fact common to other[s]'.[104] An examination of the small body of literature relating to the exercise of command in the armies of Canada and New Zealand, two nations with a similar military heritage to Australia, reveals this to be the case. Geoffrey Hayes, in his analysis of officer selection in the Canadian Army, quotes an officer who served with both the British and Canadian armies: 'Knowledge and ability in infantry skills were particularly important in the Canadian Army, where egalitarian principles were so firmly entrenched that officers received little automatic respect by virtue of their rank alone. Respect was accorded when officers were considered to deserve it on the basis of knowledge, performance and leadership.'[105] Similarly, Glyn Harper's search for a New Zealand style of command emphasises professional competence, informality in inter-rank relations, delegation of authority and obedience based on 'willing submission and respect rather than on fear'.[106] His remarks are not dissimilar to those made by various authors about the philosophy underlying command in the German Army.[107] Harper eventually concludes that it may be a 'New World' command style that he is defining as opposed to a distinctly New Zealand one.[108]

What this book reveals is that there was no single distinctive manner in which command was exercised over Australian infantry battalions during the Second World War. In their backgrounds, conduct and the challenges they faced, Australian COs shared much in common with their British, Canadian and New Zealand counterparts, despite the fact that they often served in vastly different theatres. The exercise of command over Australian battalions, like leadership, was situational in nature; it changed in response to battlefield conditions, available technology and the capabilities of COs, their staffs and their soldiers. Although a critical command function, leadership was inseparable from the other three functions previously defined. Australian COs of the Second World War, bequeathed a military ethos of heroism and aggression, had to resolve the age-old military dilemma of balancing the imperative of personal example with the necessity of being positioned to best comprehend the wider battle to facilitate decision-making and control.

Demonstrating that there is usually a kernel of truth at the heart of most stereotypes, the most successful COs were characterised by leadership founded on expert and referent power, although, contradicting traditional perceptions of discipline and inter-rank relations, many also liberally exercised their coercive power. Tactical and administrative competence was also critical to effective command but, again contrary to popular stereotypes, it was not the product of natural ability but of experience and universalised instruction, both of which grew exponentially among the AMF's officers as the war progressed. It was for these very reasons that Australian command proficiency reached its zenith in 1945: command relationships were founded on mutual trust and shared experience, and weapons, equipment and the tactics to employ them were the product of five years of victory and defeat on the battlefield. The exercise of command in Australian battalions had come a long way since the COs of 1939, relying on First World War experience and patchy militia training, appointed as much for whom they knew as opposed to what they knew, led the first battalions of the 2nd AIF overseas. The AMF, however, was never the pure meritocracy Anzac folklore would have us believe.

The figure of the battalion CO, the fighting commander, the 'old man', has slipped through the gaps in Australian Second World War historiography. He has been regarded as not quite a fighting soldier but not quite a commander either. Nevertheless, during the Second World War he was the most critical individual in Australian infantry battalions and, as such, his influence on the conduct of Australian operations was profound. As Sir David Hay, an officer of the 2/6th Battalion and later its historian, has written: 'An infantry battalion depends heavily for its performance in action, indeed for its whole spirit and character, on how it is led and trained.'[109]

CHAPTER 1

'COMPLETELY UNTRAINED FOR WAR'
Battalion command in the pre-war army

On the eve of the Second World War the AMF was an obsolescent organisation unfitted for modern war. The experience of the First World War dominated its doctrine and ethos, and its structure and organisation replicated that of the Australian Imperial Force (AIF), which had been formed in 1914, specifically for war service, and disbanded in 1921. Yet the AMF of 1939 fell far short of the AIF's standards of proficiency.[1] The AMF was predominantly a part-time force, and its regular component, known as the Permanent Military Forces (PMF), comprised only 7 per cent of its strength. The vast majority of its combatant units, including all infantry battalions, were drawn from the part-time component, known as the militia. The militia consistently failed to meet recruiting targets between 1930 and 1938, and part-time training, coupled with poor retention, produced skeleton units of partly trained soldiers. The financial stringency of the Depression years also resulted in a force devoid of modern weapons, transport and signals equipment.[2]

In 1939 most of the AMF's commanders, down to battalion level, were ageing veterans of the First World War. They represented the only significant pool of operational experience in the AMF and accordingly were valued highly. Even in the late 1930s, active service in the First World War remained a prerequisite for an appointment to a battalion command. This might have been a sound practice in the 1920s but, in the light of developments in military tactics and technology occurring overseas, in the mid- to late 1930s it served only to perpetuate the AMF's obsolescence. It also served to create a generation of battalion commanders whose physical and mental fitness to command in time of war was questionable and to block the development of a larger pool of younger, more adaptable officers.

The militia in the 1930s did have one strength, however, and this was the commitment of its long-serving officers, NCOs and soldiers, who gave up much of their leisure time to military training, despite the 'rather discouraging circumstances'[3] in which they served. The pressing need to foster recruiting and local support for under-resourced militia battalions meant that the ability to command a battalion on operations was not the sole basis on which commanders were appointed or assessed. Battalion commanders of the 1930s were products of both the society from which they were drawn and the part-time army in which they served.

The AMF's obsolescence had not gone unnoticed. Various senior officers had identified its weaknesses and urged change on the government; the AMF on the eve of the Second World War was a force on the verge of transformation. In March 1938 defence expenditure was doubled, and in September a recruiting campaign was initiated to double the strength of the militia. It was also recognised that the militia could not hope to mobilise quickly enough to deal with defence emergencies of which there was little warning. The creation of the AMF's first ever force of regular infantry soldiers, the Darwin Mobile Force, was announced in October 1938. Before the year was out, the report of the AMF's inspector general, Major General Ernest Squires, recommended, among a wide range of overdue reforms, the formation of two brigade groups of regular soldiers.[4]

The uneven and often poor knowledge of tactics and command procedures among the AMF's senior militia officers prompted the establishment of a command and staff school in August 1938. There was also a realisation that senior command was perhaps beyond all but the most committed part-time officer. Hence the training of the regular officers of the PMF, hitherto principally confined to staff functions, was broadened to prepare them for senior command appointments. In the militia battalions the first members of a new generation of COs, who were to benefit from such innovations as the Command and Staff School, were beginning to break the monopoly of the First World War veterans. It was this generation that would carry the AMF through the Second World War.

The measures described above, however, had only a limited effect on the standards of the AMF before it found itself at war. As Jeffrey Grey has written, 'The demands of modern warfare required well-trained, equipped and balanced armed forces, and these are not created overnight.'[5] It was all too little, too late.

The Influence of the First World War

In the interwar years, AMF regulations stipulated that COs be appointed for a period of four years; appointments would be extended beyond this period only in exceptional circumstances.[6] The main purpose of this policy was to spread command experience widely. In the resource-poor environment of the 1930s the preparation of units for war was largely abandoned as an unattainable goal and replaced with the more pragmatic aim of training a cadre of officers and NCOs on which an expanded wartime army could be founded. Ostensibly, the process by which COs were selected was thorough and fair. When making a new appointment, a divisional commander was required to appraise the available field of officers and forward a recommendation to the Promotion and Selection Committee of the Military Board. This committee would review the recommendation, then either confirm it or seek further justification. The committee could overrule a recommendation, and on occasion it did so.

For the appointment and promotion of officers the militia operated on a regimental system loosely based on the British Army model. The preferred practice was to appoint a new CO from within a battalion, usually the most senior major if he was suitably qualified. Twenty-one of the incumbent battalion commanders in 1939 had at one time commanded the battalions in which they had originally been commissioned or had joined following the First World War;[7] at least one commanded the battalion he had joined as a private.[8] The Military Board was most likely to question a recommendation that would result in the senior major in a battalion being superseded by a more junior officer from within or an officer of equal or junior standing from another unit.

First World War service was a highly valued commodity in the interwar AMF. In 1918 the *Defence Act 1903* had been amended to give preference for promotions and appointments to officers with active service experience.[9] Although there was a sound practical imperative for this policy, the moral significance of First World War service should not be underestimated. It served as a character reference, marking an officer as a man who had done his duty when the nation called. In 1939 there was no AMF CO old enough to have enlisted in the First World War without his parents' consent who had not done so.[10] Qualified officers, of all ages, who had not served during the war were regularly passed over for promotion to command appointments.[11] At the outbreak of the Second World War, veterans of the First World War commanded all but eleven of the fifty-four standard infantry battalions in the AMF.[12] Two

Melbourne battalions, the 22/29th and 39th, serve as examples of how command positions in militia battalions were monopolised by First World War veterans. Every officer appointed to command in both units between 1921 and 1939 had seen active service in France.[13]

By the late 1930s, however, the First World War was 20 years distant and its veterans were well into middle age. In 1939 the average age of the veteran COs was 51. The oldest, John Grant, commanding the newly reformed 61st Battalion, was 56 and the youngest, Frederick Burrows of the 36th, was 42. The retiring age for lieutenant colonels in the militia was 56 and, in the PMF, 60.[14] As Grant's appointment demonstrates, however, the Defence Act made provision for officers to be retained for a maximum of two years beyond the prescribed retirement age.[15]

The suitability of men in their early fifties to command an infantry battalion on operations is questionable. Whatever the mental ability of such men, there can be little doubt that their physical stamina would not have compared with that of younger officers. Drawing on his experience as a staff officer in the First World War, J. F. C. Fuller wrote in 1932 that the three pillars of generalship were courage, creative intelligence and physical fitness, all of which were the 'attributes of youth rather than middle age'.[16] This insight was even more applicable to the command of a battalion. In August 1918 the average age of AIF COs was less than 33.[17] In 1938 Squires reported that the retirement ages for senior officers in both the militia and the PMF were too high and, noting the physical demands placed on lower-ranking officers, that many were retained beyond the age at which they were 'fitted to carry out efficiently the duties of their rank'. He also believed that the retention of such officers impeded the promotion of younger men, limiting the development of a pool of command experience.[18] This was particularly critical in the militia, the primary purpose of which was to train a cadre of commanders, officers and NCOs to expand the AMF rapidly in time of war. Squires recommended that the retirement age for lieutenant colonels in both the militia and the PMF be dropped to 53.[19]

The utility of the First World War experience found among potential and incumbent COs in the late 1930s is doubtful. By September 1939 there was only a single CO still serving who had commanded a battalion during the First World War.[20] Another eleven had held subunit command or senior staff positions, but the bulk of COs in 1939 with First World War experience had gained it as junior officers and NCOs, principally on the Western Front. This experience lacked doctrinal relevance because of changes in organisation and tactics; it was specific to a single theatre, the conditions of which were unlikely to be replicated; and it did not embrace the challenges posed by command at a senior level. The only real relevance of First World War experience to COs in the late 1930s was the insights it provided into the general nature of the battlefield and the ways men reacted to it.

By peacetime criteria, however, the cohort of veteran COs did represent considerable command experience. In September 1939, quite to the contrary of the AMF's regulations, several officers had been commanding battalions for twice the duration of the First World War,[21] and more than a third of the AMF's COs were on their second or third posting as such. The longevity of these officers as COs supports Squires' comments about older officers blocking the progression of their younger colleagues. The other two-thirds of the 1939 cohort do, however, demonstrate that there was a steady turnover of battalion COs in the late 1930s, but at the outbreak of war almost half of this group had held their commands for a year or less.

The COs without First World War experience in September 1939 were characterised principally by their relative youth. Their average age was 36, although the youngest, Ivan Dougherty, commanding the 33rd Battalion, was 32. The command experience of some of the younger COs compared favourably with the First World War veterans, but many were among the group that had not yet attained a year in command. The most experienced was Theo Walker of the 24/39th Battalion, who first assumed command in June 1935. At this time, Walker was one of only three COs not to have served in the First World War.[22] He was reappointed in early 1939 partly on the strength of improvements in training and administration that he had overseen but also to give him further opportunities to develop as a 'sound battalion commander'. Despite Walker's success, there is still a hint that something was missing: the report recommending his reappointment noted that he was 'one of the most successful Commanding Officers who were too young for service in the AIF'.[23]

THE TRAINING OF PRE-WAR COS

Given their paucity of operational command experience, the likely success of militia COs depended on the quality of their training. Between the wars, AMF officers could be promoted to lieutenant colonel only by successfully completing a course of instruction that culminated in written and practical examinations.[24] The extent and scope of these examinations varied for PMF and militia officers. As it was not usual practice for PMF officers to command militia battalions, this discussion will concentrate only on the experience of the latter. The militia 21A course[25] for promotion to lieutenant colonel was six days long and consisted of a series of lessons and tactical exercises without troops (TEWTs). Although the basic qualification for a unit-level command, the 21A course actually instructed and assessed officers on their ability to conduct brigade operations;[26] an officer was tested in the basics of battalion operations before promotion from captain to major. This implemented section 21A (2) of the Defence Act, which prescribed that an officer of the AMF should not be promoted beyond major until he had passed a course of instruction that demonstrated he was 'fitted to command in the field a force of all arms'.[27] In addition to the six-day course, militia officers were also required to pass written examinations that tested their knowledge of doctrine, staff procedures, military law and military history before gaining the 21A qualification. Having completed the course and the examinations, an officer was awarded a certificate stating that he had 'attained the necessary standard' to assume a 'lieutenant-colonel's command in the field'.[28]

The 21A course and examinations, although intellectually demanding, were only a basic test of aptitude for command, in addition to confirming an officer's commitment to the militia, as large amounts of leisure time had to be devoted to study. This promotion system had several faults. First, it failed to test officers in the practical application of their knowledge to a command in the field. The gentlemanly environment of the TEWT could never replicate the stresses imposed by the command of real troops and the battlefield environment. There is no indication that command post exercises, in which a commander and his staff are put through their paces in a mock headquarters, were used as a means of assessment. Second, the examinations and courses were conducted on a regional basis by formation headquarters and therefore failed to attain uniformity in either standards of instruction and assessment or the interpretation of the AMF's doctrine.[29]

The doctrine that formed the basis of the training for aspiring lieutenant colonels was obsolete. Although the notional units they were manoeuvring on TEWTs represented a combined arms force, these were the combined arms of 1914: infantry, artillery, machine guns and horsed cavalry. The tactical problems set at a 21A course conducted at Camden in May 1930 included no tanks or armoured cars, nor the means to oppose them.[30] Although discussing his training as a staff cadet at RMC Duntroon between 1933 and 1937, Donald Jackson has provided an insight into the AMF's approach to warfare at the time. In his memoirs Jackson reflected that the training at Duntroon reflected an 'obsolescence' in the 'Army at large' and, while military theorists overseas discussed armoured warfare, air bombardment and amphibious operations, his course 'persevered with 1918 theories and equipment': 'My class certainly could have done its stuff in France or with the Light Horse but we saw a tank and aeroplane almost in passing, did not study motorised logistics, armoured operations or close air support, and had no acquaintance at all with mines, anti-tank and anti-aircraft weapons . . . Not for us was the slit trench and weapon pit . . . rather, we spent an inordinate time on the finer points of engineering for trench warfare.'[31]

A further indication of the state of the AMF's tactical thinking, as it affected the training of future COs, can be gauged from the subjects covered in the military history exam. Displaying some practicality, both of the campaigns studied were examples of mobile warfare, but not necessarily those of the most relevance to an officer of the 1930s. One was the operations of the BEF in October 1914, and the other the operations that began with the third battle of Gaza and ended with the fall of Jerusalem. These case studies were probably selected because they reflected the AMF's extant force structure but, given developments in combined arms warfare, a more relevant First World War case study would have been the Allied Western Front offensive of 1918. Alternatively, with Japan looming as a threat, case studies of unconventional operations in close country, such as the campaign in East Africa, may also have been relevant.[32]

There is evidence that light armoured units were incorporated into TEWTs conducted by Army Headquarters towards the end of the decade,[33] so it can be assumed that they also found their way into the problems presented on 21A courses. Few militia officers, however, would have had any real knowledge of the capabilities of armour or of the conduct of operations in cooperation with it. Indeed, the knowledge of their instructors and assessors would have been similarly lacking. Before 1937 the AMF possessed only four tanks and a very small cadre of regular officers with knowledge of their employment; it was not until 1939 that demonstrations of their capabilities were staged for infantry and artillery units.[34] Similarly, the 21A course required officers to prepare artillery fire plans, but there is little sign of infantry–artillery cooperation in the unit exercises that would have prepared infantry officers for this work.

In any case, it is doubtful that a thorough knowledge of combined arms operations could have been achieved in a mere week. Major General John Lavarack, the CGS, commented that the 21A course was too short for a satisfactory standard to be attained, particularly when many militia officers took several days to 'get away from the atmosphere and outlook of their civil employment'.[35] The British Army's course for regular infantry COs in the 1930s was three months long, and even that was judged as being too short.[36] Ultimately, the standards achieved by the 21A courses were low and uneven. There was a tendency to assess a student's results relative to his past experience rather than against an absolute standard.[37] Lavarack thought that few officers who completed the course were 'fitted for the immediate exercise of command' in war.[38]

Some officers did seek to further their military knowledge in their own time. All of the former militia officers whom I interviewed recalled studying works of military history and theory borrowed from the Returned Soldiers and Sailors Imperial League (RSSILA, or RSL), the United Services Institute (USI) or state and municipal libraries. In 2000 Frank Sublet still had a copy of Sun Tzu's *Art of War* that he had 'borrowed' from the USI's Perth library in 1938.[39] Indeed, such study seems to have been the main source of exposure to emerging military thought and new equipment. John Field was an ardent student of mechanisation during the 1930s – an essay on the subject won him first prize in a USI competition in 1932.[40] The inherent weakness of such education, however, was that it was self-directed and theoretical. Field could read about the latest thinking on mechanisation, but when he deployed on exercise with the 40th Battalion it was still on trains or civilian buses, and in the field it was still boots that moved men and horses the supplies.

The shortcomings of the militia training and promotion system were compounded by the state of its under-resourced and under-strength battalions. The CO was more of a resource manager and, as will be soon discussed, a community liaison officer than a battlefield commander. There was little, if any, opportunity for him to practise and refine his own skills. Battalions were generally too weak to provide any useful experience for officers.[41] Their peacetime establishments were pegged at about half their wartime strength, and even these meagre numbers could not be maintained during battalion exercises. At their annual training camps in 1938 the 24th and 37/39th Battalions, respectively, were able to field only 135 and 168 men – barely enough to form a company.[42] There was little chance to exercise company commanders, let alone the CO.[43] In addition, the need to keep repeating elementary training because of high turnover in the ranks largely prevented the conduct of tactical exercises that would have given senior officers experience of handling units in the field.[44]

Further, battalion resources rarely permitted the regular conduct of TEWTs. In the 1937–38 training year the 57/60th Battalion conducted only two TEWTs and the 29/22nd and 59th Battalions one each.[45] Some units and formations conducted night classes for their officers but, as with more formal training, standards varied widely depending on the instructors. The net result was that the practical command experience of officers undertaking the examinations for promotion to lieutenant colonel was patchy at best, and there was little chance of it being remedied once they were appointed to a command.

The AMF's leadership was well aware of the deficiencies in the training and experience of its COs. As has been noted, Lavarack harboured grave doubts about their abilities, which he first took to the Military Board in 1935, his first year as CGS. To remedy the situation he suggested PMF officers be trained and employed as COs for militia units.[46] Such a practice would also prepare them to assume more senior command. Lavarack was convinced that the complexities of modern warfare meant that it was unlikely that 'highly efficient commanders for higher formations' could be produced from 'part-time officers'.[47] Lavarack's suggested reforms envisaged a much-enlarged role for PMF officers. Hitherto they had largely been confined to training and staff positions, and there were few opportunities for subunit or unit command beyond the small number of permanent artillery and engineer units; to gain regimental experience newly graduated PMF officers were sent on exchange with units in the British and Indian armies.

Well aware of the ingrained cultural attachment to the citizen soldier in the Australian military establishment and government,[48] Lavarack sought to introduce change gradually. He recommended that initially PMF officers be appointed to command militia battalions only in remote areas where a suitable local officer could not be found.[49] Lavarack's major training innovation was the establishment of the Command and Staff School (C&SS), first proposed in 1936, to train both militia and PMF officers. Addressing the Military Board on 12 October 1936, Lavarack wrote that the establishment of the school was an 'essential step in the achievement of one of the first aims of the Board's policy i.e. training of Commanders and Staffs' and that the existing state of tactical training could not be allowed to persist'.[50]

Recognising that most militia officers could not be expected to increase substantially the time they devoted to military training, the proposal for the school aimed to improve their tactical training by raising the standard of instruction and promoting a common understanding of doctrine. All

'The existing state of tactical training could not be allowed to persist.' Major General John Lavarack, Chief of the General Staff 1935–39, had grave doubts about the abilities of militia officers and instituted various reforms, including the establishment of a Command and Staff School, to improve standards. Lavarack would later command at division, corps and army level during the Second World War. (AWM Neg. 100129)

21A courses would be conducted at the school and their length increased by a fortnight.[51] As the concept of the school was further developed, a new two-week course for lieutenant colonels was also added.[52] It was also envisaged that the school would study and test British Army doctrine, the acknowledged basis of AMF tactics and procedures, to develop 'a system' more suited to the AMF's training, armament and geostrategic situation.[53]

C&SS opened at Randwick Barracks, Sydney, in August 1938. Fred Chilton, then a major in the Sydney University Regiment, attended the school in mid-1939 and recalls being 'fully extended' by the problems that it set.[54] The influence of the school's establishment on the training of the AMF's command cohort before the outbreak of war, however, was negligible. In the year to September 1939 it conducted four courses for militia officers.[55] The first year of the school's operation was considered as a changeover period, and officers were still able to qualify for promotion under the old system.

In the late 1930s other efforts were being made, both by Army Headquarters and the various formations, to improve the tactical ability of senior officers. A TEWT conducted in May 1937 for senior officers of the 4th Division involved the employment of armour, anti-tank and anti-aircraft weapons, and reconnaissance aircraft.[56] In the same year, 500 officers and men from units in the Sydney area participated in an amphibious landing at Jervis Bay, and in Western Australia the 16th Battalion conducted an exercise in which it opposed an amphibious landing by an enemy force actually put ashore from an RAN vessel.[57] Army Headquarters training instructions for 1938 suggested that formations should consider conducting command post exercises but advised that resources would permit only a limited number of such exercises to be held.[58]

Major John Lloyd, second in command (2iC) of the 37/39th and then 24/39th Battalions from 1937 onwards, has succinctly summed up the state of training of the militia's senior officers on the eve of the Second World War. The militia, he recalled, had 'old First World War transport, arms and ideas . . . I could say I was as well trained as any of them, but in the same breath I could say I was completely untrained for war.'[59] As Albert Palazzo has pointed out, this situation undermined 'the entire rationale for the militia's existence'. The militia's purpose was to train a cadre of leaders on which a wartime army could be founded, but its units lacked the structures, personnel and equipment to give such leaders any significant experience in 'real tasks and responsibilities'.[60]

The difficulties confronting militia COs as war approached were not unique. Their experience and ability was remarkably similar to that of the COs of Britain's interwar citizens' force, the Territorial Army.[61] In 1936 the Chief of the Imperial General Staff, Field Marshal Sir Cyril Deverell, commented that there were too many Territorial Army COs 'who through no fault of their own [had] no tactical experience'.[62] Like the militia, the Territorial Army of the 1930s had been starved of funds and resources and lacked modern equipment. A high turnover of personnel had denied most COs the opportunity to conduct large-scale exercises and complex tactical manoeuvres with their units, impeding both their own professional development and that of their subordinate officers.[63]

'First World War transport, arms and ideas.' Almost indistinguishable from the men of the First AIF, despite twenty years having passed, B Company of the 56th Battalion stand at ease during their annual training camp. Glenfield, NSW, March 1937. (AWM Neg. P00989.047)

Opportunities to exercise with other arms were rare – the London Rifle Brigade observed only a single tank demonstration during the interwar period[64] – and TEWTs were effectively the only means by which senior Territorial Army officers were trained in combined arms operations.[65] In one regard, however, the training of the Australian COs was superior. Territorial Army officers, like their AMF counterparts, were examined on the principles of battalion operations as part of their promotion examinations for the rank of major.[66] This, however, was the last formal training they received unless they were seeking command of a brigade, when they were expected to attend the three-month course at the Regular Army's Senior Officers School.[67]

COs in a citizens' army

Military prowess alone, however, was not the sole basis for appointing militia COs between the wars. Reflecting the society of which it was a product and traditional notions of the officer class, the AMF sought men with a wide range of middle-class virtues to command its battalions. As spokesmen and role models for the AMF they were also expected to have integrity and be temperate in their habits. A survey of confidential reports and command recommendations from the 1930s reveal the character traits the AMF found desirable in its COs. Among the most commonly used phrases in these reports were 'of sober' or 'temperate' habits; 'tactful'; 'strong character'; 'self-reliant'; 'keen'; 'zealous'; 'hard-working'; and 'unquestioned' or 'undoubted integrity'.[68] A CO was expected to be a man of imagination and drive, devoted to his part-time military career. Indeed, in a citizens' army starved of funding and resources, and struggling to find and retain recruits, a battalion was likely to flounder unless its CO was thus committed. Such commitment required COs to be involved in their local community, promoting the interests of their battalion as well as a wider interest in defence issues. In modern parlance, COs in the 1930s were expected to be good community citizens.

One of the most highly valued non-military attributes of a potential CO was contacts with the local community that could be exploited to encourage material support and foster recruiting. The significance of the relationship between the battalions and their local communities was summed up by Major General Thomas Blamey, 3rd Division's General Officer Commanding (GOC), in 1935: 'The army of Australia rests more on the people than the army of any other country in the world, and with the interest of the people behind it will grow into a very live force.'[69] Arguing for an extension of Claude Cameron's appointment as CO of the 18th Battalion, the GOC of the 1st Division noted that Cameron had recently stepped down as mayor of the municipality of Ku-ring-gai. Hence he would be a continuing asset to the battalion if he stayed in command. When the 37th was reraised in East Gippsland in August 1939 Albert Stewart, the former CO of a light horse regiment in the area, was given command. Recommending Stewart to the Military Board, the GOC of the 3rd Division wrote that he held the 'complete confidence of the citizens of East Gippsland'.[70]

A potential CO was also regarded favourably if he lived in the same area his battalion recruited from. Not only did this add to the local identity of the unit but also, at a time before the advent of modern communications, the widespread use of private motor vehicles and rapid public transport, a local CO facilitated the most effective administration and oversight of the battalion. One of the unsuccessful candidates for command of the 18th Battalion at the time of Cameron's reappointment was the senior company commander. He was qualified for promotion and had operational experience in the First World War, but lived at Gosford and

Conveniently placed: a portrait of Claude Cameron taken while mayor of Ku-ring-gai. Demonstrating the importance of local connections to the well-being of Australia's citizen army, Cameron received an extension of his appointment as CO of the 18th Battalion due to links he had established during his time as mayor. Cameron was CO from 1936 to 1940, and would later command the 8th Brigade. (Courtesy Ku-ring-gai Municipal Council.)

was therefore considered 'not conveniently placed' to command the battalion, which had its headquarters in the Sydney suburb of Willoughby.[71] Officers suitably qualified to take command of battalions in the country were sometimes difficult to find. When a qualified lieutenant colonel was not available, a more junior officer could be appointed. In late 1938 Lavarack's policy of using PMF officers to fill vacant battalion commands in remote areas was also implemented with the appointment of two such officers.[72] Both situations, however, were still considered less than ideal.

The significance attached to having a local militia CO, particularly during the expansion of 1938–39, can be gauged by the example of Duncan Maxwell of the 56th Battalion whose promotion to lieutenant colonel was accelerated to replace the PMF officer who had been commanding the unit since 1937. Maxwell was a doctor in Cootamundra and, despite a distinguished record as an infantry officer in the First World War,[73] had not been active in the militia until he became the 56th's medical officer in February 1938. In late 1938 Maxwell was approached by Major General Iven Mackay, then GOC of the 2nd Division, and asked whether he would be interested in resuming his career as a combatant officer. Although expressing doubts that his commitment as a doctor to saving life would conflict with having to order men into battle – doubts that, as we shall see in chapter 5, he should have heeded – Maxwell assented to McKay's request.[74] He was promoted to major and appointed the 56th's 2iC in August 1939, and by November was both a lieutenant colonel and CO.

Imaginative approaches to maintaining battalion morale, thereby aiding retention, were also highly valued. Patrick McCormack was recalled to duty from the Unattached List in April 1937 to command the 37th Battalion, having previously commanded an artillery unit. His appointment seems to have been largely due to his reputation as a 'keen and enthusiastic organiser of sporting and social activities' within units as a means of increasing the interest and *esprit de corps* of all ranks. An insight into how such qualities were balanced against military skills is provided by a remark that described McCormack's judgement as being deliberate but sound.[75] Other COs were granted a second period of command as a means of rewarding efforts to increase a battalion's morale and public profile. Another factor in Cameron's reappointment as CO of the 18th Battalion was a plan he initiated to have it outfitted in a new and distinctive ceremonial uniform. The new uniforms had not arrived by the time Cameron's tenure expired, and the GOC of the 1st Division recommended he stay in command until the battalion could parade in their new uniforms.[76] Similarly, the 27th Battalion adopted Scottish dress in 1938 to foster *esprit de corps* and create a highly visible and distinct local identity. Its CO, Francis Best, also remained in command to see through the changes and allow the battalion in its new form to be 'soundly established before a change [was] made in the command'.[77]

Best was also considered a valuable CO because of his civilian employment as a lecturer in mechanical engineering. It was a position of responsibility and respect, it provided access to a large body of potential recruits, and it left Best with at least one free day each week that he devoted to his military work.[78] Les Peterson studied under Best at the Adelaide School of Mines and recalled Best recruiting from his lectern: 'It was obvious to me if I didn't go down and join his Army I wouldn't pass his subject.'[79] Best's recruiting technique also highlights the manner in which links of patronage from the civilian world were transferred into the AMF. Best was not the only CO to recruit from his workplace or the other circles in which he moved in his leisure time. Jack Clarebrough of the 39th Battalion sought to mine his own social circles for recruits for the 39th Battalion, then based in Melbourne's eastern suburbs. In 1936 he targeted Scotch College alumni and, in 1938, members of Melbourne's rowing fraternity.[80]

Clarebrough's recruiting strategies demonstrate the dominant middle-class ethos that underlay command in the militia units. The recruits from Scotch College, one of Melbourne's premier private schools, were intended to form a 'crack' platoon from which 'candidates for commissions could be drawn'.[81] Likewise, the rowers were to be formed into a special company because 'rowing men', Clarebrough wrote, 'are . . . superior to all other athletes [Clarebrough was a rower] and of the type that should and could lead in war'.[82] Claude Neumann's examination of the militia between the wars argues that its officers were drawn from a narrow social elite.[83] This holds true for the COs. A simple survey of the background of the pre-war command cohort reveals that more than 75 per cent of the COs serving in 1939 were engaged in white-collar work. Of this large majority, the single most numerous professional group were men of the professions: lawyers, accountants, engineers, architects and so on. Imbued with an ethos of public service integral to middle-class masculinity, service – and leadership – in the militia was a respectable social obligation for men of their position. They possessed the requisite education to cope with the intellectual aspects of command, and their civilian positions were evidence of a demonstrated ability to organise, to set and achieve goals and to express their opinions lucidly and have them acted on – all skills essential for command.

Furthermore, it was only men of the middle or upper classes who possessed the income to sustain a part-time military career as an officer. Although the AMF's commissioning and promotions policy in the inter-war years was notionally promotion by merit and experience, the system never functioned in the egalitarian manner intended. Officers were required to purchase their own uniform. Although they were given an allowance to do so, it was barely adequate.[84] Officers were still regarded as gentlemen and as such expected to pay regular dues for membership of the officers mess, attend various regimental functions, the frequency of which increased in proportion to rank, and maintain mess attire for them. All of this entailed expenditure that made part-time soldiering an expensive pursuit.[85] Financial constraints often determined whether a soldier would apply for a commission, and a class distinction between officers and the men they commanded was often evident.[86]

When considering the influence of class and patronage on promotion in the interwar militia, one other network has to be considered. During the 1930s many militia officers, from the most senior to the very junior, were involved in secret armies set up across Australia in response to fears of a left-wing takeover of government. Among them were two divisional commanders, several battalion and regimental COs and 2iCs, and numerous other lower-ranking officers.[87] Those involved guarded the secrecy of their organisations zealously. Although several historians have managed to piece together the details of several of these organisations, the full extent of their membership and influence is a secret that has largely been taken to the grave by the men involved. Hence it would be wrong to overestimate the influence of the links of patronage fostered by the secret armies. There is no basis for suggesting that the militia was ruled by a secret extra-regimental cabal. It is useful, however, to speculate how the links between these men, forged on the basis of mutual and implicit trust, might have influenced their decisions when it came to appointing subordinates. Was it just coincidence that two successive COs and the 2iC of the 39th Battalion, based in the League of National Security (LNS) heartland of Melbourne's inner east, were all high-ranking LNS functionaries?

The militia in these years has often been described as an amateur army, but such a tag is problematic. The common definition of an amateur is one who 'engages in a pursuit as a pastime rather than a profession'.[88] Although part-time soldiering was not the means by which militia COs made a living, the commitment that it entailed and its central position within Australia's defence planning made it much more than a simple pastime. The militia exhibited many of the characteristics that modern social theorists use to define a professional organisation: a distinct code of conduct; the practice of specific skills developed through systematic education and training; and a group ethos emphasising the utilisation of those skills for the benefit of the wider community. Inherent in this definition of a professional organisation, however, is the notion that the skills practised must be both efficient and appropriate, and represent the most recent developments in the theoretical principles underlying its activities.

The standards of militia COs fell short of this benchmark. Their doctrinal knowledge was limited and outdated, their tactical ability mediocre, and their age meant that their capacity to withstand the physical and mental rigours of modern warfare was doubtful. As Lavarack commented, this was 'not due to any inherent defect in the individuals concerned'.[89] These weaknesses reflected the amateur characteristics of the militia: its lack of universal standards of instruction and assessment; the absence of an institutionalised means to study and develop tactics and procedures; and a system of promotions and appointments that was only partly based on merit. As a group, militia COs of the interwar period can be characterised as having a professional ethos but amateur standards.

CHAPTER 2

THE FOUNDATIONS OF BATTALION COMMAND
Forming the 2nd AIF, 1939–40

On 15 September 1939 Australian Prime Minister Robert Menzies announced the formation of a special volunteer force of one infantry division and auxiliary units for service overseas. As the Defence Act prevented militia troops from serving outside Australian territory the special force would be raised in isolation from the existing military structure, although it was planned that it would draw heavily on its resources of manpower. The special force soon became known as the 2nd AIF, and its single infantry division, known as the 6th, was only the first of four that would be formed in the next 18 months.

Major General Sir Thomas Blamey was appointed to raise and command the new division, and would subsequently become GOC of the 2nd AIF when the decision was made to expand it into a two-division corps in April 1940. The 2nd AIF would become Blamey's force. Raising it from scratch, he faced a considerable challenge. The AMF was in a parlous state in September 1939, and the pool of command talent available had little depth and a very limited range of experience. Blamey personally involved himself in most aspects of the 2nd AIF's formation, and his policies would continue to be a significant determinant in the operational experience of its infantry battalions throughout the war.

In 1939–40 Blamey oversaw the selection of a cohort of solid and dependable, although not necessarily spectacular, battalion COs. As products of the militia, they represented many of its faults, but in most cases they were also the best available. In raising a completely

Blamey's force: Major General Thomas Blamey, GOC of the 6th Division and later GOC of the 2nd AIF, inspects new recruits, Melbourne Showgrounds, Victoria, December 1939. Blamey exerted considerable influence over the selection of the 2nd AIF's original COs and would continue to do so for the rest of the war. (AWM Neg. 000315)

new force to fight in an unknown theatre of war, it would seem that great significance was placed on the qualities of trust and predictability essential to the efficient functioning of command relationships. There are also strong indications, however, that Blamey selected these men only as caretakers, to raise and train the battalions while a younger generation of officers were prepared to take command on operations.

The formation of the 2nd AIF's infantry battalions set the tone and standards for command practice in the AMF throughout the rest of the war. The infantry battalion was much more than a collection of companies and platoons. It was a network of interdependent roles and relationships that the CO had to guide and foster, but also allow himself to be drawn into. AMF doctrine and regulations provided a sound foundation for the development of leadership and therefore cohesive military units ready to begin their training for war.

Personal politics and the formation of the 2nd AIF

The selection of the first COs for the 2nd AIF was heavily influenced by existing militia command relationships. Those militia officers who aspired to a battalion command in the 2nd AIF were required to nominate themselves via the headquarters of the military district in which they served. The district commandants were, in turn, required to recommend whom should be selected for senior commands. Brigadiers made their selections from these lists, although they were not bound to heed the commandants' recommendations. The brigadiers' selections ultimately had to be approved by Blamey, first in his capacity as GOC of the 6th Division and later as GOC of the AIF.

There is no consolidated record of potential COs who volunteered for the AIF at the outbreak of the war, although the commanders of existing militia battalions constitute a rough starting field. It cannot be assumed, however, that all militia COs offered to serve in the 2nd AIF. Some withheld their services because they wanted to transfer their entire battalions to the 2nd AIF.[1] Others were reluctant to volunteer because, as senior lieutenant colonels, promotion to command a militia brigade was imminent, and at least one ambitious and far-sighted officer held himself back in the hope of a command with an armoured formation, which he was sure would soon be formed.[2] No PMF officers were initially considered for the command of AIF battalions.

There is little archival trace of either an official set of criteria or a formalised process used to select COs. The cornerstone of the selection procedure seems to have been personal acquaintance with the candidates' character, prior experience and performance. Blamey's direction to the newly appointed brigadiers to begin considering 'suitable' COs gave no guidance as to what prerequisites made an officer suitable.[3] Their selections were discussed at a meeting chaired by Blamey in Melbourne on 13 October 1939, the main purpose of which was for him to approve each candidate. A flavour of the resulting discussions, illustrating both the influence of personal relationships on the selections and the lack of a formalised selection process, is provided by the recollections of Jack Stevens, who had been appointed to command the 6th Division's signals unit. Stevens listened with amusement as Brigadier Arthur 'Tubby' Allen, the newly appointed commander of the 16th Brigade, justified his selection of George Wootten to command the 2/2nd Battalion. Wootten weighed more than 18 stone (approximately 115 kg), and Blamey declared him 'far too heavy for a Battalion Commander'. Allen, who owed his nickname to his own portly stature, countered that he had seen Wootten recently and he weighed only about 16 stone (approximately 100 kg). Blamey, also quite rotund, relented: 'If he's only 16 stone I suppose it's all right.' Thus, Stevens concluded, '. . . one who proved to be one of the most able and distinguished leaders of the 2nd AIF won his "weight-for-rank" race!!'[4]

An examination of the likely candidates for the battalion commands and the eventual selections confirmed by Blamey on 13 October reveals the limited field from which the brigadiers could chose and that implicit criteria were applied. These criteria emphasised active service experience in the First World War and demonstrated competence in command of a militia battalion. Close working relationships with the brigadiers in the militia also seem to have contributed to the selection of several candidates.

The 16th Brigade offers a case study. The brigade was to be raised in New South Wales, and hence 'Tubby' Allen was free to select COs from any of the officers who offered their services in that state. In 1939 there were 19 infantry battalions in New South Wales,[5] and the COs of 11 of these are unlikely to have been considered as potential AIF COs because four were several years over-age, two were regular soldiers, three had held their commands for less than a year, one withheld his services and another, although having commanded a battalion for two years, had extensive experience in the Australian Army Service Corps and was appointed as a staff officer in the Australian Overseas Base. Thus Allen's field was effectively eight officers for four commands: six First World War veterans in their early forties and two younger officers in their late thirties who lacked such experience. All had commanded militia infantry battalions for periods ranging from one to seven years.[6]

Allen's first shortlist reveals that officers other than incumbent militia battalion commanders were also considered. It included three officers from the field deduced above – Frederick Burrows, Claude Cameron and Victor Windeyer, COs, respectively, of the 36th and 18th Battalions and Sydney University Regiment (SUR) – and three from outside it – Kenneth Eather, who was on the Unattached List in 1939 but had commanded the 56th and 3rd Battalions between 1933 and 1938, John Crawford, a former CO of SUR, and George Wootten, CO of the 21st Light Horse Regiment. Ultimately, Allen selected only Eather and Windeyer from his first shortlist, which points to a process of discussion and deliberation carried on right into the meeting on 13 October. Allen's other two selections were Vivian England, CO of the 55/53rd Battalion, and Percival Parsons of the 4th; Allen had little personal knowledge of Parsons but decided to include him on his list, almost at the last minute, on the basis of a recommendation from the commander-designate of the 18th Brigade, Brigadier Leslie Morshead.[7] Allen's selections display a preference for command experience. Eather, England and Parsons were among the longest-serving infantry commanders in New South Wales, and both Eather and England had commanded battalions in Allen's militia brigade. England and Parsons had both also served in the First World War as junior officers. Windeyer is the exception among Allen's selections. Although being certified fit for command in 1932, he did not assume command of SUR until 1937 and had no experience outside the regiment. He had, however, earned a particularly high reputation during his time as CO. Perhaps Allen was seeking to achieve a balance between age and experience. England and Parsons represented considerable experience, but they were both 45 and Allen must have realised that they would not have been able to stay in command for much longer than 18 months. Eather provided a balance, having five years command experience as well as being relatively young at 38.

Windeyer ultimately never served under Allen. He failed his preliminary medical examination for enlistment in the 2nd AIF and hence Allen had to find a replacement. He turned away from the incumbent militia infantry COs and selected George Wootten. Wootten was quite a contrast to Windeyer. Trained as a regular officer at Duntroon, he had seen extensive service with the AIF during the First World War. He attended the British Army's staff college in 1919 and returned to Australia, and a succession of staff postings in the PMF, the following year. Disgruntled with the poor conditions offered to regular soldiers in the 1920s, Wootten left the army in 1923 to practise law. He returned to service in the CMF in 1937 as CO of the 21st Light Horse Regiment. In Wootten, Allen was obviously banking on military experience. At 46, Wootten was over-age for a lieutenant colonel in the 2nd AIF, and his hefty build left several in the AMF doubting whether he had the stamina for battalion command.

Allen's selections were also limited by direction from Blamey that the 'militia must be maintained' and not stripped of all its best officers.[8] Both Claude Cameron and Kenneth Ward had experience comparable to those selected yet were never appointed to AIF commands. Cameron was a well-respected and successful militia CO with a distinguished record in the First World War and would later command the 8th Brigade. Ward was one of the new generation of militia COs. Aged 36, he had not seen service in the First World War but had commanded various linked incarnations of the 20th Battalion since July 1937. If a 1941 confidential report is any guide, he had exercised his commands with both judgement and zeal.[9]

Frederick Galleghan was another apparently well-qualified officer who was initially overlooked for an AIF command. His case was different from that of Cameron and Ward, and his treatment further demonstrates the significance of existing command relationships in the appointment of the 2nd AIF's first battalion commanders. Galleghan volunteered for overseas service the day immediately following Australia's declaration of war. He was one of the most senior lieutenant colonels in the militia at the time, although not, as his biographer Stan Arneil states, the most senior.[10] Galleghan was an experienced soldier, having served as a non-commissioned officer (NCO) on the Western Front, and had commanded a succession of militia battalions since 1932, but he was overlooked for a command not only in the 6th Division but subsequently in the 7th as well. Although recognised as a capable CO, his abrasive personality and tendency to disregard procedure had won him few friends in the AMF's hierarchy.[11] Allen's handwritten notes on his letter from Blamey reveal him to have been considering Galleghan early on, but he never made it to his shortlist. We can only speculate as to whether Galleghan's reputation caused Allen to disregard him in the interests of harmonious command relationships or whether he had been told discretely that Galleghan was not to be considered.

The selections made by the 6th Division's brigadiers are remarkably consistent, which suggests clear verbal guidance from Blamey. The majority of the COs had First World War experience, and there were at least two with such experience in each brigade. The 16th and 17th Brigades were almost carbon copies of each other with three veteran COs, all in their mid- to late forties, and one younger commander, without First World War experience, in his late thirties.[12]

Several figures stand out for the further insight they provide into the way personal relationships influenced the selection process. Theo Walker was selected by Brigadier Stan Savige to command the 2/7th Battalion in the 17th Brigade. He had served as Savige's 2iC when he was a battalion commander and subsequently as a CO in his militia brigade. John Mitchell, the oldest of the group, was appointed to the 2/8th Battalion at Blamey's behest, although Savige did value the diversity of tactical experience offered by his prior command of the 20th Light Horse Regiment.[13] In Tasmania, John Field was appointed to command the 2/12th Battalion in the 18th Brigade over the heads of the two incumbent militia COs there, despite having never commanded a battalion, due to a strong recommendation from the District Commandant on whose headquarters he served.[14] Field's appointment also reveals the limitations of the regional model being used to raise the AIF, which dictated that a CO had to be a resident of the state in which his battalion was raised. An officer as lacking in command experience as Field would not have been appointed CO of a battalion raised in New South Wales or Victoria.

The presence of younger officers among the COs is evidence that succession planning was being considered. With the maximum age for a lieutenant colonel in the 2nd AIF having been set at 45, it is hard to believe many of the older officers, some already over-age, were expected to last long in command. Similarly, it is unlikely that they were regarded as future brigade commanders. Their value to the 2nd AIF lay in their familiarity with the AMF's systems and procedures. In forming new battalions from scratch, Blamey was not taking chances with younger, inexperienced commanders. He wanted men who could coordinate the training and administration of a battalion and who had a good idea of the demands of active service. The role of these officers was to form the battalions and prepare them for action. At the same time, they had to select promising and reliable subordinates and groom them to take over. The importance placed on pre-existing relationships and personal recommendations in the selection of the COs also demonstrates that what was wanted were officers whose abilities were known and whose actions could be predicted.

The formation of the two brigades needed to complete the 7th Division (the brigades of the 6th Division had been reduced from four to three battalions, thus forming the 19th Brigade) in April 1940 perpetuated the command selection practices established with the 6th. When the COs selected for the 20th and 21st Brigades are examined the limited numbers of suitably qualified officers is immediately apparent. The two remaining available officers from Allen's original shortlist were appointed to command the two New South Wales Battalions, and in Western Australia a brigadier was demoted to provide a suitable CO.[15]

Although Major General John Lavarack was the commander of the 7th Division, Blamey continued to exercise tight control over the appointment of its officers. Whether this was Blamey seeking to exercise the same quality control over his corps as he had over his division, or whether it was a usurpation of Lavarack's authority, characteristic of the continuing feud that would blight their relations throughout the war, is unclear. The recollections of one of Lavarack's brigadiers vividly illustrate both Blamey's involvement and the dwindling pool of command talent. Promoted from command of the 6th Division's signals unit, Jack Stevens was appointed to raise the 21st Brigade, comprising a battalion each from Victoria, Western Australia and South Australia. He recalled that the biggest problem he faced was the appointment of his COs as the best officers from the militia had already been selected for the 6th Division. Ultimately, Blamey took the selection of two of his COs out of his hands: 'He told me there was only one suitable man in [Perth and Adelaide] and I could scarcely argue.'[16] By personally directing Stevens's choices, Blamey was operating outside the chain of command and subverting Lavarack's influence.

Stevens was permitted a free hand only in his home state of Victoria. From a field that he described as 'woefully weak' Stevens selected William Cannon, CO of the 58th Battalion, to raise the 2/14th. Cannon was 42 and had been wounded during the First World War. He had commanded militia battalions since 1928. Stevens had his doubts about Cannon's fitness and ability but, as a long-term friend, he did have the advantage that he was a man he could trust: 'He was not my ideal of a Battalion Commander in 1940, but I believed he could weld a new unit together and take it into its first fight.'[17]

Although the average age of the COs for the new division was much lower than those of the 6th, in terms of experience they conform to the same pattern. Both the 20th and 21st Brigades had two experienced COs and one novice. Each brigade also had two First World War veterans, although this experience was not always relevant – Cannon, for example, had served as an artillery signaller in the First World War. Blamey's restrictions on Stevens's appointments, his own reflections on the field in Victoria, and the appointment of the two remaining available candidates from Allen's original shortlist, all indicate that the pool of experienced command talent in the militia was drying up. One last trip to the well, to raise another division, would find it almost empty.

Reflecting their militia origins, the new AIF COs were drawn from a relatively narrow strata of society. They were overwhelmingly urban-dwelling, middle-class, white-collar workers. Just under half were professional men in civilian life: lawyers, engineers, accountants and a single architect. Another four were clerical workers, and two worked in retail and marketing. The list of professions was rounded out by a company director, a university lecturer and a dental technician. Only one of the new COs was a PMF officer. A third had degrees, a proportion far in excess of the wider population where the rate of university education was a little over 1 per cent.[18]

As a demographic sample, the first 18 AIF COs serve as an exemplar of the wartime COs as a whole, a full demographic examination of whom appears in appendix 1. Throughout the war the professional men dominated battalion command positions, constituting 26 per cent of all appointments. Men from other white-collar professional groups, such as company directors, employers and managers, those previously employed in sales and marketing, and teachers and university lecturers collectively constituted another 51 per cent of appointments. Thus, white-collar Australia provided 77 per cent of the AMF's COs while comprising only 24 per cent of the wider male population eligible for military service. As would be expected with such an occupational profile, the majority of the COs – 75 per cent – hailed from the state capitals or major provincial cities. Again, this proportion was larger than that in the wider male community where urban-dwellers represented only 60 per cent.[19]

The preponderance of urban white-collar workers was only to be expected. In a society were only a minority of boys would matriculate from high school,[20] they represented the best-educated section of society. They dominated battalion command in the pre-war militia, and there were strong links between that force and the beginnings of the army that would fight the Second World War. This dominant socioeconomic profile among the AMF's COs could perhaps be equated with a 'command class'. It would be wrong, however, to view these men as the elite of Australian society. They were not necessarily the men who governed and directed it, but they were the men, used to responsibility, who made it function. They were men with influence, but not men with the final say. It was natural that these were the men on whom a citizens' army should rely. In the middle command strata of the AMF during the Second World War they fulfilled the same role as they had in peacetime.

THE PMF IN COMMAND

The 7th Division was unique in that it introduced the first PMF officer – Robert Marlan of the 2/15th – to a battalion command in the 2nd AIF. Marlan's appointment foreshadowed the limited role that regular officers would play as infantry COs throughout the war. Ultimately, the PMF would provide only one CO in 12. Twenty were graduates of the Royal Military College, Duntroon, and members of the Australian Staff Corps, and one began the war as a lieutenant in the Australian Instructional Corps.[21] There were never more than seven PMF officers commanding battalions at any one time throughout the war – a peak achieved in early 1941. Another smaller peak – four regular officers in command simultaneously – was achieved in late 1945, but between the start of 1942 and the beginning of 1944 only five regular officers commanded battalions and never more than two simultaneously.

The dearth of regular COs had its roots in the interwar period. Despite Lavarack's efforts in the late 1930s to expand the role of the PMF, the perception that the only place for regular officers was on headquarters staffs still enjoyed wide currency in 1939, among both the government and the militia. Deep-seated suspicions of the capability of Staff Corps officers to command troops in the field were heightened by other prejudices. The PMF had earned the ire of some politicians owing to the active opposition of some of its senior officers to the government's defence policy in the 1930s.[22] Additionally, the reputation of PMF officers had suffered as a result of the hard times faced by the AMF before the war. The poor conditions of service forced many capable regular officers out of the army, and those who remained included the 'less capable officers who enjoyed the security of employment not found in the civilian community'.[23] Their presence is quite likely to have influenced the perceptions of both the citizen soldiers they served beside and the politicians and public servants they served under.

When the 2nd AIF was formed, Prime Minister Menzies decreed that all commands in the newly raised force would go to militia officers.[24] No writer has yet been sufficiently able to separate the influence of prejudice from that of reason – Menzies is reputed to have told his cabinet that he did not 'want any funny business about regular soldiers in this war'[25] – but the main reason for holding the regular officers back would seem to be the War Cabinet's fears regarding the intentions of Japan. They wanted to retain a firm base of military experience to train the militia for home defence.[26] By the time the decision to form the 7th Division was made on 28 February 1940, the situation in the Asia-Pacific region seemed clearer and hence it was considered that more regular officers could be spared for the 2nd AIF.[27] War Cabinet policy henceforth was that regular officers would be appointed to formation, unit and subunit commands on a basis of one in three. In the infantry battalions, this quota was never met, demonstrating the stranglehold that the citizen soldiers retained on command positions throughout the war.[28]

The initial postings of PMF officers in the 6th Division largely defined their role for the rest of the war. Although they received no battalion commands, regulars were appointed adjutant in all of the 6th Division's battalions, and it was in this role that PMF officers had their most profound influence on them. Similarly, all of the 6th Division's original brigade majors were regular officers. Staff Corps officers were posted in and out of battalions to gain field and regimental experience, but their posting histories demonstrate that it was still believed their principal role in the AMF was as staff officers. There was a shortage of trained staff officers in the AMF throughout the war, and it was believed it would have been a waste of the skills and experience of PMF officers to deploy them in any other way.[29] More than two-thirds of PMF COs were appointed to command while serving in operational staff, base or lines of communication postings; close to half returned to similar postings, albeit of a higher grade, after their period of command. The average command tenure for PMF battalion commanders was only eight months. None held multiple battalion commands, and only one would go on to command a brigade during the war, yet by the beginning of 1945 a fifth of the AMF's lieutenant colonels and close to half of its brigadiers were regular officers.[30]

Across the AMF, the performance of many PMF officers would suffer as a result of their limited range of military experience, and four would be sacked from the command of infantry battalions. One of these officers was Lewis Loughran, appointed to command the 8th Battalion in June 1945 from General Headquarters South-West Pacific Area. On his departure he was hailed by General Douglas MacArthur as an outstandingly thorough and dependable officer,[31] but he was sacked within two months of taking command of his battalion on Bougainville. Loughran's commander, Brigadier Arnold Potts, subsequently wrote to the Military Secretary on his behalf: '. . . the appointment of an officer to the command of a b[attalio]n engaged in a jungle campaign and actually in contact, who had reached the rank of Lt. Col. with little or no regimental service or command in action, was most unsatisfactory and also possibly unfair to the officer concerned.'[32]

THE ROLE OF THE CO

The task faced by the new COs in the early days of the 2nd AIF was daunting. On the morning of 4 November 1939 a troop train deposited a motley band of recruits at Ingleburn Station. Bound for the 2/2nd Battalion, they hailed from all over northern New South Wales, and the battalion historian records that they were in high spirits when they arrived, some having been 'almost poured on the train as it left their home town'. The new recruits were marched the three kilometres to Ingleburn camp and, just a little worse for wear, assembled in gently swaying ranks on the 2/2nd's parade ground. Surveying his latest draft of potential soldiers, George Wootten was heard to remark: 'Great God! What have I done to deserve this?'[33]

The validity of selecting officers with command experience and proven organisational abilities was immediately demonstrated as the newly appointed COs set to raising their units. Confusion and administrative inefficiency prevailed as the rapidly expanding and increasingly active elements of both the militia and the 2nd AIF outstripped the abilities of pre-war structures to support them; roughly half of the direct enlistments into the 2nd AIF for the entire war occurred between 1939 and 1940.[34] A CO's responsibilities, as defined by Australian Military Regulations and Orders (AMR&O), were extensive. In broad terms, a CO was 'responsible for the maintenance of discipline, efficiency and proper system' in his unit and for its 'training and readiness for war'.[35] More specifically, a CO had to manage three basic commodities: equipment, personnel and information. He was responsible for ensuring that all weapons and equipment were maintained in an operable condition, that any men or animals under his command were adequately fed and their health maintained, and that 'general information' was widely disseminated to all troops.[36]

By outlining the basic responsibilities of a CO and recommending the manner in which they should be discharged, AMR&O laid the foundations for leadership within the AMF. The challenge confronting COs faced with disparate mobs such as that which stood before Wootten at Ingleburn was to assert their authority and use it to begin the process of aligning the motivations and actions of individuals to the goals of the battalion. AMR&O underpinned this process. The regulations guaranteed that the basic needs of soldiers were catered for, thereby developing a reliance on and an affinity with the battalion as a formal group. They also provided the CO with his three principal leadership power bases: legitimate power based on his status in the battalion; coercive power based on the disciplinary sanctions he could impose; and reward power based on his ability to recognise and reward achievement in the battalion. The establishment of the other power bases was dependent on the way the authority conferred by AMR&O was exercised and the degree to which the responsibilities it outlined were met.

The imposition of military discipline was seen as the first step in creating an effective battalion. The Defence Act and AMR&O provided a CO with strong coercive powers to deal summarily with a range of minor offences and thereby influence standards in the battalion. COs were enjoined to 'use every effort to prevent the commission of offences and to suppress any tendency to screen their existence'.[37] More serious offences were required to be referred to a court martial, although in some cases a soldier could still elect to have the matter dealt with by his CO. In the case of a private soldier, a CO was able to award a maximum punishment of 28 days of detention or field punishment and stoppage of pay for the same period.[38] The disciplinary powers of Australian COs were considerably stronger than those of their British counterparts.[39]

In 1939, and throughout the war, COs followed the lead of AMR&O in the way they enforced discipline. Standards were high and transgressions rigorously punished. Arthur Godfrey, the original CO of the 2/6th Battalion, believed that success on the battlefield resulted 'entirely' from 'unfailing discipline'.[40] His thoughts were echoed by other COs later in the war. Fred Chilton reflected that a unit with slack standards in barracks would carry these on to the battlefield. William Parry-Okeden of the 30th Battalion similarly remarked on the 'hundreds' of faults in dress and conduct that a CO had to be 'continually checking on'. The Australian soldier, he noted, 'can become the most untidy ill-disciplined person in the world if he is not ruled with iron discipline'.[41] COs generally dispensed justice in a swift, severe and remarkably uniform fashion. Aside from 'the old man', two of the most common nicknames for COs were 'February' and 'Leap Year' resulting from their frequent imposition of their most severe summary sanction. Paul Cullen, the last CO of the 2/1st Battalion, quipped wryly that the phrase 'human rights' could never be applied to battalion justice. Emphasising the almost arbitrary nature of a CO's authority, disciplinary sanctions were sometimes bestowed en masse. On one occasion Roy King of the 2/5th Battalion held a mass orderly room on his parade ground, fining all of the battalion's leave defaulters in one go.[42]

Although battalion justice could be severe,[43] the disciplinary ethos that emerges from the reflections of former COs is strict but fair.[44] Chilton summed up the importance of impartiality: 'If he [the CO] is seen to be fair, it doesn't matter if he's a bit hard at times.'[45] Soldiers were left with little doubt as to what to expect in the case of misbehaviour. Tom Daly of the 2/10th Battalion commented that when his troops were 'up before him' they knew the rules and they knew the penalties: 'It wasn't as if something was happening to them they hadn't been told about.'[46] Exercising coercive power in this fashion conformed to the principles of sound leadership. The actions of COs were predictable, which in turn engendered trust, and with trust came respect. Mark Johnston's study of soldiers' writing reveals that discipline meted out by COs was generally accepted with good grace and that a remarkable number of stern disciplinarians were popular with their men.[47]

AMR&O did also advise COs to exercise discretion in the use of their coercive power.[48] As both 'judge and jury', COs had considerable latitude in their enforcement of discipline, and discretion remained a characteristic of battalion justice throughout the war. Chilton recalled: 'You need to give a bloke a fair go; it was terribly important for morale.'[49] The application of coercive power was tempered by both common sense and the personality of the man behind the badges of rank. As described by Phil Rhoden, CO of the 2/14th Battalion, battalion justice was a simple, transparent process: an offender was marched in to the CO, and they either 'provided a plausible excuse or not'.[50] There are many examples of COs awarding lenient penalties to men with dependent families; turning a blind eye to illegal gambling when there was a lack of recreational facilities or the organisers were perceived as honest; or of yielding to their sense of humour and going easy on men who told 'an original plausible story'.[51]

The discretionary exercise of coercive power was also a means by which COs could reinforce loyalty to the battalion. While disciplinary transgressions against the battalion's standards and personnel were swiftly and consistently punished – 'If somebody let the battalion down I had no mercy on them, none whatever,' Daly recalled[52] – those offences committed against outsiders were frequently treated more leniently, often went unpunished and were occasionally praised. In late 1939 troops of the 2/1st Battalion rioted and burnt down a canteen hut operated by a civilian contractor who was perceived to be profiteering. Their CO, Ken Eather, is said to have delayed the response of his duty officer to let the hut properly catch alight, and no one was ever charged with its destruction.[53] Similarly, after a spate of brawls in 1940 between the men of the 2/15th Battalion and those of the Darwin Mobile Force Robert Marlan simply reprimanded his men on parade, then went on to tell them he was proud that they had not disgraced themselves.[54]

COs often reacted with hostility to accusations made against their men by the personnel of other units. An element of personal pride was probably involved in such responses, but they also served to unite the battalion behind their CO. While in camp at Greta in 1941 a guard of the 36th Battalion was admonished by an officer from another battalion who believed he had not come to attention as a column of troops passed. The 36th's CO, Muir Purser, overheard, sprang from his office and stood in the middle of the road waving his fist, yelling: 'You look after your own troops. Leave my men alone and go to blazes.' The battalion history records that as a result of this outburst, Purser earned the respect of his troops, who felt that here was a leader who stood with his men.[55]

Although many soldiers simply regarded 'criming' by the CO as the price of a good time on leave, their expectations of a CO's conduct were generally much higher.[56] This attitude reflects one of the central tenets of leadership discussed earlier: that a leader's personal actions need to match his expectations – COs had to set an example – but also hints at prevailing social norms expected to be upheld by figures of authority.[57] This was in keeping with the direction that a CO was responsible for maintaining 'unblemished the honour and character of the corps'. Lambert's CO in *Twenty Thousand Thieves* is both a socially powerful grazier and a drunkard. A continuing theme of the book is the declining respect for 'Groggy Orme' as a result of his licentious ways, despite many vivid descriptions of the troops' own quests for sex and beer.[58] Phil Rhoden's reflections on the qualities of a good commander authenticate the attitudes portrayed by Lambert: '. . . one shouldn't be a flamboyant type that shows off and drinks too much in the mess [and] should be a solid citizen.'[59] The model CO was 'courteous, detached, authoritative'.[60] He was friendly with the troops, but not familiar; officers who tried to be too familiar raised the troops' suspicion and lost their respect.[61] In any case, as Daly reflected, there were so many comings and goings in a battalion that a CO was not always able to learn the names of all his troops. He definitely knew the 'good ones' and the 'bad ones', but there were 'a few in the middle' he was a bit hazy on.[62]

By enforcing discipline in a fair and consistent manner and behaving in accordance with the troops' expectations of an authority figure, COs were not only setting the standards required for the efficient operation of the battalion but also building trust and respect. Thus, they broadened the foundations of their leadership to embrace referent power. Other activities directed by COs in this formation period had a similar dual role of developing the strength and cohesion of the new battalions and strengthening the CO's leadership.[63] Many of these activities were as old as organised armies and included intra- and inter-unit sports, collective endurance tests and subunit competitions. These activities raised essential standards of fitness, cleanliness and equipment serviceability as well as providing the recognition of effort and achievement essential for individuals to identify themselves with a formal group. They provided the COs with the scope to embrace the exercise of reward power to further broaden the foundations of their leadership. Many of the AIF's initial

Broadening the foundations of leadership. Men of the 2/6th Battalion take part in a Lewis Gun drills race during a unit sports competition, Seymour, Vic, 2 February 1940. The image illustrates not only one of the means COs used to build subunit cohesion and *esprit de corps* but also the obsolescent equipment with which the first infantry battalions of the 2nd AIF trained. (AWM Neg. P02006.016)

COs introduced intra-unit competitions that remained a feature of camp routine for the rest of the war. Rewards varied and provide an insight into the personalities of the various COs. In Theo Walker's 2/7th competition winners were announced in the battalion's routine orders, and in the 2/12th the occupants of the best-maintained hut were awarded a silk pennant sewn by John Field's wife; theatre tickets were the go in Robert Marlan's 2/15th, while in the 2/8th John Mitchell appealed to purely mercenary interests, awarding a £5 prize each week to the company that mounted the best turned-out guard.

THE CO'S STAFF AND OFFICERS

There is little obvious sign in the archival record of the activities of COs during the formative months of the 2nd AIF. They flit in and out of the pages of war diaries and official reports, and are accorded a paragraph here and there in battalion histories. These disparate references, however, do allow a composite picture of the activities of COs during the formation of the 2nd AIF to be assembled. They were interviewing new recruits and candidates for promotion; attending planning conferences; making representations to higher headquarters for more supplies and equipment and improvements in camp facilities; liaising with philanthropic organisations and other civilian authorities; writing training guidelines and reporting on standards achieved; inspecting and addressing parades; giving lectures; and dispensing justice in their orderly room. These tasks were only a small component of a CO's overall responsibilities – a 27th Brigade circular in November 1940 pointed out that a CO was responsible 'for everything'[64] – the complexity and extent of which required a capable staff and a well-organised command system to fulfil them.[65] *Infantry Training* (1937) stated that 'an efficient headquarters [was] essential in every unit, and, without such efficiency, the efforts of the commander will often be nullified'.[66] The headquarters of the infantry battalion provided the CO's principal advisers, who were also often his confidants, and each was responsible for particular aspects of the battalion's administration.

The officers closest to the CO were his second in command (2iC), often known to the troops as 'Twic',[67] and the adjutant, who was his personal staff officer. The most senior of two majors in the battalion, the 2iC was responsible for administrative and logistical matters in the battalion but was also the CO's understudy and had to be ready to assume command at 'a moment's notice'.[68] The adjutant, usually a captain, oversaw the operation of the orderly room – the battalion's administration office – and handled the bulk of the CO's personal administrative work, such as correspondence and the preparation and distribution of orders, directives and the reports and returns required by higher headquarters. Don Jackson, the original adjutant of the 2/1st Battalion, recalled that he was both the CO's 'executive spokesman' and 'manager'.[69]

The close working relationship of the CO and the 2iC and the adjutant often resulted in similarly close friendships. As adjutant, Don Jackson was also a 'confidant' and 'devoted friend through thick and thin' to his CO, Ken Eather, and he believed that such an accord was 'essential' between a CO and all of his principal staff officers.[70] The 2iC's personal experience as a senior officer meant that he also often became the CO's chief confidant and the sounding board for his plans. In some battalions the exercise of command was almost a double act. The history of the 2/6th Battalion reveals that Arthur Godfrey, never particularly fond of administrative work,[71] devolved a great deal of his disciplinary responsibilities to his 2iC, Major Hugh Wrigley.[72] In the 2/16th, Alfred Baxter-Cox left the selection of new recruits and officers and their reception to his 2iC, Major Arnold Potts,[73] so that he could concentrate on the detailed planning of the battalion's training. Similarly, in the 2/27th Murray Moten relied heavily on his 2iC, Major Alexander 'Bandy' MacDonald, to perform the vital function of training his officers.[74] At the heart of some of these double acts were complementary personalities. MacDonald, down-to-earth and slightly eccentric, seems to have compensated for the quiet, almost austere manner of Moten and was a popular figure with the troops. In the 2/6th the relationship was reversed. Godfrey was loud and convivial, Wrigley proper and stern.

The command relationship between Arthur Verrier, the original CO of the 2/10th Battalion, and his adjutant, Captain Tom Daly, demonstrates both the dynamics of such command arrangements and their potential to undermine a CO's authority. Daly has reflected on the way their relationship developed after Brigadier Morshead advised Verrier to 'leave the thinking to Daly, and you do the barking': '[Verrier said] "I thought about that, and I don't think it's a bad idea . . . if you want me to do some barking, you tell me." And that's the way it happened. If I said a company commander wasn't sparking on all six, Verrier would have him in and bellow at him . . . We had a great relationship.' Verrier's actions demonstrated an astute assessment of his own capabilities and the sense to recognise and employ the talent under his command. It earned him the enduring respect and loyalty of Daly: 'He wasn't a clever chap, he didn't profess to be. But he was a good commander.'[75]

Verrier's close relationship with Daly, however, seems to have lessened his authority with some of his officers and senior NCOs. His reliance on Daly depleted his stocks of expert and referent power:[76] 'In my opinion Tom Daly made and ran the B[attalio]n. As for A[rthur]V[errier] . . . he was the "old man", a figurehead who rarely, if ever, had any influence on our activities. Should he have initiated a pronouncement it would have come to us through Tom Daly.'[77] There is a hint that some felt ignored or excluded. Lieutenant Geoff Matthews wrote of a TEWT the 2/10th's officers participated in: 'Most of the afternoon spent lying under a tree yarning while CO & Adjutant fought the war.'[78] There seems to have been little question about Verrier's authority among the troops, however. One former sergeant remembered Verrier as 'a thoroughly lazy bastard' but 'a great bloke in front of the troops'.[79] Although they may have been enforced by Daly, the standards of behaviour and performance expected in the 2/10th were regularly enunciated by Verrier in lectures to his troops and monitored by personal inspections.[80]

The CO's two principal advisers regarding the physical and psychological well-being of his troops were the regimental medical officer (RMO), commonly known as 'the Doc', and the chaplain, known to the more pious troops as the padre and to the less so as the 'sky pilot'. As attached specialist officers, usually with minimal military training, these men occupied a unique position in the battalion. Speaking of his RMO, Fred Chilton recalled: 'He was of us, but not of us.'[81] The maintenance of health standards in the battalion was delegated to the RMO[82] and AMR&O warned COs that they would 'incur grave responsibility' if they ignored the advice of their RMOs without sufficient grounds.[83] Given the potentially different backgrounds and outlooks of a CO and his RMO a considerable degree of trust and mutual respect was needed to ensure an effective working relationship. One RMO wrote that a CO should have so much trust in his RMO that 'he gives him a free hand to issue what orders he considers necessary'.[84]

Although ostensibly attached to the battalion to minister to the spiritual well-being of the troops, the chaplain's role was often more akin to that of a morale officer, and one of his main jobs was ensuring the provision of comforts. He was often assisted in this task by representatives of philanthropic organisations, such as the Salvation Army or Australian Comforts Fund (ACF), who were attached to the battalion. A set of 'Notes for Commanding Officers' emphasised the link between the men's 'worldly needs' and their 'spiritual welfare', noting that a padre had to be a 'good scrounger above the average' and that 'the best way to a man's heart is through his stomach'.[85]

Both the RMO and the chaplain, outside their official functions, were nodes in the CO's own informal information network. As 'Notes for Commanding Officers' advised, it was possible for a CO to 'learn a great deal of the "inner mind" of the B[attalio]n from both of them'. Their duties brought them into regular contact with both officers and other ranks (ORs), and their position outside the chain of command often engendered an openness that was not to be found among the formal command relationships in the sections, platoons and companies. Some COs also embraced this spirit of openness, sharing thoughts and feelings that they felt could not be discussed with other officers.[86] The RMO and the chaplain, however, also represented one of the main threats to a CO's authority. Both served an ethos incongruent with the battalion's primary aim of inflicting death and destruction, and both were also bestowed with authority from outside the battalion owing to the respect accorded their callings by Australian society at the time. Conflict between COs, RMOs and chaplains did occur, and when it did the former generally acted swiftly to remove the threat to their authority.[87]

The regimental sergeant major (RSM) provided the CO with another insight into the 'inner mind' of the battalion. He was the battalion's most senior other rank, and his principal responsibility was the maintenance of discipline among the troops. He was assisted in this role by a small regimental police section, but worked principally through the battalion's sergeants. Fred Chilton maintained that the sergeants often knew the men better than their platoon commanders, which made the RSM a valuable source of information.[88] Phil Rhoden agreed: '[The RSM is] moving in places you don't move in, he hears things that you don't hear. If he's a good RSM he'll tuck aside the trivial and report to you the guts. Having reported it, he expects you to deal with it; he expects you to support his view. If his view's the right view, it's the battalion's view.'[89]

The principal agents of a CO's authority, however, were his subordinate officers. It was they who implemented his policies and orders – as one commentator in the AMF Christmas book *Jungle Warfare* wryly observed, their 'chief job' was 'passing down to the other ranks kicks in the pants'.[90] Some former ORs struggle to assess the influence of their CO, which indicates the extent to which he worked through his subordinate officers. At least one CO, John Mitchell of the 2/8th, also used his subordinates to enforce an official barrier between him and his men. The battalion routine orders for 4 February 1940 stated that no soldier had the right to complain directly to the CO. Mitchell obviously felt the need to reinforce the hierarchical structure of the battalion, a structure that would have been anathema to many of his men with no previous military service. Inherent in this order is also an attempt to undermine the creation of any informal group structure in the battalion. It concluded that no complaints from groups of men would be considered, even if directed through the chain of command.[91]

When the AIF battalions were formed each CO was allocated a group of militia units, usually including his former command, from which to recruit his officers. This practice was obviously designed to allow the new COs to select a proportion of their officers from men with whom they already had a command relationship, to facilitate the rapid establishment of the structures and procedures of the new battalion. It provided the elements of predictability and trust vital to the effective exercise of command. The practice also had the effect of extending the patronage networks that had already influenced the selection of senior officers. Personal loyalty would appear to have been an important motivating factor for many of the militia officers who volunteered for the AIF. For example, in Victoria the militia battalions that produced the largest number of officer volunteers for the AIF were those whose COs had also volunteered.[92] This led to a degree of nepotism in at least one battalion where the CO favoured officers from his previous militia command.[93] Others, however, were consciously even-handed, taking a 'reasonable share' of officers from each allocated militia unit to avoid creating problematic command relationships in their new battalions.[94]

Revealing a gentlemanly, slightly patriarchal ethos behind the authority structures in the battalion, the way a CO was expected to discipline his officers was somewhat different from the treatment accorded ORs. AMR&O directed COs to adopt an almost fatherly persona and 'give advice to the young and inexperienced'; make themselves the 'arbiter of any disputes'; and when an officer's conduct was likely to 'interrupt the harmony of the corps' explain 'in the most forcible manner' the consequences of his actions.[95] It was customary for disciplinary transgressions by officers to be punished informally, often by ostracism in the mess.[96] Possibly referring to their subordinate status to 'the old man', battalion lieutenants were often known as 'junior'.[97]

Although AMR&O said very little specifically about the training of other ranks in a battalion, it did lay down several guidelines regarding a CO's responsibilities for the training of his officers. He was to provide 'systematic and efficient instruction' for his officers in all of their 'professional duties' and prepare them for promotion. Additionally, he was to provide opportunities for his more experienced officers to act in appointments senior to those they held.[98] Here again AMR&O laid down the foundations for effective leadership. Such training was also part of the preparation of the battalion command system for war. Training subordinates to act in higher capacities fortified the chain of command against the inevitable casualties of operations.

Training battalions, training COs

Despite their responsibility for preparing their battalions for war, the AIF's initial battalion commanders had little control over the direction of their training. Brigade commanders specified subjects to be covered, and often the way these were to be taught, as well as the degree of emphasis that each was to be given. Although at liberty to formulate their own training syllabi within these guidelines, the COs still had to submit them to their brigade headquarters for approval. Hence, the most direct influence over battalion training was exercised by the brigadiers.

For instance, the first training directive produced by the headquarters of the 17th Brigade clearly stated that 'COs [were] to make [their] own syllabus' but also specified that, among other activities, 25 hours were to be spent on drill, 5 hours on bayonet training, 50 hours on weapon training and 6 hours on anti-gas training. Instructions were also laid down regarding the type and frequency of physical training to be conducted; that night training was to take place one night per week; and that the specialists of the intelligence and signals sections were to be trained under brigade auspices. Brigade headquarters even directed COs as to when they should train their officers and NCOs and which members of their staff they should appoint to do so.[99]

The individual battalion syllabi were structured on conventional military lines that first sought to train the individual soldier, both intellectually and physically, before progressing through the training of each successive subunit. This process culminated with the training of the battalion as a whole and its testing by brigade staff. The tactical training of the battalions in Australia was unremarkable. It concentrated on the traditional phases of war – advance, attack, defence and withdrawal – in a very pedestrian fashion. The battalions had few vehicles or heavy weapons; tanks were usually represented only as a threat to be countered and not a supporting arm, as were aircraft. The nature of this training, compared to the type of warfare that had been witnessed in Europe, worried at least one CO. John Field had made a point of studying the latest developments in military thought when a militia officer. He noted in his diary that both he and Tom Louch of the 2/11th Battalion were concerned by the principles underlying Morshead's training of the 18th Brigade.[100] They felt they were preparing for a different war from the one they would be required to fight. Field found Morshead's continued insistence that his battalions train with horses and limbers particularly 'disturbing' because it was envisaged that they would be employed overseas as motorised units and therefore had to train 'individually and tactically with that in view'.[101]

It could be argued that the training of the battalions was shaped to match the resources on hand as few motor vehicles, tanks or other mechanised transport were available in Australia at the time. The training that was conducted specifically for the AIF's first COs, however, did seek to address some of the challenges of both mobile and mechanised warfare. Field attended a TEWT in the Ingleburn–Moorebank area in early 1940 that was based on countering the German tactics employed in Poland. Stevens conducted TEWTs for the 21st Brigade's COs based on similar scenarios. Field's response to the exercise in which he was involved, however, demonstrates that his grasp of mechanised warfare was based more on awe than understanding. He noted that the day had 'really demonstrated the little influence which the infantry can have on the battlefield'. Reflecting on the general level of ability witnessed among his peers, he continued that none of them, including himself, were 'tactical "tigers"'. They were all 'school boys together', and 'high and low' had much to learn.[102] Wootten of the 2/2nd would eventually command the 9th Division, but Brigadier Allen recalled that in the early days he had much to learn about infantry tactics.[103]

The brigade-administered TEWT was the main form of training provided specifically for the new COs, although the demands of raising and training units meant they were conducted only on an irregular basis. TEWTs sought to develop not only the COs' tactical abilities but also their grasp of administrative matters and the practical skills of command. A TEWT conducted by Brigadier Stevens for his COs in August 1940 provides a particularly thorough example. The COs were tenaciously quizzed regarding the tactical and administrative situation of their notional battalions: what arrangements had they made for feeding their men? Had the counter-attack force conducted its reconnaissance? How many stretcher-bearers had been allocated for the day's advance?[104] Stevens also used the TEWT as an opportunity to have his COs think about the principles underlying their tactics. They were asked to reflect on such issues as whether forward infantry should always expect to be overrun by armour and whether a successful defence should be based on a spirit of revenge. Practical skills, such as giving and receiving orders by radio, which was still largely a novelty, were also practised. The brigade war diary also reflected that the TEWT had helped foster the working relationship between the three COs and their brigadier, which was as critical as the command relationships within the battalions.[105]

The newly established Command and Staff School (C&SS) in Sydney played a limited role in the training of the new AIF COs. Only Eather of the 2/1st Battalion is recorded as having attended a course there following his appointment. Following the declaration of war the school's resources were devoted primarily to the training of company commanders, and this emphasis could be seen as the beginnings of command succession planning in the 2nd AIF. Of the 18 infantry captains and majors who attended a special company commanders' course held for AIF officers at C&SS in December 1939, 11 would eventually command battalions.[106] Sixteen infantry officers attended the next course, held in January 1940, and seven were subsequently appointed to battalion commands.[107] The general standard of tactical ability of the AIF's officers at this stage can be extrapolated from the report on the December course: 'On the whole the standard was good, but somewhat uneven owing to the fact that a number of officers started with very little tactical background.'[108]

The battalions of the 6th and 7th Divisions generally embarked for service within three to six months of their formation. Training had usually progressed to company-level exercises and, in some cases, rudimentary battalion exercises, but as an 18th Brigade report from April 1940 demonstrates, battalions were far from battle ready when they left Australia's shores. The major weakness observed on the 18th Brigade's exercise related to command in its battalions. Control was often lost, leading to 'subunits getting out of hand, delays and stickiness': 'As soon as a shot was fired the whole B[attalio]n at once went to ground.'[109]

This situation was only to be expected. The efforts of a CO and his staff in these early months had been focused on the establishment of 'discipline, efficiency and proper system', and their own training would follow now that they had an efficient, responsive organisation to train with. The consistency of the formation narratives in the battalion histories indicates a similar approach by COs, which must be attributed to their common backgrounds in the militia. It was unspectacularly effective. All of the original COs still held their commands at embarkation. They oversaw the transformation of varied groups of men into cohesive units – exemplified at one extreme by the disciplined march-pasts staged in Sydney and Melbourne in early 1940 and at the other by the troops' penchant to fight the members of other AIF units or the militia. As the historian of the 2/4th Battalion concluded his story of its formation, 'so a battalion was born, reared and presented as a complete, well-dressed and orderly unit',[110] but much training was needed.

The required training, particularly the development of effective commanders, could only occur overseas. Although the AIF battalions had been structured in accordance with the most recent British organisation, which will be discussed in the following chapter, they were still essentially 'hollow' units, lacking specialist equipment, modern support weapons and motorised transport. This equipment would be provided from British sources once the battalions were deployed overseas and it would only be then that their COs would have the chance to fully prepare themselves and their battalions for modern war.

PREPARING FOR OPERATIONS

Once the AIF began arriving in the Middle East and, in the case of the 18th Brigade, Great Britain, battalion training returned to basics to refresh individual skills after the long sea voyage and to introduce troops to new weapons and equipment as they became available. From this point, battalion training again progressed through various of levels of collective training until finally the battalions were exercising as part of a division and their COs were being tested in all of the facets of their command.

During this period COs were being tutored and tested by their superiors and attempts were being made to redress their weaknesses. For instance, the 16th Brigade war diary notes that in one exercise Brigadier Allen had to reposition a company of the 2/1st Battalion that Ken Eather had deployed along a crest and left silhouetted against the sky and thus vulnerable to observation and fire.[111] A month later the same war diary noted that Allen was pleased with the increasing confidence of his COs.[112] This training also served as a basic test of aptitude for command on the battlefield. By the end of 1940 two COs had been replaced. Leaving little doubt as to what his superiors thought of his abilities, Percival Parsons – described as a 'pleasant, hard-working trier'[113] – was posted from the 2/4th Battalion to become inspector of the AIF's canteens.[114] In December, Thomas Cook of the 2/5th was replaced after being captured on several occasions and having his badly sited headquarters overrun during a divisional exercise.[115] Wootten also relinquished command of the 2/2nd to oversee the establishment of an organisation to train reinforcement soldiers and officers arriving from Australia. There is no indication of doubts about his tactical ability, but later recollections of his lack of mobility as a brigade commander would suggest that his weight might still have been a concern.

Immediately on the arrival of the first battalions of the AIF in the Middle East in February 1940, places at British Army schools had been allocated to instruct the Australians in the most recent practices and doctrine of their principal ally. Most of the newly arrived COs were identified for additional training. Between October and November 1940, six COs were dispatched to the Senior Officers Wing of the Middle East Tactical School (METS) at Helwan, south of Cairo. It is notable that five of these officers – England, Godfrey, Louch, Mitchell and Wootten – were First World War veterans, presumably sent to update their tactical knowledge. The sixth CO sent to METS at this time was Eather, and earlier remarks regarding his performance in June might provide an insight as to why.

British Army schools in the Middle East, METS foremost among them, were to become an important element in the regeneration of the AIF's

'A pleasant, hard-working trier.' Lieutenant Colonel Percival Parsons, original commander of the 2/4th Battalion, was the first 2nd AIF infantry CO to be replaced as part of the process of preparing the battalions for battle. He continued to serve throughout the war in a variety of administrative positions. (AWM Neg. 001599/10)

command structure, and their influence can be traced even beyond the departure of the last of the Australian battalions from the Middle East in February 1943. A quarter of all of the AMF's infantry COs, including ten future brigadiers, completed either the company commanders or senior officers courses at METS. Another six future COs attended the Middle East Staff School (MESS) at Haifa in Palestine and 19 the Combined Operations Training Centre (CTC) at Kabrit in Egypt.[116] Demonstrating how the British Army schools formed part of a progressive training system for AIF officers, at least 11 of the officers previously listed as having completed the December 1939 and January 1940 company commanders courses at C&SS also completed courses at METS, MESS and CTC.

The senior officers syllabus at METS was an intensive unit command course and demanded a high standard of application from its students. In addition to teaching the latest British tactical doctrine it also dealt with leadership and discipline, battalion administration, the formulation and conduct of unit training programs, the tactics and organisation of enemy and allied forces, and such practical skills as navigation, the delivery of orders, the use of radio equipment and the conduct of reconnaissance under tactical conditions. The course's emphasis was on the conduct of combined arms operations in open terrain, although warfare in urban, mountainous and forested areas was also addressed.[117] Methods of tuition were varied and included lectures, discussions, TEWTs, and small and large-scale demonstrations.[118] Initially five weeks long, it was the most extensive military course that most Australian officers had completed in their careers. The few extant Australian assessments of METS rate it highly; Paul Cullen, who attended the company commanders course, described the tuition as 'excellent'.[119] The courses, however, were only as good as the doctrine on which they were founded. Précis from the senior officers courses conducted in November–December 1940 and July–August 1941 contain many of the weaknesses and misconceptions that bedevilled British tactics and operational procedures throughout 1941 and early 1942.[120]

The two sets of précis also demonstrate that METS courses developed as experience grew. The tactics of trench warfare featured in the courses conducted at the end of 1940, but had gone from the syllabus by mid-1941. The outline of the air force cooperation lecture presented in August of that year reflected the experience of recent operations, including those in Syria and Lebanon, which had concluded only in the previous month. During 1941 the length of the course was extended from five to seven weeks and, as the level of operational experience of the students attending METS rose, greater scope was provided for them to share the insights of that experience with other students.[121] The instructors were similarly experienced. The chief instructor of the senior officers course at METS at the end of 1941 was Adrian Hamilton, who had commanded the 2nd Battalion, the Black Watch with distinction on Crete. John Field, who by this time had commanded the 2/12th during operations at Tobruk, was impressed by Hamilton's tactical knowledge and his abilities as an instructor.[122] Although there are no extant notes or précis from Australian students attending the senior officers course in 1942, improvements in the syllabus and methods of instruction presumably continued as during early 1942 a concerted effort was made to integrate operational analysis,

An important element in the regeneration of the AIF's command structure: four Australian officers (from left to right), Captains Walter Hiscock, Donald Cleland and Arthur Oldham and Major Rudolph Bierwirth, attend a lecture at the Middle East Staff School, Haifa, Palestine, 17 April 1940. British Army schools such as the Middle East Staff School and the Middle East Tactical School were critical to the development of command ability in the 2nd AIF. Bierwirth was one of seven future battalion COs who attended the Staff School. (AWM Neg. 001494)

doctrine and training within the British and Dominion forces in the Middle East.[123]

As 1940 drew to a close, this centralised system of command and tactical training had yet to have a significant influence on the COs of the AIF but, as we have seen, replacements were already needed. The officers appointed to replace Parsons, Cook and Wootten represented the beginnings of a new, younger, more thoroughly trained, command cohort. Ivan Dougherty, the 33-year-old 2IC of the 2/2nd Battalion, took command of the 2/4th, and Fred Chilton, another of the 2/2nd's officers assumed command of that battalion. Chilton was 35 and a graduate of C&SS and MESS. In the 2/5th Cook was replaced by the austere 2IC of the 2/6th Battalion, Hugh Wrigley. A First World War veteran, Wrigley had a little more in common with his predecessor than did Dougherty or Chilton, but at 42 he was still younger than the average and took command fresh from the senior officers course at METS.

By December 1940 the battalions of the 6th Division had undergone a year of training and were fit and enthusiastic. Many officers subsequently reflected that this was the time when the AIF was at the 'top of its form' and had a 'dash and keenness' it would never regain.[124] At the end of 1940 Savige was obviously confident of the abilities of his brigade. An entry in the brigade war diary in late November noted that the work carried out by the battalions on exercise was 'of a very high order' and that their commanders acted with 'skill and confidence'.[125]

Such assessments of the battalions and their COs need to be tempered by more sober judgements. Eather later judged his battalion to have been 'fully trained', even by the standards of 1944, but conceded that it would have improved with a further six months training. He also reflected that it took the experience of their first campaign to round off the battalion.[126] After serving throughout the war as a battalion adjutant, 2iC and CO, and as a staff officer at brigade and divisional levels, Don Jackson reflected that the battalions of the 6th Division had much of their field training 'once over lightly'. He continued that 'in general the basic ploys of night operations, attack, defence and withdrawal had been exercised but once or twice, and battalions went into formation exercises tactically raw'.[127] Jackson's comments also raise the question of just how much opportunity the COs had had to hone their skills. In December 1940 the 6th Division had yet to experience battle and the sober proficiency that this would produce. Just as it would for their soldiers, battle would prove to be the final test for the COs. It would demonstrate whether the correct balance had been struck between youth and experience and how efficient were the methods put in place for the AIF to regenerate its command structure.

CHAPTER 3

'WE WERE LEARNING THEN'
The AIF's Mediterranean campaigns, 1941

The Mediterranean campaigns of 1941 were the AMF's command nursery in the Second World War, and it was here that its operational prowess in later years was founded. Seven of the Australian brigadiers involved would go on to become divisional commanders and of the 68 battalion COs, 25 would subsequently command brigades. All up, 150 COs or future COs served in the Mediterranean theatre – 93 per cent of the officers who would command battalions on operations throughout the war.

The commencement of operations in January 1941 marked the emergence of a meritocracy among Australian COs. Effectively cut off from the rest of the AMF, the AIF had to develop the ability to regenerate its command structure from within in order to withstand the inevitable casualties of battle. We have already seen how the AIF, following its arrival in the Middle East, had begun to identify and replace those unfitted to the rigours of operational command and to train and promote those with demonstrated ability. Each fortnight lists of officers observed to be ready for command were forwarded to AIF HQ; formation commanders had to certify that they would accept the officers listed as COs in their own formations.[1] Although there was still room for patronage networks to operate, the experience of battle would become the principal arbiter of an officer's fitness to command. Indeed, by the time they embarked for Greece in April 1941 only three of the 6th Division's battalions were commanded by the officers who had overseen their formation.

Map 3.1: The Mediterranean theatre

Three AIF divisions fought during 1941: the 6th, the 7th and the 9th, which had been raised in the United Kingdom in October 1940. (The organisation of these three divisions, which was not finalised until February 1941, is shown in figure 3.1.) They did so in five very distinct campaigns: the 6th Division in the ad hoc conquest and pursuit of the Italian Army in Libya in January and February, and the ill-conceived and poorly resourced defence of Greece and Crete in April and May; the 9th Division in the harried retreat to Tobruk where it was subsequently besieged from April to October, reinforced for several months by the 18th Brigade; and the 7th wresting the craggy hilltops of Lebanon from the Vichy French in June and July. Yet despite the diverse nature of these operations, the challenges facing the battalion COs were quite similar. In action they found the existing command system, and in particular its communications, inadequate for maintaining control over increasingly fast-moving, dispersed and complex operations. They had to develop means to manage this situation, as well as adapt their tactical training to the realities of the battlefield. Battalions were never at full-strength nor fully equipped, and battle quickly revealed the shortcomings both in the

THE AIF'S MEDITERRANEAN CAMPAIGNS, 1941 79

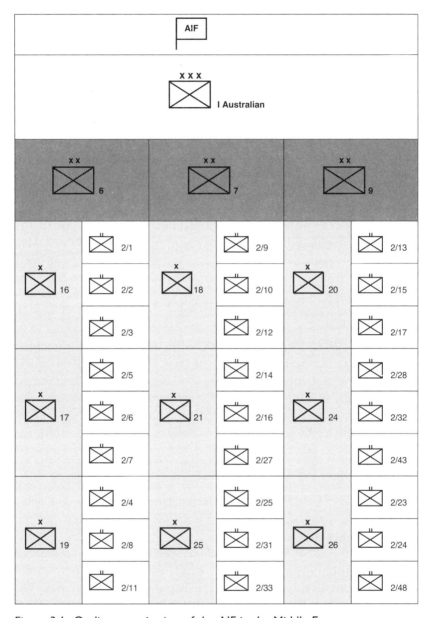

Figure 3.1: Outline organisation of the AIF in the Middle East

battalion organisation and in the tactics that had been developed for its employment.

THE AIF INFANTRY BATTALION

In *Twenty Thousand Thieves*, Eric Lambert described the thoughts of his fictional AIF CO, Lieutenant Colonel Fitzroy, as he watched his battalion move forward along a desert road: 'For some reason, the sight of the convoy creeping past in the early dawn had given him a perception of their enormous strength, and some foreknowledge of the destruction or defiance of which they would be capable. Under his command these men had been turned from a mass of raw civilians into a formidable machine.'[2] Lambert's description of the battalion as a machine is appropriate. Its subunits, each with different characteristics, capabilities and roles, were its constituent parts. These parts were mutually dependent, and their actions had to be directed to complement each other and achieve the unity of purpose inherent in the battalion's mission. This function was fulfilled by the battalion's command system of which the CO was an integral part.

The 2nd AIF was originally organised in accordance with interwar establishments that were little different from those of the First World War. The British Army, however, had implemented a new organisation for the infantry division and its constituent battalions in 1938. The new pattern division was completely motorised, and its infantry component consisted of three brigades of three battalions. When first formed, the 6th Division was still dependent on horsed transport and consisted of three brigades of four battalions, radically different from their British equivalents. Recognising the difficulties that would be presented by trying to integrate the Australian division into a larger British force, Blamey recommended to the Military Board that the 6th Division be restructured to reflect the new British organisation.[3]

Henceforth, all AIF infantry battalions were organised in accordance with the British 1938 establishment, which would ultimately form the basic model for all the AMF's front-line battalions during the war. The new organisation reduced the strength of the battalion from 902 to 792, a loss of manpower supposedly offset by an increase in firepower.[4] As will be shown, however, this increase was more theoretical than real. In outline, the 1938 pattern infantry battalion consisted of four rifle companies,[5] a headquarters company of six specialist platoons and a battalion headquarters (the complete organisation is depicted in figure 3.2).

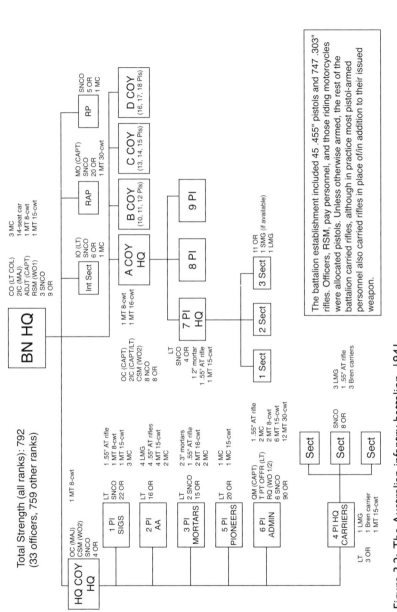

Figure 3.2: The Australian infantry battalion, 1941

Foundation of the battalion: a Bren light machine gun in action in a forward position on the Tobruk perimeter, Libya, August 1941. As the principal weapon of the battalion's 36 rifle sections, as well as the carrier and anti-aircraft platoons, the Bren was the foundation of its combat power. The Bren was a robust, reliable weapon, but lacked the range, sustained rate of fire and suppressive capability of larger machine guns like the Vickers. (AWM Neg. 009510)

The main fighting strength of the battalion was its four rifle companies. These were the smallest subunits in the battalion considered capable of conducting independent operations for short periods and were the CO's principal manoeuvre elements. Each rifle company was commanded by a captain and consisted of three platoons. The platoon, in turn, was made up of three sections. It was commanded by a lieutenant with a five-man headquarters: a sergeant as 2iC, a two-man 2-inch mortar team, a runner and a batman. Every officer in the battalion was provided with a batman. This was partially a legacy of the notion that an officer was also a gentleman and hence needed a servant, but was also based on the principle that officers looking after their own needs could not devote their attentions to those of their men.[6] The 2-inch mortar was a new addition to the infantry's arsenal and was intended to provide the platoon with its own indirect fire support.[7] Each platoon was also equipped with a Boys anti-tank rifle. Although the platoon was still expected to march into battle, it was allocated a small truck to carry its crew-served weapons, ammunition, entrenching equipment and the soldiers' packs and greatcoats.

The section was the battalion's most basic tactical element and consisted of between 9 and 11 soldiers.[8] Its primary weapon was the Bren light machine gun. British doctrine envisaged the section as being indivisible – it either fired its weapons or it moved – and it was essentially a support system for the Bren; rifles were regarded principally as short-range defensive weapons.[9] The platoon formed the basis for fire and movement, the practice whereby one element provides supporting fire to enable another to move, and the essence of all infantry tactics. It has been argued that this concept placed the British section – and hence the British battalion – at a significant disadvantage compared to those of the German Army, which, although still built around a single light machine gun, were trained to conduct fire and movement within the section.[10] British doctrine, however, did make limited provision for fire and movement to be conducted within sections, and there is ample anecdotal evidence that Australian sections were trained in this practice and employed it on operations.[11]

Three platoons of Headquarters Company provided the CO with his own fire support. The Anti-Aircraft Platoon, equipped with four Brens mounted on trucks, was responsible for the defence of the battalion from air attack while moving and for the defence of battalion headquarters when stationary. The platoon also had a secondary role to provide the CO with a mobile reserve.[12] The Carrier Platoon had a similar role. Equipped with ten light armoured vehicles with mounted Bren guns, it provided fire support, roughly equivalent to a rifle company, which could be moved rapidly across the battlefield with some degree of protection. The Bren gun carriers, however, were not light tanks, and it was expected that the Brens would be fired from the ground. The mobility of the Bren gun carrier, however, suited it to employment in the protection of the battalion's flanks and rear, and as a means to conduct local reconnaissance.[13] Two 3-inch mortars, operated by the Mortar Platoon, were the CO's third means of fire support.

A further two platoons of Headquarters Company facilitated the operations of the fighting elements. The Pioneer Platoon provided a light engineering capability and, if required, could also fight as conventional infantry. The Administrative Platoon – with more than 90 personnel, the largest platoon in the battalion – was responsible for the battalion's transport, supplies and personnel administration.[14] The only personnel-related function not controlled by the Administrative Platoon was the battalion's medical support, the Regimental Aid Post (RAP), which was part of battalion headquarters. The combination of administration and combat support elements in the Headquarters Company made it a large, complex and often unwieldy command, and it was therefore entrusted to the second major in the battalion.[15]

In battle, the administrative elements of the battalion were split into three echelons: 'F', 'A' and 'B'. 'F' – fighting – consisted of a small number of vehicles essential for communications, liaison and the carriage of support weapons, and hence required at the very forefront of battle. 'A' Echelon was composed of the remainder of personnel and transport required for the immediate sustenance of the battalion in action, such as the platoon, RAP and ammunition vehicles. It was usually deployed out of the immediate battle area, but the CO had to balance its safety against his ability to protect it and the need for a rapid response from it. 'B' Echelon consisted of non-essential personnel, stores and vehicles, and a left out of battle (LOB) group from which the battalion could be rebuilt if it suffered heavy casualties.[16] 'B' Echelons were deployed away from the battle area for their protection and were often concentrated under brigade control.

On operations, the CO's principal role was to assimilate information and on the basis of it make and implement timely decisions as to the employment of all of these subunits; that is, the command functions of decision-making and control. This process was outlined in *Field Service Regulations*, Volume 2, the keystone document of British operational doctrine, which stated that a CO's first responsibility on the battlefield was to make certain he had a thorough understanding of his battalion's mission and its object. As the quality of their decisions depended on the information on which they were based, COs were reminded that it was their 'essential duty' to make the 'best possible arrangements' to provide themselves with information relating to the enemy, conditions in the area of operations and the position and state of friendly troops, including his own. Noting the preoccupation of forward troops with the fighting on their front, such arrangements needed to be implemented parallel to the chain of command.[17] Within battalion headquarters there was an intelligence section with the role of collecting and analysing information for the CO.[18] The intelligence officer, a lieutenant, was described as the CO's 'extra pair of eyes'[19] and was expected to deliver a candid assessment of the battlefield situation. Intelligence officers were advised: '. . . any encouragement given to the wishful thinking of your Commander out of mistaken loyalty is the greatest disservice you can render him.'[20] The liaison officers whom battalions exchanged with brigade headquarters, and often supporting units as well, were another important component of a CO's independent information network.[21]

Field Service Regulations also emphasised the value of personal reconnaissance by the CO.[22] 'Notes for Commanding Officers' warned, however, that personal reconnaissance should not be conducted 'at the expense of tactical handling'.[23] A much later précis issued at Australia's Land Headquarters (LHQ) Tactical School was more blunt. A CO, it observed, could not command his battalion 'when dead, or even when pinned down by hostile SA fire'.[24] Ultimately, *Field Service Regulations* left the CO to determine the best place for himself during the course of a battle and directed he place himself where he could 'best control the course of the action, remembering that at the crisis personal example and leadership are the best means to ensure success'.[25] It was emphasised that wherever he ventured a CO needed to be readily available and hence have access to his own means of communication. He also needed to ensure that a suitably qualified officer was left to make decisions and manoeuvre the battalion in his absence.[26] Although the 2iC was the reserve commander and thus earmarked to fulfil this role, his administrative role often saw him deployed with B Echelon as the commander of the LOB group. Thus, in practice, the battle 2iC was often the OC of Headquarters Company – the junior major in the battalion.[27]

Emphasising that the CO was just one part of the battalion's command system, several forums were convened in which he could exchange information and receive advice from his staff and subordinate commanders before issuing his orders. The first of these groups was the Commanders or Reconnaissance (R) Group, comprising the officers who accompanied the CO on his personal reconnaissance once the battalion had been issued orders and hence was also the CO's primary advisory group during the formulation of his plan.[28] Doctrine suggested that it comprise the CO, the intelligence officer or the adjutant, possibly the signals officer and the carrier platoon commander, and the commanders of any supporting units.[29] In practice, COs would often include one or more of their company commanders in their R Group.[30] This saved the company commander(s) from having to conduct their own reconnaissance later, provided a sounding board for the CO's ideas and ensured that another senior officer had knowledge of the battalion's area of operations. The CO's other major forum in the battalion was the Subordinate Commanders' Group, more commonly known as the Orders, or 'O', Group. Having formulated his plan, it was here that a CO passed on the orders to make it happen.

The O Group included the company commanders, the commanders of the specialist platoons, the intelligence officer, the regimental medical officer and any attached liaison officers. Doctrine emphasised that orders within a battalion should always be passed verbally, principally in the interests of speed, and that there should be few occasions when detailed written orders were required.[31]

When making plans, COs were advised to keep things simple. Simplicity facilitated the understanding of subordinates, made command of the operations easier and lessened the chances that something would go wrong. Orders were expected to be 'clear and definite' and passed to all ranks 'so that every senior leader and man [knew] exactly what part he [was] to play'.[32] COs were advised to make maximum use of surprise in their plans and use the ground to its best advantage. Offensive action was emphasised as the only way to ultimately defeat the enemy, which, combined with Australian First World War folklore, resulted in an often blind adherence to aggressive tactics. As a result of the experience of the First World War, the application of fire support was at the heart of British doctrine, and *Field Service Regulations* urged that any plan, whether offensive or defensive, 'should be made in terms of fire power rather than of men'.[33] Thus, COs were expected to employ any available supporting arms to their full potential. Early in the war, however, as will become apparent, the structure of the British and Dominion armies, and in particular that of the battalion, was not necessarily in keeping with this basic tactical doctrine.

THE BATTALION'S SUITABILITY FOR BATTLE

Although Australian COs in the Mediterranean theatre had considerable destructive power at their disposal, their battalions had several critical weaknesses. These were incongruous with a tactical doctrine that emphasised the employment of firepower and left the battalions vulnerable on a combined arms battlefield. They were revealed quickly by the AIF's operational experience, just as they had been earlier apparent during the British Army's operations in France.[34] The AMF, however, was surprisingly slow to modify the structure of the battalion, even though commanders seem to have been regularly consulted as to its strengths and its shortcomings.

The greatest weakness of the 1938 pattern battalion was its paucity of heavy-calibre supporting weapons, which meant a CO was beholden to supporting units he did not always directly control when his troops encountered opposition that could not be readily dealt with by small arms. With only two 3-inch mortars, a CO could provide simultaneous fire support for only two of his companies and was considerably outgunned by German battalions equipped with six mortars of similar calibre. Particularly in the mountainous terrain of Greece and Lebanon, where the movement and employment of field artillery was limited, the dearth of 3-inch mortars was sorely felt and COs recommended an immediate increase in their numbers.[35]

The investigations of the Bartholomew Committee convened to examine the experience of the BEF in France had already recommended that the number of 3-inch mortars in a battalion be increased, but this change was delayed awaiting the production of sufficient stockpiles of ammunition.[36] The number of 3-inch barrels in the 1938 pattern battalion had been reduced due to introduction of the 2-inch mortar at platoon level, but this weapon did not have the same range or destructive power and the CO did not exercise direct command over platoon mortars.[37] The siege of Tobruk also revealed shortcomings in the 3-inch mortar. Its range was half that of the 3000 metres of which the Italian and German 81mm mortars were capable. This meant battalion mortars had to be positioned among the most forward positions, making them vulnerable to attack and drawing unnecessary fire down on the posts.[38]

In a similar fashion to the mortar, the Vickers medium machine gun had been removed from the new battalion organisation because of the increase in automatic firepower provided at platoon level by the Bren. Although eminently suitable for employment by infantry on the move, the Bren was simply not capable of delivering the sustained volume of fire, and suppressing the same large area, that the belt-fed Vickers could.[39] This was critical both when supporting an attack and when holding defensive positions. The 4000-metre range of the Vickers, compared to 1500 metres of which the Bren was capable, made it particularly useful in defence. Under the new organisation, Vickers guns were concentrated in special machine gun battalions, the subunits of which would be available to support battalion operations.[40] The priority of fire support, however, was allocated by the superior commander, which meant it would not always be available to the battalion CO. After the campaign in Greece, Ivan Dougherty reported that machine gun support was 'not satisfactory' and that 'the old system of machine gun companies should be re-established'.[41]

The requirement for longer-ranging automatic weapons in the battalion is readily apparent in the numerous stories recounted in battalion histories of the large quantities of captured automatic weapons employed by Australian battalions from the battle of Bardia (3–5 January 1941) onwards. In Greece, where battalions were often strung out along hillsides with the aim of engaging the Germans as they crossed valleys, the lack of Vickers guns, with their long range and high volume of fire, was sorely felt. Fred Chilton's 2/2nd Battalion was thus handicapped at Pinios Gorge on 18 April 1941. Although German troops could be observed advancing down the opposite slope, they could not be engaged forward of the river that was the battalion's last defensive barrier because they were beyond Bren gun range.[42]

The .5 inch Boys anti-tank rifle was the most inadequate weapon in the battalion's armoury. It was heavy, cumbersome and effective only at the shortest of ranges against light tanks and armoured cars.[43] On occasion it was employed in a harassing role over long ranges, and in some battalions it was to experience a renaissance during the Pacific War as an effective means of dealing with Japanese snipers, but in its primary role it was effectively useless. Hence the battalion CO was also without a means to deal with enemy tanks. Much has been made of the tactics of the infantry of the 2/17th Battalion during the German attack on Tobruk on 13 April 1941. When tanks penetrated between their perimeter posts the Australians sat tight and waited for the following infantry. Devoid of an effective anti-tank weapon, there is little else that they could have done.

The presence of Italian tanks had persuaded some COs as early as the battle of Bardia of the need for their own effective anti-tank capability. After that battle, Arthur Godfrey formed an anti-tank platoon within the 2/6th Battalion employing captured Italian 47mm Breda anti-tank guns.[44] Godfrey might have envisaged a dual role for these guns because at Bardia he had personally directed the fire of a 2-pounder gun from the brigade anti-tank company against Italian emplacements.[45] This experience highlights another weakness in the supporting weapons of the 1938 pattern battalion: the absence of a 'bunker buster'.[46]

COMMUNICATIONS AND CONTROL

The 'great weakness' of Australian battalions in 1941, according to Tom Daly, who as brigade major of the 18th Brigade was in a good position to observe, was their 'poor' communications, which 'by modern standards were almost primitive'.[47] The infantry battalion's signals platoon provided the CO with the principal means by which he exercised his command in the form of runners, dispatch riders, field telephones and visual signalling equipment. Wireless sets, with a range of approximately five kilometres, had been added to the equipment of Australian battalions in 1940, but these sets were not issued to the battalions of the 6th Division until after the Greek campaign and were among the extensive list of equipment shortages that plagued the 9th Division during the siege of Tobruk. It was a deficiency sorely felt.[48]

The dearth of wireless communications made the exercise of command difficult. This was particularly the case during the 6th Division's advance into Libya where the wide expanse of the desert, combined with the relatively small number of troops employed, dictated that COs disperse their companies over wide frontages. Field telephone lines were vulnerable to both artillery fire and the movement of vehicles and could not be laid or repaired fast enough to maintain pace with advancing infantry.[49] COs had to balance their need for knowledge of the situation as a whole and their responsibilities as part of the larger command chain against their ability to personally supervise the execution of orders and intervene at critical junctures.

A variety of means were employed to try to overcome these command difficulties. Some COs anticipated where the friction of the battlefield had the greatest potential to derail their plans and sought to position themselves where they could take immediate action if required. On the evening before the attack on Bardia, the divisional transport that was to move the 2/2nd Battalion to its assembly area failed to arrive, which had the potential to 'abort the whole attack'. Fred Chilton, however, had anticipated difficulties with this rendezvous and made sure he was there to supervise. When the required transport did not arrive, he was able to order all stores be offloaded from his own vehicles, which shuttled the battalion to the assembly area with little time to spare.[50] The next morning, Ken Eather, who had similarly positioned himself at the site of the 2/1st's break in to the Italian perimeter because it was he where he anticipated most difficulties, also intervened to keep the attack running to plan. Confusion and casualties prevented the engineers from blowing gaps in the wire. Eather, being on the spot, was able to reorganise the engineers, have the wire blown, then stand in the cover of the anti-tank ditch directing his platoons towards the gaps in the wire with his walking stick.[51]

At times, COs opted to remain mobile, roaming the battlefield in motor vehicles. In some instances, particularly in flat open areas where a battalion was reasonably close together and moving quickly, it allowed COs to react immediately to the changing tactical situation. The use of

Keep the attack running to plan: Lieutenant Colonel Fred Chilton, CO 2/2nd Battalion (centre), consulting Brigadier Cyril Lomax, commander of the 16th British Brigade, in the lead-up to the battle of Bardia, Libya, 27 December 1940. To try to minimise the friction of battle Chilton briefed his men with a large sand model, then sought to position himself at the points on the battlefield where the plan was most likely to come under strain. (AWM Neg. 005265)

Bren gun carriers by Ivan Dougherty and John Mitchell during the 19th Brigade's rapid thrust into the heart of the Italian defences at Tobruk on 21 January was a prime example of this.[52] Remaining mobile, however, also had the potential to lessen a CO's control. Without wireless equipment they could only communicate with – and therefore command – the troops whom they could see. At Bardia, shelling caused Tom Louch to abandon his car, leaving him stranded on the battlefield. He later reflected that if anything had gone wrong he would have been uncontactable and 'powerless to help': 'My place was at B[attalio]n HQ somewhere in the rear: and that is where I should have been.'[53]

Other COs exercised more restraint, remaining in a central location to the rear of their forward troops. This allowed them to maximise the effectiveness of their intelligence staff, liaison officers and field telephones to maintain their situational awareness and exercise their command. They were still able to make trips forward when required, usually to resolve confusion or contradictions in information received, or to solve an immediate tactical crisis.[54] The strengths of such an approach are readily apparent in Theo Walker's efforts to regain control of the 17th Brigade's floundering attack south of Bardia on the evening of 3 January. Walker established his headquarters in an abandoned Italian bunker, then employed his staff to start making sense of the confusion. Although two companies remained lost for much of the night, Walker's envoys were always able to return directly to his position, and were readily able to contact the two companies most heavily engaged.[55] Writing of a conference he had with Walker before issuing orders to renew the attack, Savige noted, 'I . . . obtained from Walker the present situation on his front and his dispositions which, as would be expected of Walker, were known, and he had a good grasp of the situation.'[56]

Static command posts were particularly suited to the defensive operations mounted at Tobruk and in Greece, although a 24th Brigade report did note that they needed to be sited well forward.[57] At Tobruk, battalion command posts were generally connected with field telephones to all their platoons, and often to posts of section strength or smaller, including observation and listening posts outside the perimeter, providing COs with a good degree of situational awareness and a ready response from supporting artillery.[58] These communications depended heavily on captured Italian signal wire, however, as the battalions' standard allocation of 13 kilometres proved barely adequate.[59] Similar difficulties were encountered in Greece. The only way Chilton was able to keep in touch with his widely dispersed companies at Pinios Gorge was via a 'long hoarded' cache of Italian line.[60] Without his own purloined stash, Donald Lamb, whose 2/3rd Battalion was deployed to Chilton's left, had to rely solely on runners for his communications.[61]

The stable command environment that prevailed through most of the siege of Tobruk seems to have fostered a 'dugout' mentality that, as will be discussed later, served to alienate at least one CO from his troops. The command system was only as strong as a D5 phone cable, as was demonstrated by the confusion that reigned in the 2/24th Battalion's headquarters during the German attack on Ras el Medauuar on 30 April– 1 May,[62] or the supposed loss of 2/12th Battalion posts on 16 May, which, it was subsequently discovered, never left Australian hands.[63]

When large-scale attacks were mounted beyond the perimeter COs generally sought to command them from their command posts inside it and did not venture forward.[64] This left them with a limited capacity to exercise any influence over events once their troops had crossed the start

A limited capacity to exercise any influence over events once the troops had crossed the start line: Lieutenant Colonel John Crawford, CO of the 2/17th Battalion, monitoring the progress of a patrol outside the Tobruk perimeter from a forward observation post near Post 32. (AWM Neg. 020207)

line.[65] A 2/12th Battalion report noted that during the battalion's attack on the German salient at Tobruk on the night of 3 May the information reaching the CO was 'meagre' despite line parties having advanced with the assaulting troops.[66] Field eventually resorted to relocating himself to the headquarters of the 2/23rd Battalion, which was holding the sector of the perimeter adjacent to where his attack was being mounted. From here he could glean details of his battalion's progress from reports phoned in from the 2/23rd's forward posts.[67] Some COs occupied observation posts as an alternative means of staying in touch with the unfolding battle, but visibility was usually impeded by dust, smoke, mist, darkness or the landscape.[68]

Therefore, with electronic means of communication unreliable, if not non-existent, and personal observation problematic, COs relied on their runners and liaison officers to pass orders and information. The potential for losing control was vividly demonstrated by the attack made on the Tobruk salient by the 2/9th and 2/12th Battalions on the night of 3 May. When heavy German fire forced the assaulting troops to ground, the momentum of the attack was lost and the programmed artillery barrage crept away. With little knowledge of where their troops were, neither CO could safely adjust the artillery. The forward troops were to have indicated their position by the use of flares but, as John Field noted in his diary, 'the Boche' also 'filled the air with flares of every kind'.[69] The 9th Division report on the Tobruk siege clearly states that both the COs and the company commanders were unable to exercise command of their troops and 'the battle became a bitter hand to hand struggle'.[70] A recurring feature of the battles for the German salient at Tobruk was the exploits of isolated parties of Australian infantry going unnoticed by battalion headquarters and suffering for want of reinforcement and artillery support.[71]

Although much more mobile owing to their offensive nature, battalion operations in Lebanon suffered from similar control problems.[72] As the temporary loss of two companies of the 2/27th Battalion on their approach march to Miyeoumiye on 13 June demonstrated,[73] in Lebanon it was precipitous terrain rather than wide open spaces that dispersed battalion subunits. Wide valleys and wadis made the use of not only line but also of liaison personnel problematic. In one instance a battalion that was only three kilometres across a valley from brigade headquarters required a dispatch rider run of 58 kilometres to maintain communications.[74] The practical result of these difficulties was well illustrated by a report on air cooperation during the battle of Damour, which noted that the average time required for the passage of information from a battalion to a brigade headquarters was 90 minutes – far too long for the timely provision of air support to forward troops.[75]

The battalions of the 7th Division embarked on the Syrian campaign equipped with four portable wireless sets, two short of the authorised battalion establishment at the time.[76] The results were mediocre. There was still not enough wireless equipment for it to be issued to all companies and specialist platoons. The 108 Mk 1 sets were found to be finicky and unreliable, and in at least one battalion all four were inoperable throughout the campaign.[77] Inexperienced operators, the effects of mountainous terrain on reception, and the 108 set's relatively short range all complicated its usage.[78]

Where functional equipment was combined with well-trained personnel, however, the beginnings of a much more responsive battalion command system can be seen. Selwyn Porter, CO of the 2/31st Battalion, attributed a great deal of his unit's success in Lebanon to its use of wireless communications. As one of the battalions raised in the United Kingdom, it had been issued at an early date with enough British-manufactured

No. 18 radio sets, widely considered to be superior to the Australian-made 108 sets, to equip each company, and its signallers were well trained in their use. In Lebanon the battalion instituted a series of procedures to maintain control by wireless.[79] Hence, Porter wrote: '. . . control did not suffer by reason of subunits moving beyond range of telephone lines, as was so often necessary in the mountainous country surrounding Jezzine. Company commanders could hold personal conversations with each other and with me; and this meant a great deal when wide gaps occurred in the course of operations.'[80]

THE NOVICE ARMY

With COs' capacity to influence battles in progress often limited, the most extensive opportunity for them to affect their outcome was before they began. The AIF's preparations for battle in 1941 were of uneven quality. At times they were careful and methodical and at others hasty and slipshod, betraying an army that was still developing and refining its skills. Battalion war diaries and reports show COs' personal battle drill to have been reasonably good and generally in accordance with the procedures laid out in *Field Service Regulations*: they took the time to conduct their own reconnaissance, consult with supporting arms commanders and formulate their plans. The time spent in such preparation varied with the tactical situation. For the counter-attacks mounted against the German salient at Tobruk COs had no more than a few hours for their reconnaissance,[81] whereas before the deliberate attack on Damour, Murray Moten of the 2/27th Battalion was able to spend at least two days closely observing the Vichy positions he was to assault.[82]

Once the start line was crossed, the conduct of the battle essentially passed into the hands of subordinate commanders, and therefore the provision of adequate time and resources for them to conduct their own battle procedure was critical. At Bardia, Chilton had his intelligence staff construct a sand model of the 2/2nd Battalion's objectives, which he used to walk his company commanders through 'every step of the operation'. A timetable was also set in train so that the company commanders and platoon commanders, in succession, could complete their reconnaissance and plans, and brief their subordinates.[83]

The hurried battle procedure that preceded the costly defeats of the counter-attacks on the Tobruk salient compared unfavourably to the 2/2nd's thorough preparation at Bardia. Before both the 2/48th's attack on 1 May and that of the 18th Brigade on 3 May, no subunit commanders were given time to reconnoitre the ground over which they would advance nor the objectives they would assault, yet they were expected to do so in the dark with no prior rehearsal.[84] As Field mused afterwards, 'a night attack in this desert country with its lack of directing features is a hard task even if there had of [sic] been plenty of time for daylight observation'.[85] In both attacks companies were unable to maintain direction, faulty maps resulted in critical navigation errors, platoons could not locate objectives, and casualties mounted owing to other positions that had not been identified.[86]

The performance of COs cannot be isolated from the larger command system of which they were a part. Company commanders were also developing their skills at the same time as their COs.[87] Although much of the 2/9th Battalion's troubles during the 18th Brigade's attack on 3 May can be attributed to a lack of time for subunit battle procedure, its CO, James Martin, noted that the inexperience of his subordinate officers contributed to the loss of control during the battalion's attack: 'At night the hardest part of an off[ice]r's job is to keep control of his com[man]d. In this case, some of our officers require much more experience yet, and are inclined to "lead" small attacks instead of controlling the whole force.'[88] The Australian COs were just one component of a novice army. Just as their success depended on the ability of their subordinates, it also depended on the judgement and skill of their superiors, and the nature of command relationships that were still evolving. The experience of the 2/3rd Battalion at Bardia demonstrates the effect of inexperience at several levels of command. Divisional orders had specified the battalion occupy a linear position on its objectives, requiring all four rifle companies[89] and thus allowing none of the depth on which a sound defence, in the case of counter-attack, could have been based. On the afternoon of 3 January Italian tanks pierced the 2/3rd's thin line and proceeded to threaten its headquarters. This counter-attack also revealed Vivien England's poor control of his battalion and its supporting arms. England had to race the tanks across the desert in his Bren gun carrier to improvise the defence of battalion headquarters, which was saved only by the chance arrival of a platoon of anti-tank guns.[90] It was later realised that the divisional orders for the attack on Bardia were far too prescriptive; in a very telling comment Brigadier 'Tubby' Allen later reflected, 'It didn't matter – we were *learning* then.'[91]

Likewise, during the siege of Tobruk, unrealistic timings and faulty tactical appreciations at brigade and divisional levels were as much to blame for the failure of the counter-attacks against the German salient as were shortcomings in the battalions. Too much was expected of the battalions in these attacks,[92] but the almost naive enthusiasm of the COs also influenced the results. There is a definite sense that some COs felt they had something to prove in their first battles. Although Field was concerned about the short time he had to prepare for the 3 May attack on the German salient, and was asked by Major General Morshead if there would 'be time to get it all down to the men', he did not argue strongly for a postponement of the attack. He subsequently wrote, 'the decision to go on with it was taken and Martin and I were eager for the chance'.[93] Similarly, there is no sign that Bernard Evans sought further time to plan and rehearse the 2/23rd's attack on the morning of 17 May after information was received that the posts it was intended to recapture were still in Australian hands. As Barton Maughan noted, Evans task had changed from 'primarily recapturing lost ground' to 'attacking a fully developed enemy position held since 1st May'.[94] Five hours before H-hour, with his final orders issued, his troops on the move to their forming-up place, and his headquarters closed, Evans switched objectives and modified artillery programs. An hour before the attack began two new posts were also added to its objectives.[95]

It is difficult to separate enthusiasm from the aggressive spirit Morshead sought to engender in his COs at Tobruk, and that, more broadly, lay at heart of Australian tactical doctrine. Evans was an aggressive and forthright character, described as both a man of drive and as being cocky and arrogant.[96] On 30 April he had sought permission to counter-attack the burgeoning German salient in the sector of the neighbouring 2/24th Battalion. The 2/24th's own CO, Alan 'Jiggy' Spowers, had also exhibited a potentially dangerous mix of aggression, desperation and tactical naivety – seeking to combat German armour with Bren gun carriers and Boys rifles.[97]

Perhaps the most disastrous example of a CO seeking to make his mark was the muddle-headed attack mounted across the Wadi Muatered, on the southern edge of the Bardia defences, by the 2/6th Battalion under the command of Arthur Godfrey. Ordered only to stage a demonstration by fire, Godfrey, with the acquiescence of Brigadier Savige, mounted a full-scale battalion attack. Savige has intimated that Godfrey's soldierly pride was offended by the relatively passive role he was ordered to play in the 2nd AIF's first battle.[98] The assault was pure folly, and was described by one 2/6th platoon commander as neither well planned nor well executed.[99] B, C and D Companies attacked Italian posts perched on

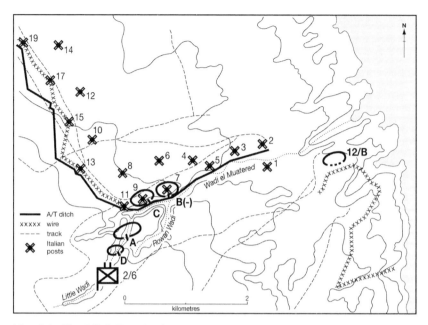

Map 3.2: The 2/6th Battalion's positions on the Wadi Muatered at nightfall on 3 January 1941.

the far bank of the steep wadi, and all succeeded in forcing an entry. B and C clung to their gains; D was heavily counter-attacked and forced to withdraw into the wadi, which was covered by well-sited machine guns and registered by artillery. Forty-eight men from D Company crossed the start line; only four returned to the battalion that day.[100] Godfrey was left with two companies marooned and vulnerable across the fire-swept wadi. There was no depth to their position, no artillery support was available after the initial assault and, although Godfrey still had a company in reserve, there was little chance of getting it across.

The 2/6th remained pinned down along the Muatered until the final collapse of Italian resistance at Bardia on the morning of 5 January. In keeping with his impulsive conduct, Godfrey had planned to renew the attack on Post 11 with A Company on the night of 4 January,[101] despite having no artillery support and having long since lost the element of surprise. Fortunately, after further reconnaissance, the plan was abandoned. His attitude is illuminated by his refusal to try to outflank the Italian positions after a platoon, operating independently to the east, had found a clear and practicable route to do so. Godfrey told his battle 2iC, Major

George Smith, that 'he had no —— intention of giving up what he'd gained'. Smith was told to lie to Brigadier Savige that it was impossible to extricate the companies in contact when this could have been easily achieved under cover of darkness.[102] The fight along the Wadi Muatered, the intended demonstration by fire, resulted in the heaviest casualties sustained by any unit at Bardia.[103]

Godfrey's insubordination was unusual. Implicit in the command culture of the AIF in the first campaigns was a strict adherence to orders. We have already seen how Field and Martin were reluctant to argue at Tobruk. Although there was no boyish enthusiasm apparent in his actions, Victor Windeyer had similarly fallen into line a few days earlier, despite believing the 2/48th's counter-attack against the German salient to be 'impossible'.[104] Those who questioned orders were rapidly put in their place. Raising doubts about the chances of success of a renewed assault on the stubborn post R7 after the failure of his battalion's attack on 17 May, Evans was told by Brigadier Wootten: 'Listen, laddie, when I want your advice I'll ask for it.'[105] Although not commanding a battalion at the time, Tom Daly was similarly censured after writing an appreciation critical of Morshead's plans to renew the attack on the salient. He was told: 'If that's the way you feel, Daly, then your best place is back in Australia.'[106] Such attitudes were not confined to the 9th Division. Brigadier Stevens recalled some 'holy rows' that he had in Lebanon with the acting CO of the 2/16th Battalion, Major Arnold Potts, who frequently queried his orders. Potts was eventually told that there would be no discussion of his orders and he was simply 'to obey them'. Stevens's conclusion to this anecdote illuminates what was expected of a good CO: 'And he did [obey orders]! He was a grand soldier.'[107]

The role confidence, trust and personality played in the command relationships discussed above is difficult to quantify. It could be surmised that there was a distinct lack of both trust and confidence in the command decisions relating to the salient attacks at Tobruk. Morshead and Wootten seem to have been reluctant to accept the judgements of their subordinates, and they, in turn, lacked the confidence to stand by them. A degree of professional arrogance could also have been involved. Wootten and Morshead had seen service at relatively high rank during the First World War; Evans, Field, Martin and Windeyer had not. Personality clashes might also have played a part. Wootten and Evans were both confident and forthright, Potts was 'pigheaded'[108] and Stevens 'highly strung'.[109]

There is evidence of more flexible and productive command relationships. During the initial phase of the Lebanon operations, when a rapid advance, destroying Vichy opposition in encounter battles, was still central to the Australian plan, the 2/31st Battalion encountered a 'strong and cleverly designed system of defences' south of Khirbe.[110] With his companies unable to move forward, then forced to yield ground, Porter initially vacillated, perhaps reluctant to communicate to higher headquarters his growing belief that the 'encounter phase of our advance had ended almost as soon as it had begun',[111] and that a deliberate attack would be needed. He ordered his forward companies to mount a quick attack, but cancelled it after the troops had crossed their start line.[112] Porter then sought the time and artillery support needed to conduct a deliberate attack. The recollection of one company commander that 'hot words' then passed along the chain of command indicate that Porter's request was not popular.[113] He stood his ground, and eventually the Commander Royal Artillery (CRA) of the 7th Division came forward to discuss his artillery requirements. The two officers conducted a joint reconnaissance, allowing Porter to justify his attack plans; two regiments of artillery were subsequently provided in support, and a deliberate attack was approved.[114] It took place before dawn on 11 June and, with an overwhelming volume of artillery fire, succeeded with little difficulty and few casualties.

Trust and mutual respect were at the heart of productive command relationships. In the 19th Brigade, Brigadier Horace Robertson trusted Iven Dougherty absolutely,[115] and the nature of their command relationship is evident in the latter's recollections of the second day of the battle for Tobruk:

> I told the Bde Comdr I was not happy about moving across the open ground until the Battery at 40834342, and this position generally, was silenced . . . The Brigadier agreed with me and, as 8 Bn could not get to this position in time, he agreed to my suggestion that I take the position in my objective. Then I found the 'I' Tanks could not get up to me by 1000 hrs, & told the Brigadier I'd have to wait for them – he agreed.[116]

Dougherty clearly had sufficient confidence in his own judgement to query his orders and Robertson an equal amount of trust in his subordinate's abilities to engage in a dialogue and revise the orders.

Trust, however, could also be abused. Savige believed he and Godfrey had 'mutual faith and trust in each other' and that the latter would not deliberately disobey orders.[117] As a result Savige gave Godfrey considerable freedom of action at Bardia, resulting in the Wadi Muatered fiasco – a case of pride undermining trust and obscuring a CO's prime responsibility to understand his mission.

Leadership in battle

In addition to the intellectual demands of the battlefield, each CO had to face his own personal challenges as a leader, a follower and an individual on the battlefield. Some had much to prove, and sometimes sought to do so at the expense of their troops. For others, responsibility for the lives of their men was a heavy burden. COs had to control their own doubts and fears and reassure and motivate troops who were often inexperienced or facing seemingly impossible odds. What was valued most by both a CO's subordinates and his superiors was a calm and capable hand amidst the chaos of battle. These qualities, however, were almost contradictory to the other defining characteristics of Australian command culture inherited from the ethos of the original AIF: vigorous leadership based on aggression and personal example.

The pithy citations for the 18 DSOs awarded to Australian infantry COs during the campaigns of 1941 highlight some of the dominant characteristics of the battlefield leadership they exhibited.[118] Embodied in these citations – originally drafted at brigade level and ultimately approved by Blamey – are the institutional expectations of how a CO should behave. Fifteen commend their recipients for personal courage or gallantry, implying that they were in a position far enough forward to be at serious risk from enemy fire; three are specifically mentioned for regularly moving among their most forward troops and several others for daring personal reconnaissance. This personal bravery, however, was not recognised simply for its own sake. Thirteen COs were lauded for inspiring or setting an example for their men. The language of the citations – 'dash', 'determination', 'energy', 'willpower', 'fearlessness', 'indefatigable' – speaks of a command culture founded on vigorous and aggressive personal leadership.

The commitment to such a leadership style often led COs to take risks that could be seen as imprudent for men of their responsibility. The casualty figures – only one CO was killed, after his vehicle struck a mine, during the campaigns of 1941[119] – belie the risks they often took. On the second day of the battle for Bardia, for instance, Eather responded to a halt in the advance of one of the 2/1st Battalion's companies by taking up a rifle and leading a Bren gun team forward under fire, 'thus restoring the confidence of his men' and allowing the advance to continue.[120]

Often such acts of leadership were inspired by genuine camaraderie. On 11 April 1941, at Vevi, in Greece, Dougherty set off in a Bren gun carrier to personally deliver revised withdrawal orders he was worried had not been received by one of his company commanders. Later, on Crete, Dougherty would remain ashore waiting for a missing section while the destroyer evacuating the 2/4th Battalion from Heraklion slipped its hawsers and began to pull away from the wharf.[121] As the 2/7th Battalion prepared for its wild bayonet charge at 42nd Street, also on Crete, Theo Walker sought to borrow a rifle and bayonet to take part, telling one of his soldiers, 'I know I should not be here, but I must be in this with you boys.'[122] He was dissuaded only by the wise counsel of his 2iC.[123] Walker's loyalty to his men was further demonstrated while they waited for evacuation from Crete. Having boarded a barge, he was told it was the last that would depart that night, which he already knew was the last of the evacuation. Walker jumped into the water and waded ashore to rejoin his men.

In some cases there was a fine line between personal example and pride or bravado. Soon after the capture of Merdjayoun in Lebanon, Porter was shot in the buttocks during an O Group he conducted on an exposed hilltop; he had scoffed at a warning from one of his company commanders that it was dangerous for them to gather there.[124] In Greece, Hugh Wrigley clung tenaciously to command of the 2/6th Battalion's rearguard operations around Domokos and Brallos, despite suffering badly from a wound sustained at Bardia. He was found at Brallos 'ashen, shivering, his arm clenched tightly over his chest', beneath his shirt a 'ghastly looking wound, seeping blood and pus'.[125] Wrigley had persevered for fear of losing his battalion. It was a brave action, but not a wise one. As one of his company commanders recalled: 'no wonder he made a few very stupid decisions. Sick men should not lead battalions when razor sharp decisions are to be made.'[126]

Vigorous personal example was not the only component of successful battlefield leadership. Lurking among the superlatives of the DSO citations there are also such phrases as 'sound planning', 'steadying influence' and 'coolness under fire'. They hint at another style of leadership, that of quiet, unshakeable proficiency, exemplified by Fred Chilton. In the midst of the 2/2nd Battalion's desperate fight for Pinios Gorge on 18 April 1941, the New Zealand general Bernard Freyberg spoke to Chilton by field telephone. Brigadier Allen recalled that the conversation was 'just like a business call', and after hanging up, Freyberg reported, 'You have a fine man up there, Allen. He's as [cool] as a cucumber.'[127] Paul Cullen, one of Chilton's company commanders in both Libya and Greece, recalled that he was a 'gentle fellow', idealistic, intelligent and diligent.[128] These characteristics were the foundation of his leadership on the battlefield. The battalion history noted that Chilton's 'skill as a careful planner, his modest unassuming bearing and his personal endurance and steadfastness' earned him the 'respect of every man in the battalion'.[129] This passage demonstrates that Chilton's tactical and organisational skills were trusted by the men to give them the best possible chance in battle.

Dougherty exhibited a similar demeanour and was likewise highly respected by his battalion. It was observed that he was 'quiet' and 'almost boyish' but that his ability, judgement and integrity had the confidence of the men.[130] The calm, predictable facade of a CO proved a steadying influence, and some unlikely characters excelled in the role. Walker was known to the 2/7th Battalion as 'Myrtle' because he was a 'rather mild looking man with a high voice and a delicate way of standing';[131] the battalion history recalled he was 'short, dapper, [and] precisely spoken'.[132] Initially, it seems that some of the AIF's senior officers doubted Walker's ability to command.[133] His reputation grew through the Libyan, Greece and Crete campaigns: so much so that following his capture in May 1941 Blamey remarked on his loss in a letter to Lieutenant General Vernon Sturdee, the CGS, describing him as a 'fine officer' and a 'serious loss' to the AIF.[134]

Accounts of the 2/7th's operations in Greece and Crete are replete with descriptions of Walker's stolid leadership and his insistence on maintaining standards of dress, bearing and soldierly behaviour despite the increasingly desperate situations the battalion encountered.[135] Walker rarely displayed his emotions, but his troops recall him visiting every section to exchange a few quiet words of encouragement the night before they joined battle on Crete.[136] Walker's leadership was validated by the behaviour of his battalion. The 2/7th participated in the withdrawal through Greece and was shipwrecked following its evacuation, but still remained a viable fighting unit throughout the campaign on Crete; its last action was a three-day rearguard stand, with little ammunition, rations or water, in the hills above Sfakia. Ordered once again to prepare for evacuation, Walker marched his battalion on to Sfakia beach, 'pushed and jostled' by the 'seething mass of disorganised, hysterical men' seeking to make their own escape.[137] There it stood, quiet and orderly in ranks, waiting to be called to embark; the battalion was seen to be still standing thus when the last ship sailed.[138] Afterwards, given the option to escape in small parties or attempt to fight their way out as a formed unit, the men of

'Myrtle': Lieutenant Colonel Theo Walker, CO of the 2/7th Battalion. Walker was the original CO of the 2/7th, and it seems his mild manners and high-pitched voice led some to doubt his suitability to command infantry soldiers. A quietly capable performance throughout the Western Desert and Greece campaigns, however, led to General Blamey describing his capture in the latter as a serious loss to the AIF. (AWM Neg. 001441)

the 2/7th opted for the latter.[139] In the robustly masculine environment of the infantry battalion, the acceptance of Myrtle Walker speaks highly of the effectiveness of his leadership. Writing of the last day at Sfakia, the 2/7th's intelligence officer remarked: 'Our stocks slumped, but the CO was a wonderful leader and gave us all that little extra "punch" we needed.'[140]

Just as out of the line men expected higher standards of conduct from their COs than they observed themselves, in battle they expected them to be in control, and extreme displays of emotion were viewed suspiciously. Although only a company commander in the 2/6th Battalion at the time, Frederick Wood had a dubious reputation for being 'excitable' and occasionally erratic in the conduct of his command.[141] Donald Lamb, CO of the 2/3rd Battalion, acquired a similar reputation when he stopped a New Zealand anti-tank unit fleeing Pinios Gorge at pistol point; he was thereafter remembered for 'having lost his block'.[142] The calm demeanour of COs like Chilton, Dougherty and Walker could perhaps be seen as a mask of command. One of Walker's soldiers recognised the inner strength that was required to wear it: 'What a load that man carried when I am sure that he could see little to be cheerful about as he came to give us a few words of cheer.' He reflected further on Walker's calm demeanour: 'I think of Lt. Colonel Walker as the example of courage of a different kind.'[143]

COs had their own doubts and experienced their own fears. In common with most former COs, Chilton maintained that he was generally too busy to be scared, but it is likely that he sought to suppress his emotions as part of his job. He recalled feeling 'not too good' when the first men under his command were killed and 'bloody awful' as his battalion was dismembered around him at Pinios Gorge, a day he described as the worst of his life. Casualties and defeat were 'part of the game'; the emotions associated with them, however, had to be controlled: 'If people can't take it, then that's the finish of them.'[144] Godfrey's diary suggests that after the 2/6th's experience at Bardia he was much more sensitive to casualties. After the first day at Tobruk he wrote: 'Thank God, casualties light.' He followed this, the next day, with: 'No more casualties – wonderful. This is my main theme song. I hate to lose one of these boys.'[145] Potts, writing to his wife, was not reticent about admitting the fear he felt in Lebanon: 'To say I was windy is to grossly understate the situation and I only had to take the mob down the hill and start them off up the other side.' A First World War veteran, Potts was familiar with war and wrote, even before he had left Australia, that he was 'a hell of a coward about some things away down at the bottom'. He wondered if the happiness of married life had not made him 'too soft for soldiering', and his subsequent letters home show him striving to keep Potts the man and Potts the commander separate – longing to be 'cuddled' and being 'down in the dumps' as a result of casualties, yet still having to pick himself up and lead troops into battle.[146]

What is immediately apparent in the leadership exercised by Australian COs on the battlefields of the Mediterranean theatre is the reliance on referent and expert power. Leadership based on coercive, legitimate or reward power seems to have had little direct influence, supporting the notion of a situational, or contingency, theory of leadership in the AIF. It could perhaps even be described as an evolutionary theory whereby COs were forced initially to depend on external power bases until they had the opportunity to build alternative power bases that depended on their own skill and personality through demonstrated behaviour. Although the personalities and actions of COs varied widely, the essential elements of successful battlefield leadership in 1941 were a willingness to share the same risks as their men, genuine camaraderie with them and demonstrated tactical ability. These elements were not always held or exercised in equal measure, and strength in one could compensate for weakness in another. Shared experience also served to strengthen the bonds between leader and led, unifying the formal group. Ultimately, trust was the key. Asked to comment on how the morale of the 2/2nd Battalion stood up to the onslaught at Pinios Gorge, Chilton replied that the battalion had been together a long time and endured a 'grim, physical campaign': 'They knew it wasn't my fault.'[147]

The demise of the First World War generation

Writing after the Greek campaign, D. R. C. Boileau, who commanded the 1st Battalion, the Sherwood Rangers, beside the 2/4th and 2/8th Battalions during the battle for Vevi Pass, reflected that physical fitness was 'the first essential of a competent officer, and that a leader who cannot function 100% when dog tired is not only useless, but a definite menace to all concerned'.[148] The fighting of 1941 revealed that many of the First World War veterans commanding Australian battalions lacked such physical and mental stamina, and they rapidly succumbed to the strain of prolonged campaigning. Sixteen of the original COs of the 27 battalions that saw service in the Mediterranean theatre were First World War veterans older than 40; 11 were replaced for failing health, a lack of physical fitness or doubts about their ability to adapt to contemporary tactical thinking. The head of the 6th Division's medical services, Colonel Harold Disher, reported that the best officers in the fighting in Libya and Greece were those younger than 40 'who had not been handicapped by experience in the last war'.[149] Several COs seem to have been conscious of their fitness levels. In Palestine, Alfred Baxter-Cox and one of his company commanders endured 'liquid days' in an attempt to shed a few pounds,[150] and on campaign in Libya Godfrey confided in his diary: 'My girth has decreased some considerable amount which is all to the good.'[151]

John Mitchell of the 2/8th provides a case study of the deterioration of the performance of the veteran COs. As CO of the 8th Battalion during the First World War, Mitchell had been awarded a DSO and bar, and lauded for his energy, tactical nous and courage.[152] Mitchell was a contradictory character. He was regarded as difficult by some of his peers and superiors – Brigadier George Vasey described him as a 'nasty

'Too old for the wear and tear of modern war': Lieutenant Colonel John Mitchell, CO of the 2/8th Battalion. Mitchell was the oldest of the 2nd AIF's original infantry COs and one of its more controversial officers, being variously regarded as a 'nasty piece of work' and a 'kindly gentleman'. The pressure of the withdrawal in Greece proved too much for Mitchell, and he was returned to Australia for 'special duties'. (AWM Neg. 006125)

piece of work' – but well regarded by his troops, one of whom described him as a 'deep-voiced, kindly gentleman', a 'considerate' commander and 'father' to the battalion.[153] As early as October 1940 Brigadier Robertson had sought his removal, believing him too old and too set in his ways to command a battalion in a modern war. This was just one factor in the mutual antagonism that prevailed between the two. Both had strong personalities, and Mitchell was also a critic of the PMF of which Robertson was a product. Mitchell, presumably, remained in command thanks to the support of Savige and Blamey.[154]

Mitchell commanded the 2/8th competently, although not spectacularly, during its advance along the Libyan coast in early 1941.[155] One of his officers lauded his aggression and dash during the advance out of Derna on 31 January 1941,[156] and on one occasion Mitchell launched a silent night attack to capture an Italian rearguard position that dominated the route of the advance. This action anticipated later orders from Robertson and 'ensured a footing on the high ground from which to advance next day' that was 'most valuable'.[157] At the next occupied escarpment, however, there are signs that Mitchell was slowing down and that his aggression was waning, giving the Italians time to fire several demolitions on the road.[158]

In Greece, the 2/8th was one of the first Australian units to fight. It met the Germans at Vevi on 12 April 1941 in mountain-top positions blasted by snow and sleet. It was a testing situation for Mitchell. His troops had had few opportunities for rest in the week preceding the battle, and the frontage he was required to defend was approximately 4000 metres long, three times longer than that usually allocated to a battalion.[159] Mitchell's actions at Vevi betray a CO yielding to the strain of command. He located his forward companies too far down the forward slopes of the battalion position and his headquarters too far back on the reverse slope.[160] The headquarters was well protected, but Mitchell had no direct observation of his front, and a visit to the companies required an 'arduous scramble' of up to an hour and a half, which he seems to have made on only one occasion to visit a single company.[161] There is no sign of the aggression he displayed in Libya. He failed to order patrols forward to secure the battalion position while his troops dug in, allowing German patrols to move right up to – and in some cases between – section posts; his counter-infiltration plan at night was to order his men to remain in their posts until first light and fire on any movement; and his piecemeal commitment of his reserve company to cover a gap between the 2/8th and the neighbouring 1st Battalion, the Sherwood Rangers, shows him acting in a completely reactive fashion to the German activity along his front.

The main German attack hit the 2/8th on the morning of 12 April. The battalion was able to hold its own for much of the day, and Mitchell did mount a successful counter-attack, but by late afternoon the withdrawal of both of the 2/8th's flanking units made its position untenable. Mitchell, ordered to hold his position until 1900, dithered and his control began to slip. He called a conference of his company commanders at 1700, but by this time the battalion was being heavily engaged on both flanks, their planned withdrawal route was blocked and communications with brigade headquarters had been cut. Mitchell remained at battalion headquarters to continue to monitor the situation and seek to re-establish communications with brigade and dispatched his adjutant to rendezvous with the company commanders; without the CO present, a decision was taken to withdraw immediately, each company in succession.[162] Before the withdrawal plans could be implemented, German tanks and infantry penetrated the battalion position, inter-communication broke down and companies withdrew piecemeal. Mitchell had no control over his battalion, and it was the adjutant who sought to take command.[163] Relentlessly harried throughout a

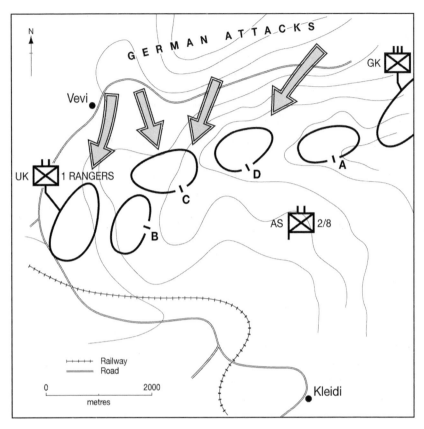

Map 3.3: The 2/8th Battalion's defence of Vevi, 12 April 1941

19-kilometre retreat, the remnants of the 2/8th reached the next rearguard position 'completely disorganised',[164] and a muster on 13 April revealed only 13 officers and 193 men present, only 50 still armed.[165] There is no account of Mitchell's actions during the withdrawal from Vevi,[166] evidence in itself of no outstanding act of leadership on his part. He appears only boarding a truck belonging to an anti-tank unit encountered along the way.[167] In a situation in which his troops were completing a gruelling withdrawal on foot and under fire, it was hardly an example of inspiring personal leadership.

At corps headquarters Blamey reacted with disbelief to reports of the destruction of the 2/8th. He remarked to Savige, 'They don't know Mitchell as you and I do.'[168] This reaction supports the notion that Mitchell owed his position to Blamey's patronage. Mitchell, however, was spent and his reputation tarnished beyond redemption. Vasey, now commanding the 19th Brigade, noted that after the withdrawal from Vevi Mitchell was 'completely exhausted'.[169] Within six weeks of returning to Palestine from Greece he was sacked at Blamey's instigation.

The arduous nature of the operations in Lebanon in June and July 1941 similarly sapped the strength and performance of several COs. In the 2/14th, Cannon struggled with a leg wound he sustained during the First World War and, with his decisiveness and aggression slipping, was eventually evacuated owing to exhaustion.[170] Lieutenant Colonel 'Bandy' MacDonald, aged 43 and appointed to replace Baxter-Cox in the 2/16th, also struggled.[171] Arnold Potts, at that time one of MacDonald's company commanders, confided to his wife that his CO 'ran himself down in his first show' and had to go out for a 'spell'. After ten days of acting command, Potts resented MacDonald's return, noting that he took a 'little nursing at times'.[172]

Efficient command relationships were also undermined by the apparent lack of physical stamina of William Crellin, CO of the 2/43rd Battalion at Tobruk. Crellin, aged 44 and a PMF officer and First World War veteran, rarely ventured beyond his headquarters and relied on his company commanders to pass on information at regular conferences. One has reflected that the 2/43rd was run 'just like a business' and that decisions were arrived at by a 'sharing of information and also suggestions, from any of us'.[173] To the men in the battalion's forward positions Crellin was an aloof and distant figure, unlike William Windeyer of the 2/48th Battalion or Bob Ogle of the 2/15th, who made regular nightly visits to the forward positions.[174] Not seriously tested at Tobruk, this laissez-faire command arrangement in the 2/43rd functioned satisfactorily but had lasting ramifications for the battalion. Without 'Daly to do the thinking', Verrier in the 2/10th also found himself out of his depth at Tobruk. As BM of the 18th Brigade, Daly was still in regular contact with his esteemed former CO: 'It was a bit much for him, he was getting too old.'[175] One of Verrier's officers was more critical: 'I don't think he knew anything. He may have been a good platoon sergeant in World War I . . . but I don't think he understood the tactics.'[176] The ultimate appraisal of the performance of Cannon, Crellin, MacDonald and Verrier is that none saw another campaign in command; all but Cannon were transferred by the end of November 1941.

Despite the unspectacular performance of such officers, there was a recognition in the upper echelons of the AMF that they had done what was expected of them. Suggesting to Blamey that it was 'time to take action to send home officers who are too old for the wear and tear of modern war', Brigadier Sydney Rowell, the Brigadier General Staff of I Corps, maintained that they had 'done a good job within their limits'.[177] Cook, sacked before his battalion even saw battle, was still awarded an OBE for his efforts in raising and training it.[178] Blamey sought to provide dignity for those removed from command and had them recalled to Australia for 'special duty' with 'all honour'.[179] Louch, Mitchell and Cannon were among those returned to Australia in this way, and it was felt they still had something to contribute to the training of troops there.[180] Other former COs were retained in the Middle East and appointed to training and base commands. Some performed outstandingly in these roles. Cook, for instance, earned a CBE commanding the Tobruk base area in the early months of the siege,[181] and Bandy MacDonald excelled as a trainer – by April 1945 he was the AMF's Deputy Director of Military Training.[182]

It should not be thought, however, that all of the veteran COs floundered. Arnold Potts succeeded Bandy MacDonald to command of the 2/16th Battalion. Described by Brigadier Stevens as a 'fine fighting soldier of World War I',[183] Potts had already proven his vigour and competence during his periods of acting command in Lebanon.[184] He would succeed Stevens to the command of the 21st Brigade and lead it into action on the Kokoda Track. Godfrey, removed from command of the 2/6th Battalion the day after he turned 45, was resurrected from the AIF Reinforcement Depot to assume command of the 24th Brigade at Tobruk and commanded it with distinction until he was killed at El Alamein.[185] Wootten likewise overcame his demise as a CO to command the 18th Brigade and then the 9th Division.

By the end of 1941 all three Australian divisions in the Middle East had seen action and were finally united as a single corps in northern Syria to ward off a possible German drive through Turkey, aimed at Arabian oil and the British base areas in Palestine. It was a time for recuperation and reflection. All three divisions had been involved in hard-fought campaigns, and the 6th and 9th had suffered particularly heavily, the first through large numbers of prisoners of war left behind in Greece and Crete, and the second through the physical privations of Tobruk.

The operational pause of the second half of 1941 saw further large batches of Australian officers dispatched to British Army schools. A third of the Australian graduates of the METS senior officers course completed it between August 1941 and February 1942, and three-quarters of Australian attendees at the Middle East Combined Operations Training Centre did so in the same period. The reflections of the Commander in Chief, Middle East, General Claude Auchinleck, after a visit to Syria in October 1941, reveal the Australians to be developing doctrine and procedures as a result of their battlefield experience. Auchinleck noted with a degree of concern that the Australians are 'tending to draw away from us in matters of principle and doctrine'. He went on, however, to comment, 'It is most obvious that they are anxious to learn all they can and also that they realise their lack of knowledge in certain directions.'[186] A further hint at the doctrinal experimentation going on in the AIF at the time is provided by the formation of an independent Bren gun carrier company in the 7th Division as a highly mobile, quick response force.[187] The benefits of such innovation would largely be denied to the 6th and 7th Divisions as in February 1942 they left the Middle East bound for a very different war against the Japanese.

By the time the 6th and 7th Divisions embarked to return to Australia, the benefits of training and experience had brought about a generational change in the AIF's COs. Fourteen new COs were appointed between July 1941 and February 1942, dropping the average age of COs across I Corps by almost five years to 39. Only six First World War veterans remained in command: only two were the original COs of their battalions, and three were destined never to see action again. The experience of the First World War veterans had been critical to raising the 2nd AIF, but it is clear that they had only ever been intended as watch-keepers until the new generation – trained in the tactics and procedures of modern warfare, blooded in battle and fit enough to resist its physical and mental strain – were ready to take command. The next chapter will show what they could achieve.

CHAPTER 4

DESERT EPILOGUE
El Alamein, 1942

The fighting at El Alamein between July and November 1942 was the culmination of the AIF's operational experience as part of the British and Dominion army in the Mediterranean theatre. With eight DSOs being awarded to the 14 COs who served in these battles, the last in the war in which Australian battalions fought as part of a full-strength infantry division employed as such,[1] an argument could be mounted that Alamein marked the high point of Australian battalion command. The Alamein battles were the only occasion on which Australian COs in the Mediterranean fought fully equipped, although not full-strength, battalions,[2] and they had more supporting arms under their direct command than they had ever had previously.

Alamein is the epilogue to the story of improvisation, learning and adaptation told in the previous chapter. The exercise of command at Alamein, and in particular the October battle, demonstrated a maturity and refinement that was absent in many of the engagements of 1941. Narratives of the battles do not indicate any of the rivalry that was sometimes apparent in earlier campaigns, and there is little evidence of COs seeking to prove themselves – there was not the need: four had already commanded their battalions in action, at least six had served as 2iCs, four had been decorated for bravery and at least six were graduates of METS. These were officers still prepared to take risks, but generally they were calculated risks firmly grounded in experience rather than the foolhardy risks of untutored enthusiasm or sheer desperation. This same experience allowed COs to judge where their presence was most necessary, and the blooming of the flexible command system that had been budding in 1941 gave them the mobility to do so. In any case, COs at Alamein tended to remain with a headquarters to the rear of their forward troops, and much greater control was facilitated by improvements in communications, more experienced troops and the development of drills to ensure that thorough battle procedure was carried out even when time was short.

THE 9TH DIVISION'S PREPARATIONS FOR BATTLE

The Australian battalions that fought at Alamein in October 1942 were the product of the AIF's cumulative experience in the Mediterranean theatres, including the most recent fighting in July and August. After the heavy casualties and defeats of July, the British Eighth Army, under the command of General Bernard Montgomery, placed an emphasis on discerning the lessons of operational experience. These were used to revise doctrine and unit organisations and to guide focused training programs.[3] At dusk on 21 October, Charles Weir addressed the 2/24th Battalion to reveal the plans for the coming battle. He had no doubt it would be equal to the trial ahead: 'We have the strength, we have the support, we have the experience, and, above all, we have the determination to succeed.'[4] Weir's words are indicative of the confidence that buoyed the Australian battalions on the eve of battle, and they also provide a succinct summary of the strengths that would see them victorious.

By the time the battalions of the 9th Division moved from Syria back to the Western Desert in mid-1942, several of the deficiencies in the organisation of the battalion had been remedied. Sufficient wireless equipment, of much improved quality, had been issued to allow for communications within the battalion between battalion headquarters, the companies and the specialist platoons.[5] The Mortar Platoon had been equipped with six 3-inch mortars early in 1942[6] and issued with ammunition capable of achieving ranges comparable to that of Axis weapons. Additionally, the mortars had been mounted in Bren gun carriers, allowing them to be rapidly deployed on the battlefield with some degree of protection.[7] An anti-tank platoon was also formed and initially equipped with a variety of non-standard ordnance.[8]

Further changes to battalion organisation resulted from the July battles at Alamein. The Australian infantry had rarely failed to capture German positions, but had often lacked the ability to hold them in the face of

A strong reserve of fire support: An Australian Vickers gun crew, El Alamein, Egypt, 26 November 1942. After the experience of Australian battalions at El Alamein in July, a machine gun platoon equipped with four Vickers medium machine guns was added to the battalion. Capable of sustained fire out to 4000 metres, the Vickers significantly increased the firepower under a CO's direct command, which would prove vital in the October battle. (AWM Neg. 013660)

determined German counter-attacks.[9] To compensate for their lack of a long-range sustained-fire machine gun all of the Australian battalions toted a diverse range of captured automatic weapons.[10] In August, four Vickers guns were issued to each battalion to form a machine gun platoon.[11] Most COs still opted to retain all of the captured weapons as well, demonstrating their preference for a strong reserve of fire support under their direct command. After the October battle, both corps and divisional reports would again emphasise the need for sustained-fire automatic weapons within a battalion, particularly in defensive operations, and recommend the machine gun platoon remain a permanent feature.[12] August also saw the issue of eight 2-pounders, made surplus when the divisional anti-tank regiment was re-equipped with 6-pounders, to each battalion to standardise the equipment of their anti-tank platoons.[13] Although the 2-pounder was a distinct improvement on the battalion's previous anti-tank armament, it struggled to do any real damage to German tanks except at close range – under 500 metres and then only against the relatively thin side armour; the 2-pounder was virtually incapable of penetrating the spaced frontal armour of German tanks.[14] After-action reports following the October battle would recommend that the 2-pounders in the battalions be replaced by the harder-hitting 6-pounders.[15] All round, the 'new broom' that Montgomery put through the Eighth Army ensured that, for perhaps the first time in the war, Australian COs would lead their units into battle with their full complement of arms and equipment.

In addition to their material strength, the battalions were further stiffened by the experience of their COs. The dominant experience of active service among the COs of the 9th Division was no longer the First World War – when the 9th Division moved south from Syria in June 1942 only two veteran COs remained[16] – but that of fighting the Germans, Italians and Vichy French. One of the most obvious manifestations of this experience was the way the COs trained their battalions. They knew the standards required of both individuals and subunits and oversaw rigorous training regimes to achieve them. As in earlier periods, COs had little influence over the broad training syllabi imposed by higher formations, but they did have a direct influence on the intensity and finesse of their implementation, and they were almost solely responsible for the training of their officers.

The COs of Alamein were characterised by their reputation as hard and realistic trainers. Bob Ogle of the 2/15th Battalion was described by his 2iC as a 'great trainer' who would have the battalion out 'all day and every day, going over sand hills doing all sorts of improbable things'.[17] Similarly, one of the 2/17th's company commanders, recalled that Noel Simpson 'tore into' the battalion like a 'shockwave' when he took command in March 1942. After six months in Tobruk and a long period on garrison duty in Syria, the battalion lacked experience in mobile offensive operations,[18] a situation that Simpson rapidly sought to rectify with a series of company and battalion exercises focusing on the advance and the attack.[19] During these exercises Simpson kept a close eye on his company commanders, routinely requiring them to justify their decisions and counselling them in a forthright manner when he considered them to have acted negligently or incorrectly.[20]

One of the AIF's most famous – or perhaps infamous – COs first earned his reputation with the training regime he imposed on the 2/48th Battalion in Syria. Heathcote Hammer, soon to earn the nicknames 'Tack' and 'Sledge', told the assembled battalion early in his tenure that its motto was to be 'hard as nails and driven by a Hammer'.[21] The nature of training in the 2/48th is exemplified by an account of a route march written by one 2/48th platoon commander: 'Chaps fell out, unconscious; all were nearly crying at the exhaustion; I was dry retching, and shivering with goose-flesh.'[22] Hammer's diary reveals that there was more to his training program than simply pushing his men to the edge of exhaustion. He was personally compiling the battalion's training schedule, delivering lectures and preparing training notes, and conducting TEWTs for his officers, which subsequently formed the basis of company and battalion exercises. Hammer also routinely conducted conferences with his officers after these exercises to discuss the lessons learned from them.[23]

The example of Hammer demonstrates how leadership should not be confused with simple popularity. Hammer and his regime were not popular. In addition to his training practices, his tongue was harsh and his discipline severe. His leadership was founded on the battlefield and the stocks of power created by his tactical ability and personal courage. One veteran recalled that after the restrained manner of Victor Windeyer, 'a lot of men' did not like Hammer's 'attitude', but he proved himself and his methods in the July fighting: 'as a battle commander he was really very good'.[24]

The sense of purpose engendered in the Eighth Army following Montgomery's arrival gave further direction to the efforts of the Australian COs. In his report for September, Bernard Evans commented that the month was 'probably the best from the tr[ainin]g point of view' in the history of the 2/23rd Battalion.[25] In the 2/48th, Hammer followed the lead of his higher headquarters and regularly promulgated discussions of the lessons learned from the exercises conducted throughout August and September,[26] which were based, in turn, on an analysis of the battalion's performance at Tel el Eisa in July.[27] The overriding emphasis of battalion – and indeed Eighth Army – training during this period was the development, refinement and inculcation of battle drills to allow the rapid mounting of operations, in a tactically sound fashion, with a minimum of preparatory orders.[28] Just as in 1941, operations in July had suffered from being hastily mounted. Evans outlined how battle drills related to the role of the CO in the introduction to a set of standing orders laid down for the 2/23rd Battalion in August: 'Battle drill must not be confused with tactics. [A] Bn Gp with good battle drill leaves its comd free to deal exclusively with the tactical situation, confident in the knowledge that all the normal precautions have already been taken by the automatic application of battle drill.'[29]

Following the heavy casualties among NCOs and junior officers in the July fighting, COs also needed to devote particular attention to the training of their replacements. Before the October battle Hammer instituted a special NCO syllabus in addition to the 2/48th's continuing training.[30] At the other end of the chain of command he adopted a battalion policy of always placing supporting weapons from both within and outside the battalion under the direct command of his company commanders. This was to make them familiar with the tactical handling of these weapons and foster close personal relationships with the supporting arms commanders.[31] Several COs took full advantage of the defensive routine of the Alamein front to develop the command abilities of their new officers.[32] This practice too was marked by the intensity engendered by the looming battle. Evans wrote that he adopted a 'very heavy patrolling policy' with 'the object of "breaking in" all ranks'. Up to 14 patrols went forward every night and 'bad patrols were ordered out again'.[33]

All of the training was ultimately tested and refined in a series of exercises that the divisional report on operations described as 'full dress rehearsals'; such rehearsals, conducted by Hammer on his own initiative, had already proved their worth in the 2/48th's successful attack on Tel el Eisa on 10 July.[34] A divisional training team conducted the rehearsal exercises in October,[35] allowing even the COs a chance to practise their role in the upcoming battle. The historian of the 2/13th Battalion concluded his narrative of this period thus: 'The Battalion had been coached along like an athlete in the hands of a wise trainer. It was battleworthy.'[36]

COs AND MOTIVATION

Envisaging the drawn-out slugging match that the second battle of El Alamein would become, both Montgomery and Morshead placed a strong emphasis on the role of battalion COs in maintaining the morale and determination of their troops. In a memorandum to Eighth Army COs, Montgomery emphasised that a unit had to be 'welded into a fighting machine, with the highest possible morale', and he left no doubt that the buck stopped with the CO. He directed that units be assembled regularly and addressed by their commanders.[37] Morshead adopted a similar tone when addressing his COs on 10 October, and emphasising their key role: 'It is on you here that so much depends, for as are the leaders so are their commands.' He urged them to force their will on their men. 'If you are enthusiastic, confident, determined, efficient', he stated, 'so will all those under you.'[38]

Several of the more charismatic COs sought to inspire their battalions on the eve of the battle with rousing orations. Demonstrating an affinity with the classics common to the well educated, Robert Turner, a barrister before the war, evoked Shakespeare's *Macbeth*, telling the 2/13th to 'Be bloody, bold and resolute'.[39] Evans of the 2/23rd and Weir of the 2/24th struck a similar note, each presaging the battle as one of the most decisive of war. Both, however, ultimately appealed to the group loyalty and pride of their men. Evans enjoined the 'old "mud and bloods"' not to let the division down, contrasting the soldiers assembled before him with the battalion's 'broken reeds' – those suffering from self-inflicted wounds and anxiety neurosis – most of whom he described as 'cast offs' from other battalions.[40] Weir similarly closed by daring his men to show weakness and face the shame that would follow: 'Finally, I say this. If there is a man among you who hasn't the guts to fight shoulder to shoulder with us, and who hopes in some way to dodge his share of the job, let him go and let no one stop him.'[41]

As Johnston and Stanley have remarked, the real influence of such orations on the troops is uncertain. Some might have been inspired while others were 'suspicious of rhetoric when action was demanded'.[42] Private Les Clothier of the 2/13th noted in his diary, without comment, that his CO had passed on a battle cry from Shakespeare. His recollections of Turner's address, and the details of the battle plan contained therein, are tinged with cynicism: 'Of course, all this is on paper & in theory. It is almost sure to be different with so much noise and confusion.'[43] Private Les Watkins recalled that Turner's invocation of Macbeth 'meant nothing' to the 'average soldier'. What spurred him and his mates on was the sense of being part of a much larger enterprise and, although they knew they would soon 'be in it right up to their necks', they appreciated 'being fully put in the picture, no bullshit, just plain hard facts'.[44]

The experience of an earlier engagement at Alamein would suggest that it was the 'no bullshit' approach that made the greatest impression on the men. Concluding his orders for the 2/28th Battalion's attack on Ruin Ridge on 26 July, Lew McCarter emphasised: 'We must get on to that ridge . . . We must not be distracted by fire from the right flank. By dawn all our troubles will be over.'[45] Sergeant John Kehoe later reflected on the influence of those words: 'One sentence in the CO's orders made a tremendous impact on me and I think motivated my actions for the rest of the night, "We must get on that ridge."'[46] The 2/28th did seize Ruin

'It is on you here that so much depends': COs of the 20th Brigade are briefed before the October battle of El Alamein. General Montgomery emphasised that the plan of the battle be systematically briefed down the chain of command so that every officer and man knew his part. (AWM Neg. P01614.018)

Ridge that night, but was surrounded and forced to surrender the next morning. The task of rebuilding the 2/28th from its LOB group and the few escapees from the disaster of 27 July fell to Jack Loughrey. The battalion history records his first battalion parade after Ruin Ridge: 'In a short, simple address he told his troops that a big responsibility rested on their shoulders. Their task was to rebuild the battalion, so that it could avenge the losses it had suffered and could play its part in helping to liberate all those who had been captured on Ruin Ridge. The sincerity of his appeal produced a spontaneous outburst of clapping.'[47] Some of the influence of the words of McCarter and Loughrey can perhaps be attributed to their congruence with the personalities of the two men. McCarter was a 'big, breezy' officer with an aggressive character[48] and a reputation as an outstanding fighter[49] – just the man to deliver the abrupt ultimatum he did. On the other hand, the battalion history recalls Loughrey as a man uncomfortable with military formality but who at the same time found it difficult to mix with the troops[50] – anything but his quiet, succinct appeal for his men's assistance would have been out of place.

COs on the battlefield at Alamein

Command practice during the October battle was quite similar from battalion to battalion, which indicates the influence of common doctrine and training. Morshead had directed that headquarters be sited as far forward as possible but warned that it was 'foolish' to site them too far forward or in tactically unsound positions.[51] In the advance, battalion headquarters, reflecting what was now standard practice at all levels of command throughout the Eighth Army, were divided into 'rear', 'main' and 'tactical' elements. The 2/48th's tactical headquarters or 'CO's Group' was typical. It comprised the CO and his principal staff officers – the adjutant, intelligence officer and his liaison officers to brigade headquarters – as well as a section strength protection party.[52] Practice varied between battalions, but the tactical headquarters also had access to a jeep and a Bren gun carrier to allow it to move rapidly across the battlefield. As Hammer noted, its position in battle was 'flexible',[53] but generally it moved to the rear of the forward companies, along the axis of advance.

This mobility, however, did not come at the expense of control as it had during the fighting of 1941. The battalion tactical headquarters was equipped with at least one No. 108/18 model wireless that could be carried by a man and used to communicate with the companies, the other elements of BHQ and the supporting arms. Depending on battalion practice, another such wireless might have been mounted in either the CO's jeep or carrier, as might a larger No. 101 set, used to speak directly to brigade headquarters.[54] Communications personnel in the 2/48th's Tactical HQ comprised the battalion signals officer, two signallers and four runners. BHQ Main moved to the rear of the reserve companies and joined the tactical group following the capture of the battalion's objectives. At Alamein, BHQ Rear remained with the battalion LOB groups and B Echelons in brigade areas outside the immediate battle area.

Although the fighting was often bitter, tactically, the engagements that comprised the October battle were relatively straightforward.[55] The troops had been well trained and briefed, and the employment of battle drills had the desired effect of speeding up the conduct of operations while at the same time reducing the amount of work required by a CO and his staff to mount them. The influence of battle drills can be seen in the abbreviated orders issued by Major George Colvin for the 2/13th Battalion's attack on the night of 24 October: 'The orders for the night were clear and simple. C and B forward, C on the right, composite company in

EL ALAMEIN, 1942   121

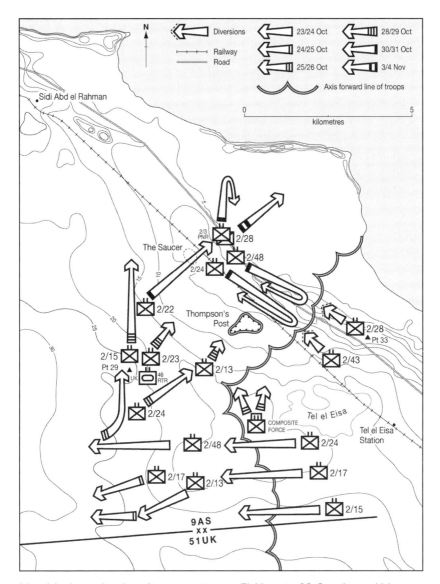

Map 4.1: Australian battalion operations at El Alamein, 23 October – 4 November 1942

reserve. Bearing of attack 271°, administration normal.'[56] The 2/48th's diarist reflected on just what the phrase 'administration normal' implied:

> It meant the preparation for the battle in detail – man loads, carrier packs. Arrangements for mines, wire, amm[unition], food, water, over head cover, sand bags, tools, arrangements for withdrawing A Tk guns, MMGs, Mortars, lifting the mines on our northern flank, and the holding of the present position while preparing to do battle in another attack. It said much of the battle drill and discipline of the unit . . . that these arrangements prior to the attack worked smoothly and without a hitch.[57]

Efficient and thorough battle drills meant there was little need for last-minute intervention on the part of COs, such as that discussed in the previous chapter. They were therefore able to exercise their commands in a mature, restrained fashion, monitoring the evolving battle but essentially leaving its conduct in the hands of their company commanders. The principal means of command was the wireless, the effectiveness of which was reported on in glowing terms in post-operational reports.[58] The 2/48th's report on the attack on Trig 29 on the night of 25 October noted that wireless 'had a marked effect on the rapidity of the attack, the employment of reserves to mop up enemy posts and permitted almost a perfect teamwork between CO and Coys in the conduct of the attack, mopping up and reorganisation'.[59] A series of lectures Hammer gave on his experience at Alamein outline the basics of a command philosophy that could be applied to many of the Australian COs there. He spoke of a CO needing to be 'hot on the trail' of information, not simply waiting for the reports from his company commanders, and using his principal staff officers to chase it.[60]

The only mention Hammer makes of a CO needing to move forward on his 'flat feet' was to check on the coordination of company positions following an attack, and the 2/28th's history has Jack Loughrey carrying out a similar role after his battalion's occupation of the Saucer on the morning of 1 November.[61] The concentration of COs on matters of overall coordination emphasises the reliance on the abilities of company commanders; providing them with support when required was one of a CO's primary roles. Hammer recommended that, if in doubt, a CO should 'stay put at Bn HQ', the position from which he could have the most influence: 'the Coy Comd howl for the old man – all kind of minor problems are cropping up and you make the Coy Comd happy if you can fix him up immediately.'[62]

Hammer offered some sound advice to commanders at all levels, counselling against rash heroics unless one wanted to 'catch the first bullet' and then not be able to command at all. 'Fight wisely', he urged, and 'keep Comds alive – they will come to the front when things are tough'.[63] COs in the October battle do seem to have fought wisely whenever able. Although the casualty rate among COs at Alamein was one of the highest of the war, the bulk of these casualties were the result not of small arms wounds sustained at the forefront of battle but of artillery fire and mines to the rear of the battalions' forward positions.[64] In accounts of the fighting up to 28 October there is little sign of dramatic acts of personal leadership by COs. Demonstrating the situational nature of battlefield leadership, however, as the operations became progressively more difficult owing to mounting casualties and strengthening opposition, COs readily stepped forward to rally their men and take hold of deteriorating tactical situations.

The same mixture of necessity, comradeship and bravado that was apparent in the actions of COs in 1941 is visible in the actions of COs on the battlefield at Alamein. On the night of 28 October the 2/23rd Battalion was ordered to make an attack in concert with the 46th Battalion, Royal Tank Regiment, on whose tanks the infantry would ride during their initial advance. The attack met with heavy enemy fire, the tanks lost their way and floundered among minefields, and the 2/23rd's platoons and companies were scattered and suffered heavy casualties – particularly those riding on the tanks. As the battle fell apart, Evans moved forward and personally directed the few remaining tanks onto enemy positions. On one occasion he led a tank onto an enemy post while standing in his jeep waving his revolver.[65] The destruction of the 2/23rd enraged and dismayed Evans. The camaraderie he shared with his men is clearly evident in his actions near the end of the attack: 'Colonel Evans threw his steel helmet on the ground and defiantly lit a cigarette without any attempt to shield the match. 'Well, boys, this looks like the end of the 2/23rd Battalion!' No one contradicted him. Calling on his men to form a single line, he led them, some 60 strong, in an attack against the main enemy position.'[66] The position, including six artillery pieces and 160 prisoners, was captured, and at that point Evans halted the advance and ordered the battalion to dig in.[67]

Charles Weir and 'Tack' Hammer conducted themselves in a similar manner on the night of 30 October when their weak battalions were committed to overly ambitious attacks against the strongest German positions on the Australian front.[68] Apparently seeking to lead by example, they

'Tack': Lieutenant Heathcote Hammer, CO 2/48th Battalion, congratulates Corporal James Hinson on the award of the Distinguished Conduct Medal, the second highest bravery decoration in the British and Dominion armies, Tel el Eisa, Egypt, 1 August 1942. The unrelenting nature of Hammer's training regime did not initially endear him to the battalion, but its experience in the July fighting at El Alamein proved his abilities and vindicated his methods. (AWM Neg. 024750)

positioned their headquarters further forward than had henceforth been the case; Hammer was between his two forward companies. The dangers inherent in this practice were borne out by their subsequent experiences. Hammer's headquarters was embroiled in the fighting for several posts. Later, contemplating a withdrawal after severe casualties, Hammer decided he needed to consult Weir, whose battalion had attacked on his right. He set off alone into the dark, armed only with a revolver. In the course of his foray, Hammer was shot through the cheek but managed to locate the 2/24th's headquarters.[69] He was, however, unable to speak to Weir because he, ordered to establish whether Thompson's Post was occupied, had departed into the night.[70] Both battalions eventually executed independent withdrawals, during which Weir was severely wounded by the explosion of two 1000-pound bombs attached to trip wires. Such behaviour was not limited to the Australians. The consolidated report on the lessons of the battle produced by HQ Middle East observed: 'There was a tendency of some commanders to go forward unescorted in order to get up-to-date information. This is foolish.'[71]

It should be noted that another common feature of the acts of leadership by direct personal example was that they occurred at night in the course of attacks. During the day, the Australians held defensive positions against concerted counter-attacks and ferocious artillery bombardments. This situation reduced the battle to a personal struggle for survival during which a CO had little direct influence on the endurance of his men. As Harry Williamson, called on to take command of the 2/43rd Battalion when his CO was wounded in the deadly confines of the Saucer, has recalled, there was very little a commander could do when everybody was huddled in roofed slit trenches to escape shellfire: 'you can't wander around talking to people'.[72]

David French has argued that the relatively small size of the British infantry battalion was a significant weakness in that unit cohesion was likely to come under strain after only relatively light casualties had been sustained.[73] The disintegration of units in combat has two principal causes: the loss of the capacity to fight, and the loss of the will to fight;[74] the latter factor obviously being influenced by the former. The experience of Australian battalions at El Alamein, however, highlights that cohesion and combat effectiveness was more than just a simple numbers game. The battalions did suffer heavily during a week of continual fighting, some being reduced to less than a hundred men, but cohesion remained strong. Although their capacity to inflict destruction on the enemy was severely reduced, they continued to respond to orders and play a role in the fighting.[75] Successful leadership might be difficult to quantify, but when the resilience of the battalions at Alamein is compared to the disintegration of the 2/8th Battalion at Vevi or that of 8th Division in Singapore, it is clear that their COs were doing something right.

EXPERIENCE AND JUDGEMENT

At Alamein the influence of experience on the 9th Division's tactics and procedures can be clearly discerned. For instance in the 2/13th Battalion the traditional dawn stand-to was abandoned because long experience in the desert had shown that the early morning was one of the clearest periods of the day, and as a result the Germans preferred to attack out of the setting sun instead.[76] Although based on experience, the ability of COs to manage risk still required considerable moral courage. On several occasions at Alamein COs chose to attack without the support of artillery barrages in order to achieve surprise. Similarly, they demonstrated a remarkable coolness in the conduct of defence, even when faced by considerable odds; resolve that almost certainly stemmed from the experience of Tobruk. Assailed by German tanks on 25 October, Major George Colvin, acting in command of the 2/13th Battalion,[77] ordered that all anti-tank guns in the battalion's area hold their fire until the armour was at the closest range possible, thereby maximising their effects. Just as the battalion's forward positions were about to be overrun, the guns opened fire 'as if by a single word of command'. Every tank in that wave was hit, and their burning hulks became the high-water mark of the attack.[78]

Similarly, when the 2/48th was set upon while defending the slopes of Trig 29 in the small hours of 28 October, the smoke and dust raised by both Australian and German artillery obscured the attackers to such a degree that they could not be accurately engaged. In a courageous decision, later commended by Morshead in a letter to Blamey,[79] Hammer requested that his supporting artillery cease firing. Under bright moonlight, visibility improved, and the battalion used its own support weapons to smash the attack at close range; the Germans had come within 400 metres of the battalion's forward positions.[80] The 2/48th's experience no doubt contributed to remarks made in the 9th Division's report on the operations, which noted that there was a marked tendency to call down defensive fire barrages, obscuring targets for the direct fire weapons that did the real damage, when observed fire would have been 'more effective and economical'.[81]

In some actions there was a degree of tactical innovation, again involving the management of risk, to create a winning advantage. One example was the attack mounted on Trig 29 by the 2/48th Battalion on the night of 25 October. Hammer ordered that two platoons mounted on Bren gun carriers, supported by a further platoon travelling on the trucks of an anti-tank troop, carry out the final assault. Given the heavy casualties suffered by the 2/48th's carriers during a counter-attack at Tel el Eisa in July, it was a potentially dangerous plan. Speed and surprise were critical to its success. The most significant risk, the presence of minefields, was discounted after careful observation and reconnaissance, and a heavy artillery barrage was to be used to prevent the employment of direct-fire weapons against the carriers and trucks. The attack was a success. The carriers crossed more than a kilometre of desert in less than nine minutes, depositing the infantry on the top of Trig 29 amidst the dust and confusion of the artillery barrage, which had ceased only a minute earlier. There was much vicious hand-to-hand fighting, but the hill fell to the Australians.[82]

There were instances, however, of rash decisions based on blind aggression and, possibly, pride. On the night of 23 October the 2/13th Battalion failed to capture all of its objectives owing to heavy opposition. The next afternoon its CO, Robert Turner, proposed to capture the battalion's objectives in daylight, against an alert enemy, with minimal artillery and armour support. Fortunately for his troops, brigade headquarters did not approve the attack. Turner was mortally wounded that afternoon while conferring with a tank commander, so we cannot be sure of his motivations. The 2/13th was the only Australian battalion not to capture its objectives the previous evening so it could be surmised that he was partly motivated by the issue of battalion – and possibly personal – pride. Additionally, the 2/13th's objectives lay on the main minefield corridor for the British 1st Armoured Division. Not until the battalion had completed its task could the mine clearing and, ultimately, the advance of the armour begin. Turner's rash decision might simply have been prompted by blind desperation.

As the failure of the 2/13th Battalion to capture its objectives on the first night of the battle highlights, the most significant factor affecting a battalion's chances of success remained the allocation of an achievable and realistic task by higher headquarters. Like the other battalions of the 9th Division, the 2/13th was well trained, and its officers and men were perhaps among the best briefed in the division owing to the diligence of its intelligence officer.[83] There is nothing, aside from differing detail, to differentiate its orders from those of other battalions, and its immediate preparations for battle proceeded smoothly. An examination of the trace of the 2/13th's objectives that night quickly confirms the judgement of the battalion historian that they were overly ambitious. It was ordered to attack on a frontage – the widest frontage of any single battalion – that in the first wave had been covered by two battalions and that would increase in width as the battalion advanced. The 2/13th suffered heavily from fire from its flanks, and the delays thus imposed resulted in its being 1500 metres from its objectives at dawn. Similarly, Morshead expected too much of the tired and depleted 2/24th and 2/48th Battalions when he ordered them into action on the night of 30 October.[84] That they survived the attacks to withdraw as formed and functioning, albeit terribly weak, units, then defend the Saucer for a day, is a credit to the abilities of Hammer and Weir.

On several occasions, both during the October battle and the earlier fighting around Alamein, COs were clearly uneasy with their orders. Don Jackson, then brigade major of the 24th Brigade, recalled William Wain's reaction to the orders for the 2/43rd to attack Miteiriya Ridge on 17 July:

'His unit had been given a particularly difficult task which involved a deep penetration of the enemy line supported by some tanks. He raised no questions with the Brigadier but engaged me in a brief exchange before going off: "Dddon, [Wain stuttered when 'excited'] we'll all be killed!"'[85] War diaries and after action reports were used as a forum in which criticisms of orders given could be officially expressed. As the 2/13th Battalion sought to consolidate after its attack on the night of 28 October, Colvin had it recorded in the war diary that he was 'not satisfied with the ground' he had been ordered to occupy.[86] Several COs similarly criticised the tactical worth and suitability for defence of positions that they had been ordered to capture and hold in July.[87] The 2/48th's war diary indicates Hammer's attitude to his orders for the night of 30 October: 'The task is a tremendous one. The bn has been fighting strenuously for 6 days during which it has launched two major attacks and withstood several strong counter attacks. The natural result has been that the bn has suffered casualties and these men cannot be replaced so that tonight the bn will be engaged in its most difficult task since the offensive was launched with the total strength of four rifle coys only 213.'[88] The question that should be posed is why these COs did not protest when given orders that were beyond the capabilities of their battalions. With few records of the conferences held at brigade headquarters, we cannot be sure that they did not. Ultimately, however, loyalty and obedience up – as opposed to down – the chain of command remained paramount. The Alamein battles revealed COs willing to disregard their orders and conduct withdrawals to preserve their battalions, but only after they had made a determined effort to carry out their often poorly conceived tasks.[89] We can only surmise their reasons. Battalion and personal pride, the aggressive ethos of the AIF, loyalty to other units and the carefully cultivated sense that the October battle was a turning point in the war, demanding that the Eighth Army needed to fight to the last man standing, all probably played a part.

By the time the battalions of the 9th Division left the Middle East in January 1943, thereby ending Australia's contribution to the land war against Germany and Italy, battalion command in the AIF had undergone a transformation. Initially, the exercise of command in Australian infantry battalions had been based on outdated doctrine, a limited base of experience and all of the idiosyncrasies of a peacetime citizens' army. In the Middle East the AIF became part of a larger British and Dominion army, allowing it to draw on a wide base of up-to-date military experience and produce a new cohort of battalion commanders. These were officers with personal experience of this new war, proven in battle, younger, fitter and more thoroughly trained than any before them.

The development of Australia's COs took place simultaneously with the development of the command system of which they were apart. When the Australians first joined the war in the Middle East the command system was ill-suited to the operations being conducted, a situation exacerbated by a lack of training and experience at all levels in Australian battalions. It was only when command systems, training and experience all harmonised with the nature of operations, as they did at Alamein, that high and consistent levels of performance on the battlefield were achieved. This progression also influenced the nature of battlefield leadership. Desperate situations elicited desperate measures from COs but, as levels of training and experience improved, instances of direct personal intervention by COs generally decreased.

Even when the AIF was split in two with the return of the 6th and 7th Divisions to Australia in February 1942, it remained Blamey's force. His influence on the selection and dismissal of COs remained strong. On the eve of his return to Australia he was still at work, directing that the new CO of the 2/17th Battalion, Maurice Ferguson, be withdrawn from the senior officers course at METS because he believed he already had enough tactical experience and needed to build a rapport with his new command.[90] Once back in Australia, Blamey continued to monitor the 9th Division's COs closely, and Morshead's signals regularly contained information as to their performance and fates.

When the COs of the 9th Division stepped ashore in Australia in February 1943 they were arguably among the best trained and experienced in the British and Dominion armies. They were the product of two years of collective experience of fighting in the Middle East and had fought, as part of a large, well-resourced, well-integrated combined arms force, in one of the turning-point battles of the war. For the past year, however, the rest of the AMF had been struggling to master another operational environment: the jungle and mountains of Malaya and New Guinea. This environment placed its own unique demands on a command system and a commander, and one of the major issues that the AMF was facing was the question of how applicable and adaptable was the experience of the Middle East to this new war.

CHAPTER 5

VICTIMS OF CIRCUMSTANCE
Battalion command in the 8th Division

The practice of command in the AMF during the Second World War reached its nadir during the fighting for Singapore in February 1942. The initial Japanese assault on the island burst on positions held by the Australian 22nd Brigade on the night of 8 February, and within two days the 8th Division, of which the 22nd Brigade was a part, had ceased to function as an effective fighting formation. Four of its battalions had disintegrated; the other two, out of communication with their brigade headquarters, were acting independently; and the animosity and mistrust that had bedevilled the exercise of command in the division since its formation was out of control – to the detriment of the confused, exhausted and increasingly demoralised soldiers on the firing line.

The Australians' month-long campaign in Malaya and Singapore had been fought at a dizzying pace. The foundation of the Japanese concept of operations in Malaya was the 'driving charge', a tactic that emphasised taking calculated risks to maintain the momentum of the advance and continually keep the enemy on the back foot. The 8th Division was deployed in southern Malaya when the Japanese landed in the north on 8 December and was held there while III Indian Corps fought the initial battles. As the Japanese pushed south without pause into January it was intended that the Australians would form the core of a force that would stop them dead in their tracks in the north-west corner of the state of Johore along a line running from Muar on the west coast to Segamat on the main road and rail corridor. This 'main battle' was never to be. After a brief flurry of resistance as the fresh Australian troops joined the fray, the British and Dominion position collapsed into an increasingly harried withdrawal. The first significant action for the Australians took place on the southern bank of the Gemencheh River on 14 January and, after a succession of what could be best termed delaying actions, they withdrew to Singapore on 31 January to prepare for its ill-fated defence.

Although the loss of Malaya and Singapore was largely brought about by a succession of poor strategic and operational decisions over which battalion COs had no control, the quality of decisions made by COs within their own sphere of influence was also lacking. The experience of the 8th Division's battalions further reinforces the impression that there was a dearth of reliable command talent in Australia at war's outbreak. The formation of four AIF infantry divisions, instead of the three originally envisaged, thinly stretched Australia's limited military resources, and at some point something had to give. That point was Singapore.

Failure at the battalion level, however, cannot simply be laid at the feet of the COs. We return again to the concept of a command system. It was not the COs as such who failed but the system of which they were a part. The means that COs had at their disposal for effecting their command were inadequate and vulnerable in the fluid, high-tempo fighting of this theatre. Direction from higher headquarters was confused, contradictory and often non-existent. This difficulty was compounded by a lack of confidence along the command chain generated by the personal conflicts in the division's higher echelons. These conflicts had a destructive virulence, and they spread throughout the brigades, undermining the mutual confidence of battalion commanders. None of these difficulties were unique to Malaya, but the calibre of many of the COs was such that they were not able to cope and adapt. Although some were more progressive in their tactical thinking than others – and the Australians were among the best prepared troops in the theatre – the tactical and logistical problems presented by the campaign in Malaya and Singapore tested them to the very limit of their abilities, and sufficient expertise, experience and time was not available to produce a new generation of COs to replace them.

The COs of the 8th Division

The 8th Division, originally comprising the 22nd, 23rd and 24th Brigades, was the only 2nd AIF division not formed under Blamey's careful stewardship. By the time the division's headquarters had opened at Victoria Barracks in Sydney on 8 July 1940, Blamey had departed for the Middle East, and responsibility for the formation of the new division was passed to Major General Vernon Sturdee, its commander designate. Sturdee was a regular soldier of 32 years experience, and for most of the decade preceding the war he had occupied senior staff positions at Army Headquarters. Sturdee's appointments within the new division reveal the continuing influence of patronage in the selection of COs as well as the exhaustion of the AMF's pool of command talent.

Sturdee wanted his battalions commanded by men who had exercised command at company level during the First World War and who had subsequently served as COs in the militia. He stipulated, however, that such men needed to be as young as possible and still mentally adaptable. Sturdee's personal ideal for a battalion commander was a young and vigorous officer: in 1938, when aged 48, he commented that he was too old for a battalion command, recalling that when he took command of the 4th Pioneer Battalion in 1917 he was only 26.[1] As a compromise between this ideal and the type of officers actually available, Sturdee further insisted that battalion 2iCs should be much younger than their COs and no older than 35.[2] Sturdee's philosophy also envisaged First World War veterans spread among the other ranks of his battalions to provide a leavening of experience to utilise in the training of new recruits and to provide a steadying influence when they embarked on their first operations.[3] The policy was directed at providing the battalions with the opportunity to develop command talent from within, through the medium of experience, both vicarious and actual. The veteran COs were expected to last 18 months at the most, before they were replaced.[4]

In several respects Sturdee's selection policy echoed that of Blamey, but his greater emphasis on succession planning can be seen as a realisation of the extremely limited nature, in both experience and capacity, of the potential COs remaining available for appointment to the AIF. It will be recalled that when the 7th Division was being formed Blamey already believed that there were few suitable men left to appoint. Ultimately, four brigades would be recruited for the 8th Division as on 23 September 1940 the War Cabinet directed that the two AIF brigades then in the United Kingdom would form the nucleus of a fourth division: the 9th. The 8th Division's 24th Brigade was allocated to complete the 9th, and one last AIF Brigade – the 27th – was raised to take its place.[5]

The extent to which the formation of the 6th and 7th Divisions had stripped the AMF of its best officers is readily apparent when the initial cohort of COs for the 8th Division are considered as a whole. None completely satisfied Sturdee's criteria. Of the 12 COs selected, seven were aged 45 or older, five had four years or less service in the interwar period and eight had no more than a year's command experience, including two PMF officers who had never commanded a battalion previously. All but two had served in the First World War, but only three had commanded anything larger than a platoon during that conflict.[6]

Given the limited field, patronage played an important role in deciding some appointments. In Victoria, Brigadier Edmund Lind opted for two officers from his own militia brigade: Leonard Roach to command the 2/21st Battalion and Howard Carr the 2/22nd. Another appointment that seems to have been made on the basis of an established working relationship was Brigadier Harold Taylor's selection of Duncan Maxwell as CO of the 22nd Brigade's 2/19th Battalion. Maxwell was a risky appointment because of his lack of command experience and the potential for conflict between his civilian medical career and the role of a combatant officer. Taylor was well aware of Maxwell's lack of command experience, but there is no indication that he had any doubts about his capabilities. Maxwell's 56th Battalion had been part of Taylor's militia brigade, and the two men had worked together closely during a three-month training camp in early 1940. 'I would have selected him again,' Taylor reflected after the war.[7] It could be suggested that Taylor had been blinded by friendship to the possible dangers of Maxwell's inexperience. Or perhaps, like Stevens before him, he was willing to take a risk, in the face of a limited field of suitable officers, to select a man whose loyalty and commitment he knew he could trust.

With the formation of the 2nd AIF's infantry divisions nearing completion, desperation to secure a command also motivated some officers to exploit their own patronage networks. The third CO in the 23rd Brigade was a political appointment over which Brigadier Lind, and perhaps even Sturdee, had no control. The 2/40th Battalion was raised as a purely Tasmanian unit as a result of lobbying by the Tasmanian government.[8] The same political forces were determined that only a Tasmanian would be appointed to command it; Geoffrey Youl, then commanding the 12/50th Battalion, became its CO. Youl was a grazier from the north of Tasmania and a member of a wealthy and influential pastoral family. He was a stalwart of the social and conservative political scene in Tasmania,[9] and it can be inferred that these factors played a considerable role in his appointment. Youl was 48, had limited infantry experience and had already been passed over for command of the 2/12th Battalion in 1939.[10]

Until the formation of the last battalion of the 2nd AIF's last brigade, Eastern Command's problem-child, Frederick Galleghan, remained available. At 43, he was still under the age limit for a lieutenant colonel in the AIF (which was 45) and had continued to command the 17th Battalion despite thoughts of resignation following his failure to secure a command in the 6th Division.[11] Galleghan's First World War experience, however, was as an NCO and fell short of Sturdee's criteria. Additionally, Sturdee, as GOC Eastern Command, was one of those whom Galleghan had earlier antagonised.[12] It was only after command of the division had passed from Sturdee to Major General Gordon Bennett,[13] and former Prime Minister William Hughes had intervened on his behalf, that Galleghan was appointed to command the 2/30th Battalion.[14] Hughes was a close friend of Galleghan, and his help had previously been enlisted in an attempt to have the 17th Battalion transferred to the 2nd AIF as a complete unit.[15] Galleghan had few friends in the AMF before seeking Hughes's assistance and even fewer afterwards. Galleghan subsequently failed to meet the rigorous medical standards applied to AIF volunteers owing to a hearing condition resulting from the First World War, but was still able to pull some strings to be passed fit for active service.

The 8th Division's original COs were an eclectic bunch. The fact that Sturdee, then the Military Board, were prepared to accept appointments that were compromises on a compromise is further evidence that the AMF was struggling to provide enough fit, qualified, experienced and relatively young officers to provide COs for four expeditionary divisions. As a cohort, there was little to bind them, perhaps with the exception of First World War experience, and even that was extremely varied and much of it of doubtful relevance. Unlike the original COs of the 6th and 7th Divisions, there was no common heritage of long and dedicated interwar service. Some of the appointments seem ill-informed, others a downright risk to the soldiers these men would command. The appointment of officers who had seen little military service in the past 20 years could be considered negligent in the extreme. Sturdee was gambling on these appointments, gambling that there would be an opportunity for new commanders to be trained and exposed to active operations before the battalions became involved in any serious fighting. It should at least be recognised that Sturdee was aware of the weaknesses inherent in his commanders and had the foresight to implement an appointment policy for other battalion officers designed to redress them.

Leadership in the 8th Division

The COs of the 8th Division exhibited a diverse range of command styles that were the product of their military experience, their civilian background and their own personality. Some demonstrated a remarkable degree of self-awareness, compensating for their own weaknesses with astute leadership, which included exploiting the talents of their subordinates. The result was the same type of complementary command relationships that emerged in many of the battalions of the 6th Division, as was discussed in chapter 2. These complementary command relationships also began the process of grooming the 8th Division's next generation of COs, validating Sturdee's selection criteria. There were also COs who remained rigid and pig-headed in their approach to command, demonstrating little appreciation of their own strengths and weaknesses, and earning little respect or trust from their subordinates. With their leadership being based principally on the exercise of coercive and legitimate power, their commands were inherently weak.

The influence on a command style of personality and experience outside the army is well illustrated by the example of Duncan Maxwell. He is remembered as having a subtle and humane approach to command that almost certainly resulted from his professional training as a doctor. Major General Bennett recalled that Maxwell 'was a doctor always looking after the welfare of his patients and his men were to some extent his patients'.[16] His command was founded on a sincere concern for the welfare of his men, who were said to worship him. Brigadier Taylor described the 2/19th as having the best camaraderie in the brigade but noted that Maxwell 'hesitated to put the boot in'.[17] Maxwell's leadership was clearly based on referent power. Roland Oakes, a 2/19th company commander, wrote that Maxwell's success lay in his ability to get others to do things for him: '. . . if something has to be done, there is seldom the need to go straight for it when there are other pleasant and devious ways to get the thing done more quickly and far more effectively'.[18]

It was widely recognised that tactics were not Maxwell's strong point, but it was felt that the efficiency resulting from the battalion's strong *esprit de corps* compensated for this weakness.[19] Maxwell seems to have been aware of his own weaknesses and relied heavily on his staff. In particular, Major Charles Anderson, the 2/19th's 2iC, directed much of the battalion's training in Malaya. Anderson was a hard, practical man, familiar with fighting in close country owing to experience in East Africa during the First World War. At the same time, he possessed the same

A spirit of humanity and mutual respect: Lieutenant Colonel Charles Anderson (far right), CO 2/19th Battalion, talking with three of his company commanders (right to left): Major Bert Bradley, OC C Company, Major Tom Vincent, OC D Company, and Major Roland Oakes, OC A Company, Seremban, Malaya, August 1941. After he took over from Duncan Maxwell in August 1941, Anderson's honest straightforward command style maintained the strong unit cohesion established by his predecessor. (AWM Neg. P00102.45)

spirit of humanity and mutual respect as Maxwell, which ensured that the two complemented each other well.[20] A similar relationship between CO and 2iC prevailed in the 2/18th Battalion. Arthur Varley is remembered by his men as being shy and retiring but, validating Sturdee's selection philosophy, was well supported by his 2iC, Major Charles Assheton. 'Full of amazing energy and drive, with a flair for thoroughness and attention to detail', Assheton ensured the smooth day-to-day administration of the battalion, leaving Varley able to concentrate on guiding the battalion's training.[21]

In the 2/19th, Maxwell commanded in a consultative fashion and keenly experimented with different tactical concepts during training. Taylor noted in his diary that this approach often resulted in a lack of decisiveness, with discussion and experimentation taking the place of a definite plan and succinct orders.[22] As with many of the AIF's original COs, Maxwell's behaviour set the tone for subsequent command relations after his departure. Bennett visited the 2/19th after Anderson had assumed command following Maxwell's promotion. He observed that Anderson was 'a bit theoretical and discussive' and that no work had yet commenced on the Jemaluang position. He concluded that there was 'Too much thinking and not enough action'.[23]

The strength of referent power as the foundation of military leadership is further demonstrated by the example of Arthur 'Sapper' Boyes of the 2/26th Battalion. Boyes was very much a product of his PMF training, which had included a period on exchange with a British guards regiment, and he always maintained the distance between himself and his men that his status demanded. As one 2/26th veteran has recalled, there was no room for doubt that he was the commander.[24] He also took his responsibility to his men seriously and conveyed this same sense of duty to his officers. They were told bluntly that they were not important in the grand scheme of things and that the welfare of their men was their main obligation. When this lecture was delivered to the officers there were several NCOs within earshot, and Boyes' message was quickly disseminated throughout the battalion. It has been suggested that it was a stage-managed performance but, whatever the case, it was said by many members of the battalion to have engendered a spirit of mutual respect and obligation.[25]

Conversely, William Jeater of the 2/20th provides a prime example of the ineffectiveness of leadership founded on legitimate and coercive power. Jeater refused to entertain ideas contrary to his own and drove rather than led; Brigadier Taylor described him as a 'slave driver'.[26] By the time the 2/20th reached Malaya in February 1941, Jeater had lost the confidence of his junior officers owing to his practice of criticising them in front of their troops, with little regard for their rank or experience, and regularly undermining their freedom of action. He would not heed the advice of any of his officers regarding training, thereby preventing any sense of trust or mutual respect developing between them.[27] Jeater demanded high standards from his troops, and exhausting tests of endurance were part of his training syllabi; he was not popular.

In the early months in Malaya Jeater's grip on his command slipped. Statistics for courts martial in the 22nd Brigade in May 1941 would seem to provide a damning comment on the state of Jeater's command. In that month courts martial were convened for one soldier from the 2/18th

Battalion, for two from the 2/19th and 11 for troops from the 2/20th.[28] Accounts of this period in the 2/20th's war diary and in the battalion history, published in 1985, are contradictory, however, which indicates a complex and contested story. When they are synthesised the image of a troubled battalion emerges. The troops were dissatisfied with their conditions in Malaya, particularly the quality of their rations, and frustrated that they were not serving beside their countrymen in the Middle East. Several incidents of collective indiscipline, significant enough to make the pages of the war diary, were the result.[29] Jeater's response was characteristically coercive. He concluded an article in the battalion newsletter: 'although we might wish to be elsewhere, as soldiers we obey. And as I said above we do it cheerfully.'[30]

At the same time as he was attempting to hold his battalion together, Jeater was struggling to hold himself together. Like many Australians new to the heat and humidity of Malaya, Jeater contracted a chronic skin rash, most likely a variant of tinea. It resulted in several long periods of hospitalisation and eventually initiated a psychological condition.[31] Jeater's ordeal revealed a complete lack of respect in his battalion. Despite his significant discomfort the battalion magazine quipped, 'We're all itching to know . . . That officers' ball . . . Why didn't the Colonel dance – was he afraid it might be too rash?'[32] and some of his troops instructed the local Malay paperboys to call, 'Read all about it – Col. Jeater's got the pox', and directed them to the officers mess. True to form, Jeater blamed his junior officers for the slight.[33] Being responsible for the discipline of their subordinates, however, they cannot be considered blameless.

Battalion discipline and good order should have been able to withstand the absence of the CO; the problems in the 2/20th were obviously rooted much deeper. A hunger strike by the rank and file protesting the quality of the rations demonstrates the level to which command relations had sunk. Jeater held his officers responsible for allowing the dissatisfaction to take hold during his absence and claimed never to have been told of the troops' complaints. His officers maintained they had sought to raise the subject on several earlier occasions but were rebuffed.[34] Given the lack of mutual respect between them, either version could be correct. The battalion history sides with the officers, but even when possible bias is taken into account its final comment on the incident is insightful: it need not have happened 'had Jeater had a better liaison with his officers'.[35]

Ultimately, Jeater proved unfit for command. His rash and attendant psychological condition led to his final evacuation to hospital in June 1941 and his replacement by the 2/18th's energetic 2iC, Charles Assheton, in

August.[36] The tragedy of the 2/20th was that Taylor rated Jeater the most knowledgeable of his COs and the battalion the most proficient in elementary skills.[37]

THE GALLEGHAN ENIGMA

The best known of the Australian COs in Malaya, and perhaps of the whole of the AMF during the Second World War, is Frederick 'Black Jack' Galleghan. Much of his reputation stems from his time as the senior Australian officer in Changi, and a great deal of the rest, as Ian Campbell has pointed out, resulted from his own self-promotion, an example of which we saw in the introduction.[38] Galleghan provides an informative case study about the influence of character on command, as well as further illustrating the factors influencing command relations discussed above.

Right from its formation, Galleghan expected the 2/30th to do everything harder, faster and better than other battalions.[39] A former 2/19th Battalion officer who was good friends with Galleghan later in life recalled that he had a 'colossal ego',[40] and it would seem that his battalion was an extension of it. The product of a difficult upbringing, Galleghan had used the army as a means to 'raise his stature' and advance his position in life.[41] Campbell is convinced that Galleghan was consciously seeking to create an elite battalion.[42] Galleghan's exceedingly high standards meant that during their early training at Bathurst the men of the 2/30th had to rise more than an hour before the official reveille time to ensure that their morning routine was carried out to his satisfaction. Galleghan's regime bordered on the absurd. During the subzero extremes of a Bathurst winter, troops discarded the minimal insulation provided by straw in their palliasses so that they could be arranged to pass inspection in the morning.[43] Galleghan was another CO who chose to upbraid his officers, like naughty schoolboys, in front of their soldiers, and he second-guessed decisions for which he had previously delegated authority.[44] Hence his relationship with his officers was strained, he was widely regarded as a 'tyrant' by his men and his reputation as a 'great ogre' spread to other battalions in the division.[45]

It is little wonder that Galleghan's troops mutinied before they had even left Australia. Following a gruelling three-day exercise, they refused to participate in a snap kit inspection.[46] Campbell has succinctly described their motivation: 'they believed themselves to be working harder than their mates from other battalions. They did not mind this, were even proud of it, but would brook no more nonsense.' Galleghan attempted to deal with the mutiny in his characteristic manner. He stormed into one section hut and physically pulled an offending soldier from his bed; he accused his padre of neglecting his duties; and, after an attempt at discussing the matter rationally with his officers, he ranted and raved at them. His efforts achieved little, and order was restored only after negotiation with elected representatives from each company the following morning.[47] No charges ensued and, seemingly conscious of his reputation and that of his battalion, Galleghan did not report the incident to brigade headquarters.[48] The mutiny did little to modify Galleghan's behaviour or the hostility of many of his troops to it. While standing on the wharf at Fremantle on passage to Singapore, Galleghan ducked a bottle thrown from the deck of the ship above.[49]

The free-booting behaviour that had kept Galleghan from an AIF command for so long also continued. He sought the assistance of Billy Hughes, again, to have the local police evict 2/30th men from Tamworth pubs at 8pm, an imposition that the police had strongly resisted. In Malaya, the 2/30th took part in a scripted exercise during which it was meant to lose so as to illustrate several tactical lessons. Galleghan felt that to lose was beneath the 2/30th and hence fought the battalion in such a way that it won. During another exercise he disputed the judgement of the appointed umpire, Anderson of the 2/19th, thus beginning an enduring antipathy.[50] These incidents highlight the hypocrisy that undermined Galleghan's command style. Arneil maintains that these incidents could be seen as signs of his unstinting loyalty to his own unit, but they can equally be portrayed as self-centredness, which has little place in a military organisation. Galleghan demanded unquestioning loyalty and obedience from his subordinates, yet he did not always accord it to supporting units or to his superiors.

The ultimate demonstration of Galleghan's hypocrisy was his reaction when he arrived in Singapore to find the freshly promoted Duncan Maxwell in command of the 27th Brigade. He greeted his new commander with: 'Well, you need not expect me to congratulate you on the red flannel you're wearing!'[51] then made a beeline to Bennett to complain. He threatened to resign his appointment and return to Australia. Bennett, revealing an insightful reading of Galleghan's character, refused to take him seriously, although at least one member of the divisional staff suggested that if Galleghan felt so strongly about the matter his return to Australia would be best.[52] Galleghan stayed.

'Black Jack': Lieutenant Colonel Frederick Galleghan, CO of the 2/30th Battalion, consults the battalion intelligence sergeant, Erwin Heckendorf. Galleghan is one of the best known of the Second World War infantry COs, but his performance in Malaya was much more patchy than his legend would suggest. (AWM Neg. 011304/04)

Overlooked in the Galleghan legend is the figure of Major George Ramsay, his 2iC. Several of the 2/30th's officers regarded Ramsay as the perfect foil for Galleghan.[53] Ramsay's letters reveal a sensitive and caring family man,[54] and as an officer he was even-tempered, unshakeable and attentive. The troops christened him 'Gentleman George', a nickname he retained until his death in 1981. One company commander recalled, 'Old Black Jack would dress us down, and George would come and cheer them up until tomorrow [when] they'd get bashed down again.'[55] Ramsay embodied a sense of loyalty common to his generation. He is remembered as displaying 'tremendous devotion' to Galleghan and the battalion, despite Galleghan's faults, which it became obvious in later years had frustrated him considerably.[56] Additionally, Galleghan's adjutant and several of his company commanders also functioned as buffers between the men and the worst of his excesses.[57]

Galleghan is an enigma, and it is difficult to separate the legend from the reality. Campbell argues that he manufactured a 'mask of command'.[58] Galleghan himself told Lionel Wigmore that 'being a commander demands hardness'[59] and, as discussed, Galleghan presented himself as the archetypal 'hard bastard'. Campbell, however, suggests that this was merely artifice, maintaining that Galleghan rarely participated in the arduous training he set his men, that he drove or rode when they walked; and that he drank, smoked and ate to the verge of excess.[60] He gained his appointment to the AIF despite being medically unfit, yet he told his troops that any man reporting sick to the MO was finished in the 2/30th and demanded the same fate for those who fainted on parade.[61] This image of a self-indulgent hypocrite is in stark contrast to the heroic figure who strides across the pages of Arneil's biography and to the command philosophy that Galleghan himself outlined to Wigmore: '[The] only way to know when soldiers are tired is to walk with them.'[62] Few of Galleghan's former men would argue that he was not a difficult character, yet he still inspires fierce loyalty. The question remains, however, did he lead or merely drive? It could be suggested that the loyalty of his men was a transference of loyalty to the battalion, born of the men's pride in their own achievements in the face of Galleghan's bullying. Galleghan's methods did contribute to the production of a cohesive battalion – the mutiny is evidence enough of its collective unity – and in Malaya the 2/30th was widely regarded as the best battalion in the 27th Brigade.[63] Galleghan had achieved this through pretence, however. His leadership was not based on respect for him as a man or as a tactician, but almost solely on the exercise of coercive power. As we will see, pretence was vulnerable on the battlefield.

COMMAND RELATIONSHIPS IN THE 8TH DIVISION

The irascible and arrogant personality of Major General Gordon Bennett is well documented.[64] An outspoken critic of the capabilities of PMF officers during the 1930s, Bennett had few friends among the AMF hierarchy and had been overlooked for divisional commands in the AIF on five separate occasions since 1939. His appointment to command the 8th Division when Sturdee became CGS in September 1940 is a further example of the degree to which the AIF had been expanded beyond the AMF's ability to provide it with competent and respected senior officers. In Malaya Bennett's very presence undermined the proficiency of his division. His reluctance to accord Brigadier Taylor any degree of independence in the training of his brigade resulted in a running feud that meant by the time of the Japanese invasion in December the two men were not even on speaking terms.[65] Bennett's personality also generated an atmosphere of disharmony and discord among his principal staff officers that rendered his headquarters almost dysfunctional.

The impact of Bennett's personality on command appointments in the division is well illustrated by Maxwell's promotion to command the 27th Brigade on its arrival in Malaya in July 1941. The appointment of Maxwell, whose weaknesses have already been discussed, was not a vindication of his tactical prowess but just another move in a political game being played out by Bennett, fighting for complete control of his division; by Sturdee, seeking to retain an influence over senior appointments in it; and Taylor, simply trying to have his opinions, as one of the division's senior officers, heard. At the time, it was generally regarded that Maxwell's appointment was due to Taylor's lobbying, an interpretation that was perpetuated by some officers after the war.[66] This interpretation, however, completely ignores the antipathy that existed between Bennett and Taylor. Maxwell was in fact low on Taylor's list of preferred candidates, which was headed by Colonel Henry Rourke, Bennett's regular army General Staff Officer Grade 1 (GSO1), and Varley of the 2/18th.[67] Ironically, Maxwell was probably appointed because he was not Taylor's preferred candidate. Galleghan, perhaps seeking to nurse his own bruised ego, maintained that Maxwell was appointed because Bennett wanted an officer with some experience in Malaya; Maxwell and the 2/19th had been in Malaya since February.[68] Wigmore insightfully observed that, given the difficult and combative relationship between Bennett and Taylor, Maxwell's appointment owed much to his benign personality.[69]

Maxwell's appointment only added to the disunity in the division. Many among Bennett's staff had no confidence in Maxwell's abilities, and his own brigade major later recalled that the brigade was not a happy one.[70] Galleghan had little respect for him, and there are also indications of a troubled relationship with Boyes. It could be surmised that Boyes, as a regular soldier, was reluctant to serve under an officer so lacking in military qualifications as Maxwell but was too imbued with respect for the chain of command to carry on in the manner Galleghan did. Maxwell took to his new job with characteristic humility, admitting at his first conference that he knew little of infantry work and that he was junior to two of his COs. He told them that he planned to consult them regularly for their ideas and opinions and that the brigade would train in accordance with the program and tactical concepts developed by the 22nd Brigade.[71] Maxwell's approach displayed a mature understanding of his own shortcomings, but it was perhaps not the best way to inspire the confidence of a long-serving professional soldier like Boyes or an officer with as much confidence in his own abilities as Galleghan. In Malaya, these two officers formed a strong friendship in, it would seem, collusion against Maxwell. When his command began to falter during the latter stages of the retreat through southern Johore in January 1942,[72] Galleghan and Boyes exercised de facto command over themselves in a consultative partnership.[73]

Other COs were also affected by the disharmony that seeped down the chain of command from Bennett. Jeater may have been partisan in the Bennett/Taylor antipathy and, at the very least, played it to his own advantage. During a visit to the 2/20th, Bennett was told by Jeater: 'I wish you w[oul]d come and spend a week with me.' Bennett further noted in his diary: 'Jeater most pleasant and full of respect etc. Kappe had spent yesterday and today with him and must have let drop something re Taylor.'[74] Bennett appears to have given Jeater favourable treatment. Although aware of the disciplinary problems in the 2/20th, he did not take any action, and when Jeater's mental state led to his removal from command he was not returned to Australia but instead appointed to command the AIF General Base Depot at Johore Bahru.[75] The gentlemanly code of honour applied to dismissed COs in the Middle East also held sway in Malaya. Blamey passed through Singapore in December 1941 and described Jeater as 'mental', yet he was still not sent home.[76]

The internecine feuding in the 8th Division had a detrimental effect on the training of the battalions, particularly those of Taylor's 22nd Brigade. In one instance Taylor sought to improve the cooperation between his battalions and their supporting artillery by including the 2/10th Field Regiment in battalion exercises and routinely attaching artillery officers to the battalions. These arrangements earned Taylor a rebuke from Bennett: 'I understand you were at 2/10th F[ield] R[egiment] last n[igh]t. I want you to understand you command 3 inf[antry] b[attalio]ns; no more. When you go into action you will be given certain supporting arms.'[77] The joint training ceased immediately.

Bennett's tendency to ignore Taylor and the chain of command often placed the latter's COs in a difficult and confusing position. In September

1941 Bennett circulated a letter to the 22nd Brigade COs regarding their defensive preparations that even stipulated tasks for small-scale patrols. Charles Assheton, who had replaced Jeater as CO of the 2/20th, was moved to discuss the letter with Taylor, who later confided in his diary: 'I have not read the letters yet but the plan has been done entirely from the map and read like it. Pity he would not give more time to his real job, which is a broad plan and the organisation of his Div HQ, which is in a parlous state, instead of poring over maps and sending out bright ideas, most of which will not work. Some day he will learn – I hope not too late.'[78]

Erratic and dangerous decisions regarding battalion command appointments continued once the 8th Division was committed to action in January 1942. During training Bennett was not impressed by the performance of John Robertson, who was subsequently found to be medically unfit. Nevertheless, Bennett still allowed Robertson to lead the 2/29th Battalion forward towards its first action at Bakri on 17 January 1942, intending to remove him after a day in command. This was presumably to allow Robertson to leave his command with honour.[79] It could be construed as a generous or thoughtful action, but any premeditated decision to remove a CO in the middle of battle must be seen as highly irresponsible.

Bennett was also complicit in two eleventh-hour changes of command as the Australian battalions prepared to meet the Japanese invasion of Singapore in the first week of February, both of which presented obvious risks and would ultimately have dire consequences. The first was Maxwell's removal of Boyes from command of the 2/26th Battalion. After the war Maxwell argued that Boyes was removed owing to exhaustion after the hard-pressed withdrawal through Johore.[80] Boyes, however, had performed with increasing competence, and would subsequently be pressed straight back into service to organise a composite battalion from reinforcements and the remnants of the 22nd Brigade's battered battalions. Given that Galleghan was also dispatched to hospital at this time, although for legitimate reasons, it is possible that Maxwell was seeking to break the Boyes–Galleghan cabal referred to earlier.[81] Despite the fact that the battalion would soon be in action, Maxwell did not appoint a new CO from within the 2/26th but opted for Major Roland Oakes, 2iC of the 2/19th, who had not seen action in Malaya.[82] Bennett did not intervene; the 8th Division's post-operational report would later describe the appointment as an 'unwise decision'. Oakes was a stranger to his officers and had little time to familiarise himself with the position; several command problems would subsequently occur during the battalion's fight for the Kranji area and its withdrawal south.[83]

In the 2/19th Charles Anderson was evacuated to hospital, leaving the battalion without a CO. Both Brigadier Taylor and the men of the 2/19th favoured Major Tom Vincent, one of the battalion's rifle company commanders, who had proved himself a capable and courageous officer during the fighting on the Malayan mainland.[84] Bennett, however, scotched the appointment, as there were more senior majors in the brigade, and Major Andrew Robertson from the 2/20th was appointed instead.[85] Thus, another battalion had a CO it did not know and did not trust foisted on it. It was not long before acrimony erupted between Robertson and his senior officers when he ordered battalion headquarters to relocate from a site carefully chosen by Anderson.[86] Robertson had had little time to acquaint himself with the ground or the battalion's existing plans, and his decision proved fateful.[87] Anderson had sited battalion headquarters to control high ground that dominated the 2/19th's line of retreat and nominated it as the battalion's rendezvous should this be ordered. When the Japanese landed they occupied the now-vacated ridge, and the 2/19th disintegrated in its efforts to batter its way out.

Tactical finesse and the fighting in Malaya and Singapore

When the Australian battalions joined the fighting in Malaya in the second week of January 1942, their COs faced the same fundamental challenge as had confronted their compatriots in the Middle East. They needed to develop a command system capable of employing a variety of arms in an integrated and timely fashion so as to match the tactics and operational tempo of their opponents. This they never quite achieved.

To varying degrees the COs of the 8th Division were brave, principled, intelligent men. Most had sought to adapt the training of their battalions to the specific conditions of Malaya, and recent scholarship on the British and Dominion operations in Malaya and Singapore rates the Australian battalions as among the best in Malaya command and describe a competent performance during the fighting retreat through southern Johore.[88] This, however, is a relative assessment. Ultimately, Australian COs in Malaya and Singapore lacked the experience and imagination required to match an enemy of the quality of the Imperial Japanese Army.

Unlike their compatriots in the Middle East, the Australians in Malaya were denied the benefits of universalised tactical instruction based on the experience derived from continuing operations. The British Army in Malaya devoted little time to the development of doctrine for fighting in the conditions encountered there, had no centralised training organisation and operated no senior tactical or staff schools, and its knowledge of Japanese operational practices was rudimentary and undermined by prejudice.[89]

As a result, Australian COs dismissed the potential of the more recent developments in warfare, such as wireless, anti-tank artillery and motorised transport, and lacked the tactical judgement – or perhaps the confidence – needed to act boldly and capitalise on their success. In post-operations reports command practices of Australian infantry battalions exhibit the same flaws that had been identified during exercises in Australia, confirming the dearth of training for senior officers in Malaya.[90] If the basic measurement of military competence is being able to minimise foreseeable risks, questions need to be asked about the imagination of officers who were incapable of even conceiving the basic, yet critical, tactical innovation of having their men's khaki uniforms dyed green for better camouflage in the jungle.

The ignorance of some Australian COs regarding their supporting arms is exemplified by their employment of anti-tank guns. Tanks were at the heart of Japanese shock action in Malaya, and Australian anti-tank detachments played a decisive role in the fighting at Gemas on 15 January 1942 and at Bakri four days later. Collectively, they destroyed 12 Japanese tanks and prevented their smashing through the middle of two battalion positions, as had happened to Indian units at Slim River only a week earlier. In the lead-up to both Gemas and Bakri the respective COs, Galleghan of the 2/30th and Robertson of the 2/29th, volubly dismissed the utility of these weapons and neglected to incorporate them into their plans, a clear failure to prepare for a likely and dangerous threat to their battalions. Galleghan proclaimed with characteristic arrogance that the Japanese would not use tanks and that anti-tank guns were a nuisance, while Robertson banned the anti-tank troop commander from his orders group and gave him only one order: not to position the guns anywhere near the infantry.[91] The troop commander disobeyed Robertson and positioned two of his guns by the road, right in the middle of the battalion position, and deployed the other two to the rear. The next morning the two forward guns destroyed eight Japanese tanks. Robertson apologised the next day as he lay dying of wounds inflicted during the battle: 'I'm

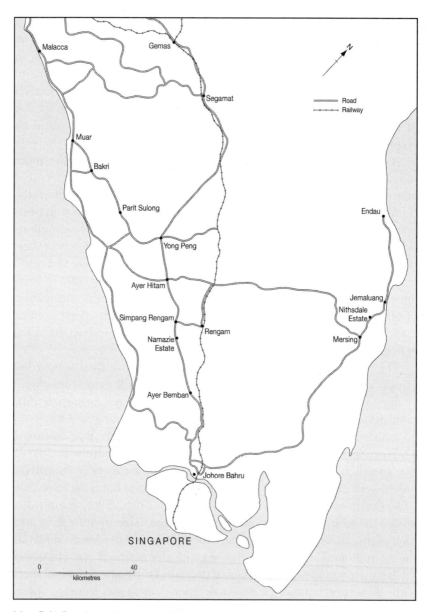

Map 5.1: Southern Malaya and Singapore

'Only for your persistence in defying my orders and positioning your guns where you did, there would have been wholesale slaughter': a 2-pounder anti-tank gun of the 2/4th Anti-tank Regiment, commanded by Sergeant James Parsons, in action against Japanese tanks, Bakri–Muar road, Malaya, 18 January 1942. This gun was positioned in the midst of the 2/29th Battalion's positions contrary to the orders of its CO, Lieutenant Colonel John Robertson. (AWM Neg. 068592)

so sorry that I acted as I did. Only for your persistence in defying my orders and positioning your guns where you did, there would have been wholesale slaughter. I'm so sorry.'[92]

Galleghan's intelligence officer later defended his CO's ignorance: 'not that the Old Man was against A/T guns, it was that he was against anything being placed anywhere easily visible, & they were. They were right on the road – they had to be!'[93] This attitude, however, betrays an inability to balance the risk of observation against the threat posed by the tanks and a lack of insight into the tactical siting of the weapons. A visit to the sites of both the Bakri and Gemas actions confirms the skill of the anti-tank gunners in their use of the ground to conceal their positions as far as the characteristics of their weapon would allow. It disproves the view of Robertson and Galleghan that these guns could not be concealed and surprise maintained. Anti-tank officers later complained that the tendency of infantry commanders to position them behind roadblocks destroyed the element of surprise.[94] Given their lack of understanding of the employment of these weapons, it was perhaps fortuitous that Galleghan and Robertson ignored the young troop commanders and left them to their own devices.

In the post-mortems of the campaign, field artillery officers also maintained that the infantry had failed to employ artillery to its full potential, largely owing to ignorance of its requirements and capabilities. They complained that the passage of information between battalions and the artillery chain of command was poor and that liaison between infantry and artillery officers was of a similar low standard owing to the lack of joint training.[95] Some COs misunderstood the command relationship between them and supporting artillery batteries and wanted to command the guns right down to 'the last wheel' instead of letting the artillery officers get on with their job. At Gemas, for example, Galleghan sited his supporting artillery in a highly vulnerable position among his forward companies.[96] Fire planning and coordination, including the integration of artillery and battalion mortars, was also poor.[97] On one occasion during the initial Japanese landings on Singapore two artillery batteries were called on to engage 32 targets simultaneously within a battalion sector; doctrinally, the largest number of targets they could engage at the once was two or, in extreme circumstances, four.[98]

The capabilities of the Bren gun carrier were also poorly understood. The absence of tank support led to the misuse of carriers by COs who regarded them as akin to light tanks, despite the fact that at close ranges their armour could be penetrated by small arms fire.[99] Carriers were employed for the purposes of shock action by Anderson during the 2/19th's fighting withdrawal from Bakri on the Malayan mainland, by his successor Andrew Robertson in an attempt to break through a Japanese force blocking the 2/19th's retreat from its initial defensive positions on Singapore Island, and by Varley in a foolhardy 'charge' to stabilise the crumbling front around Reformatory Road on 11 February.[100] In common with Spowers' action at Tobruk, it is difficult to determine whether such action was born from desperation, ignorance or a combination of both. Anderson and Varley's carrier crews were lucky not to have encountered heavy-calibre weapons – which could indicate calculated risks being taken by their COs – but Robertson's suffered heavily, and all of the vehicles involved were destroyed or disabled.

The antithesis of such dangerously aggressive use of carriers was Galleghan's attitude. He considered them to be of little use and employed the 2/30th's carriers only in limited numbers to transport machine guns and ammunition, and left the others far to the rear.[101] The type of actions fought by Galleghan's battalion, however, were well suited to the use of carriers in a more aggressive fashion. A post-operations report noted that carriers were ideally suited to fighting rearguard actions along road corridors being able to deploy heavy sustained firepower – it was common for 8th Division carriers to be equipped with a Vickers gun – then break contact at speed.[102] Both 2/26th and 2/18th Battalion carriers were used in this manner with great success to cover the respective withdrawals from Gemas on 16 January and from around Bulim on 10 February.[103] The 2/30th Battalion, like the 2/18th, acquired two British armoured cars in the course of the campaign. Galleghan made much of his acquisition of these vehicles,[104] but Campbell has argued that he never employed them to their full potential. The armoured cars were an ideal vehicle for utilising as a mobile command post. Given the communications difficulties faced by Galleghan at the time, they would have been of considerable assistance to him employed in this role, particularly if equipped with wireless as were COs' vehicles in the Middle East.[105]

The employment of battalion transport in Malaya was characterised by a lack of imagination on behalf of COs. Despite extant doctrine being relatively sophisticated, several COs seem to have been unable to come to terms with a resource with which they had little experience. On narrow Malayan roads transport was prone to bunch, and the surrounding terrain often made movement and harbouring difficult. Transport harbours were regarded as vulnerable defensive burdens. The response of COs was to reduce the number of vehicles employed in forward positions and to advocate permanent changes to the scale of issue of motor vehicles; it seems that little attempt was made to revise the operating procedures for motor transport.[106] These changes ran contrary to much of the experience of the battalions. The rapid redeployment of the 2/19th Battalion across the Malayan peninsula from Jemaluang to Bakri on 18 January would not have been possible without a full complement of battalion transport and, similarly, the 2/26th's organic transport was critical to its withdrawal following its stand at Namazie Estate on 28 January.[107] This latter move was made only with the battalion's A Echelon transport used shuttle fashion and could have proceeded more efficiently had more transport been available forward. The speed of the Japanese advance was reputed to be owing to their use of bicycles.[108] In fact, much of its momentum was derived from motor transport, and extensive use was made of the large quantity of British vehicles captured along the way.

The challenges COs encountered with the employment of their transport indicated a broader tactical weakness. Transport harbours were not inherently vulnerable nor difficult to defend: the real problem was that the battalions struggled to counter Japanese infiltration tactics. The actions at Gemas, Bakri, Ayer Hitam, Namazie and Ayer Bemban during the withdrawal on the Malayan mainland and around the causeway to Singapore Island demonstrated that Australian COs had a good eye for ground and a reasonable grasp of conventional defensive tactics.[109] Japanese accounts speak of a new tenacity on behalf of their enemy when they met the 8th Division in Johore.[110] The Australians were never surprised in their defensive positions, and Japanese frontal assaults suffered heavily. In none of these engagements were the battalions defeated or forced to withdraw owing to the Japanese penetrating or overrunning their positions, and on most occasions the Australians were able to break contact cleanly and withdraw in good order.[111] The threat was always to the rear, and the nervousness it induced led to the early abandonment of several strong positions and the failure to take advantage of at least one other: a good ambush position in the jungle defile south of Namazie.[112]

Australian COs quite readily adopted an aggressive approach to defence, but their actions were handicapped by a low level of tactical finesse, on occasion an absence of common sense and a reluctance to exploit the potential offered by the terrain for less conventional operations. Much has been made of the ambush actions by the 2/30th Battalion on the Sungei Gemencheh foward of Gemas on 14 January and the 2/18th at Nithsdale Estate on 26–27 January. Both undoubtedly surprised the Japanese and inflicted relatively heavy casualties, but both plans were also flawed. At Gemas, two platoons of the ambushing force had to fight their way across the killing zone to withdraw.[113] At Nithsdale, Arthur Varley's plan for the action required his D Company to move behind an advancing Japanese battalion and thereby hold it against the main battalion position while it was pummelled by artillery. The plan made good use of the artillery's destructive power but left D Company isolated and vulnerable to attack by the rest of the Japanese force. The company became trapped and had to be abandoned to its ultimate destruction when Varley was ordered to withdraw.

The defensive positions occupied during the Australians' withdrawal through Johore were generally fixed linear positions or tight battalion perimeters, located close to the roads and hence easily outflanked.[114] Even within these positions, proper preparations were not always carried out. In a misguided attempt to maintain an aggressive mindset in his

Much of its momentum was derived from motor transport: Australian infantry debus during the withdrawal through southern Malaya. Contrary to enduring opinion, much of the momentum of the Japanese advance was owing to motor transport. Australian COs were slow to grasp its potential despite its vital role during their withdrawal. (AWM Neg. 011303/29)

battalion, Galleghan forbade the digging of weapon pits; when shelled or bombed, his men suffered accordingly.[115] Counter-attacks were generally uninspired frontal assaults, often clumsily executed. At Gemas, Galleghan launched a company counter-attack that ran straight into the Japanese advance less than 300 metres from its start line. The troops attacked without fire support from the battalion's mortars in an effort to achieve surprise, but were required to cross a field of knee-high rubber saplings that prevented the development of any momentum.[116] Only Anderson, most likely reflecting his earlier experience in East Africa, demonstrated any real willingness to attempt to match Japanese tactics and utilise the concealment offered by jungle and rubber for tactical manoeuvre. At Bakri, presaging the tactics that would later be used to great effect by Australian battalions in New Guinea, Anderson used one platoon to attack and hold the front of a Japanese force approaching from the east while another two platoons were swung through the jungle to fall on their right flank; at the same time another company was redeployed to threaten the Japanese rear. One officer noted that the Japanese 'literally ran round in circles'; they were routed and left behind more than 140 dead.[117]

Battalion operations were further hampered by an almost Luddite suspicion of wireless communication by COs. Despite advice to the contrary from signals personnel,[118] they were dubious of the reliability of wireless and feared that its use would allow the enemy to locate their headquarters with direction finding (DF) equipment.[119] The use of wireless within battalion headquarters was discouraged and, in some cases, even banned.[120] Galleghan went so far as to abandon all of his wireless equipment.[121] The 8th Division's chief signals officer observed after the campaign that the persistent disregard of wireless communications by unit commanders resulted in their being out of communication for critical periods of the battles for both Malaya and Singapore, thus causing 'confusion and avoidable casualties'.[122] During the fighting for Singapore, once the field telephone lines were cut artillery fire orders were passed by the wholly inadequate means of flare or liaison officer.[123] The standard issue 108 portable wireless sets were criticised as being bulky and cumbersome,[124] but they were the same type of set used effectively in a mobile role in the Middle East. It must be conceded that the humid conditions in Malaya could have a debilitating effect on this equipment, but this could be managed, and was not a reason to completely disregard its use.

After the campaign, a syndicate of 8th Division company commanders noted that a much greater reliance needed to be placed on wireless communications in infantry battalions.[125] The validity of their claim is vividly illustrated by the 2/30th's ambush on the Sungei Gemencheh. B Company was deployed six kilometres forward of the 2/30th's main defensive position astride a defile on the trunk road formed by the bridge across the Gemencheh and a steep-sided cutting immediately to the east of it. The plan was for B Company to let the Japanese advanced elements cross the bridge and enter a killing zone in the cutting at which time the bridge, rigged for demolition, would be blown sky high. Cut off, much of the advance guard would be slaughtered in the cutting, with the battalion main position forming a backstop for those who got through. In the meantime, the artillery would smash the Japanese troops and vehicles banking up on the far side of the Gemencheh. The lynchpin on which it all depended was communications but, denied the use of wireless, B Company's only link with the rest of the battalion was field telephone, the lines of which had been clumsily laid along the road. Japanese scouts

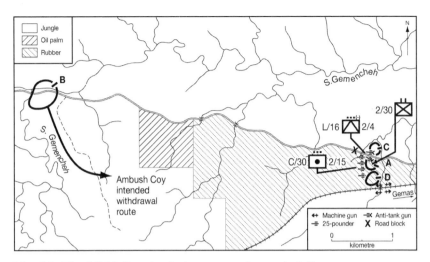

Map 5.2: The 2/30th Battalion's dispositions forward of Gemas

cut these in the first moments of the action, preventing B Company from reporting that the ambush had been sprung and the bridge across the Gemencheh blown. Thus deprived of information, Galleghan, located in battalion headquarters to the rear, hesitated to unleash the artillery. The guns never opened fire, the Japanese were spared the planned carnage and immediately began a counter-attack across the river; they were able to repair the bridge and resume their advance on the 2/30th's main position within six hours.[126] Similar difficulties in coordinating the manoeuvre of infantry and the employment of artillery, resulting from the inherent vulnerability of line communications, were encountered by Varley and the 2/18th at Nithsdale Estate.[127]

PERSONAL EXAMPLE AND CONTROL IN BATTLE

Poor communications, combined with the confined close-quarters fighting of the campaign, meant that COs in Malaya and Singapore were more often embroiled in the thick of the action than they were in the Middle East. This was sometimes by accident and sometimes by design, but the basic challenge of retaining control of the battalion remained the same. While operations in Malaya and Singapore further demonstrate how soldiers looked for and responded to the personal example of their CO, they also highlight how focusing on a small portion of an action could rapidly undermine a CO's situational awareness and hence his ability to command the rest of the battalion. Further, during the fighting for Singapore, the efforts of several COs to seek information or direction in order to improve their ability to exercise their command ironically resulted in a total loss of control and the disintegration of their battalion.

The most renowned feat of direct personal leadership during the campaign was the actions of Charles Anderson during the fighting retreat from Bakri to Parit Sulong between 20 and 22 January. Anderson and the 2/19th Battalion had been hurriedly redeployed to Bakri to reinforce the 45th Indian Brigade and the 2/29th Battalion fighting south of Muar. By the time they arrived their role was to hold open a withdrawal route for their comrades in the face of the outflanking and infiltration tactics that underpinned the Japanese driving charge. By the morning of 20 January, assailed from all directions, Anderson had no choice but to ram a column composed of his own battalion and the remnants of the 2/29th and the 45th Brigade through the parties of Japanese who had cut the road behind them. The men under Anderson's command were exhausted and frightened, and if they were to survive he had to wrench new reserves of courage and endurance from them. Anderson roamed up and down the column, stepping forward whenever example was needed. He led one bayonet charge against a Japanese roadblock, fighting with his revolver and hand grenades, and personally gave the orders for several more. A 2/19th soldier involved in an unsuccessful assault on a Japanese machine gun nest recalled how Anderson's 'calm, clear voice gave us new confidence' and that his 'own private fears abated considerably'. 'Screaming hideously', the soldier and his unwounded mates leapt to their feet and charged the position once more, killing all of its occupants.[128] Many of the survivors of the retreat have stated that without Anderson no one in the column would have survived. Anderson was awarded the Victoria Cross – the only Australian CO to be so decorated during the Second World War.

In many ways Anderson was an unlikely leader; he was myopic and a chain smoker with a terrible cough, and had a gentle whimsical sense of humour. Anderson's command persona, however, suited the retreat to Parit Sulong. Throughout the 2/19th's training he had placed a heavy emphasis on aggression, particularly the employment of the bayonet, and had conveyed the impression that he would be involved in the thick of any action encountered – he always carried several hand grenades on his person.[129] Anderson met his men's expectations in battle; hence they trusted and respected him and responded readily to his orders. Seeking to cover the withdrawal of the 2/20th Battalion to Singapore's Tengah airfield on the morning of 9 February, Charles Assheton similarly based his command on the exercise of referent power. He 'appealed to', but did not 'directly order', a group of machine-gunners to accompany him onto an exposed hill. 'Inspired by his cool approach' they readily followed.[130] COs who had not built up sufficient reserves of trust and respect could not lead in this fashion. When an irritant smoke caused Galleghan's men to break off an attack at Namazie Estate on 28 January, he drew his revolver on the retreating troops and 'threatened to shoot anybody who went farther back'.[131]

Courageous though they were, the actions of Anderson and Assheton, however, could be seen as ill-considered. In the actions mentioned both the 2/19th and the 2/20th had suffered terrible casualties among their officers, and both COs were seeking to conduct difficult fighting withdrawals while surrounded by the enemy. Anderson had to be persuaded by one of his company commanders not to personally lead any further attacks against the Japanese roadblocks.[132] Rash personal action had already killed one CO during the Bakri fighting. Despite a foot patrol having already been turned back by a Japanese road block to his rear, John Robertson of the 2/29th decided personally to report the situation confronting his battalion to brigade headquarters. He set out on the pillion of a dispatch rider's motorcycle and was mortally wounded less than a minute later.[133] The consequences for Anderson's column had he been killed can only be imagined, but Assheton's death, directing the machine-gunners who had followed him, provide an inkling of what could occur. Assheton had already let his control of the battalion slip by remaining with his rearguard instead of seeking to coordinate the rest of the withdrawal, and after his death the 2/20th ceased to function as an effective fighting unit.

The effort to clarify the confusing tactical situation on Singapore also led COs to take personal action that ultimately contributed to a loss of control and the disintegration of their battalions. Early on the morning of 10 February Samuel Pond, appointed to rebuild the 2/29th Battalion from raw recruits after the disaster at Bakri, lost the rest of his battalion while acting as a guide for one of his companies. His subsequent wanderings in search of information took him right back to 8th Division Headquarters, and he was unable to exercise command over his battalion for close to 20 hours. In the meantime his battalion mounted a fighting withdrawal along the Choa Chu Kang Road, initially under the command of the 2iC, then a company commander.[134]

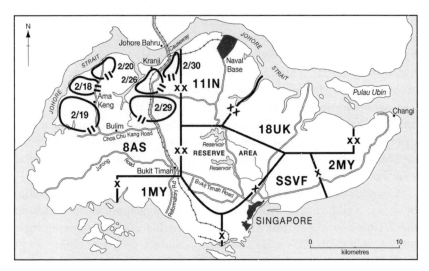

Map 5.3: Australian dispositions on Singapore, 8 February 1942

To demonstrate that the phenomenon of the wandering CO was not limited to Pond and the 2/29th, we can turn to the experience of Varley and the 2/18th on the first morning of the battle for Singapore. Realising he was in danger of losing the ability to command his battalion, he had ordered a withdrawal from its scattered outposts along the coast to a battalion perimeter at Ama Keng.[135] While seeking to consolidate this position he was ordered to report to brigade headquarters and set off on foot. While Varley was away his battalion headquarters moved more than five kilometres to the rear, followed by several parties from the rifle companies, depriving Varley of his principal means of commanding the battalion. When he eventually returned to the fragile perimeter he had to pass his orders in an ad hoc manner, and the fate of his command rested almost entirely on his personal leadership. The position was soon overwhelmed, and Varley had to fight his way out like the rest of his men.

Reflecting on the Australians' defeat in Malaya and Singapore, Galleghan condemned the command practices of some of his colleagues and subordinates: 'Unit and sub-unit comds must be taught to realise their place in battle is where they can receive and give orders and inf[ormatio]n. Mock heroics by moving about was alright last war but not in modern war where brains rather than guts is required of a leader.'[136] Although, like much of Galleghan's writing, these remarks need to be tempered with knowledge of his self-promotional tendencies and his own dependence on a leadership style founded on coercive power, they do highlight one of the major contributing factors to the rapid disintegration of several AIF battalions on Singapore. Confronted with a confusing and increasingly desperate tactical situation, and cut off from their higher headquarters, COs resorted to instinct and fell back on the basic foundation of the Australian command ethos: lead from the front. In so doing they failed to make effective use of their staff; they performed the roles of scouts, dispatch riders and liaison officers; and control of their battalions faltered and, in some cases, collapsed.

The situational factors affecting command, however, must also be considered. The COs' ability to command their units was seriously undermined by the vast frontages that they had been ordered to defend. In addition, they were poorly served by higher headquarters, which ordered them to the rear in the midst of battle rather than the basic principle of command doctrine: that a superior should go forward to consult a subordinate. On the retreat from Bakri, Anderson was lucky. Swamps on either side of the raised road limited his freedom of manoeuvre, leaving frontal attacks as the only tactical option. His battalion was well trained, not full of raw reinforcements as were several during the fighting for Singapore; and it was concentrated, not strung out among acres of tangled jungle. Despite the intensity of the action, it was relatively easy for Anderson to maintain his situational awareness and pass orders, as well as still being able to position himself to inspire and reassure the troops. These remarks are not intended to demean Anderson's courage or achievements, but simply to pose the question of whether he would have performed any differently from his comrades had he commanded on Singapore.

Moral courage

Anderson's actions during the retreat from Bakri also highlighted the moral courage that the desperate fighting in Malaya and Singapore demanded from COs. The willingness to make difficult decisions, including to risk and expend the lives of his men in pursuit of a larger objective, is a key characteristic of a successful commander. It was a strength that not all of the 8th Division's COs possessed in equal measure, and the success of operations suffered as a result. Moral courage, however, should not be an unthinking virtue. Just as operations suffered owing to an excess of concern for the welfare of those commanded or a personal moral ideal, they also suffered owing to an unhesitating obedience of orders.

Retreating from Bakri, Anderson displayed exemplary moral courage. Although a humane man, with an abiding sense of responsibility to his men, he readily ordered some of them to almost certain death to force the column through to Parit Sulong. Anderson's previous mission to hold the Bakri crossroads to allow the 2/29th Battalion and the 45th Indian Brigade to withdraw had forced him to abandon his A Echelon, positioned several kilometres to the rear, when it was attacked. It had little capacity to defend itself, but Anderson did not have the troops both to hold the crossroads and to march to its aid. He stood firm at the crossroads, and his A Echelon was overrun. After reaching Parit Sulong and being unable to wrest the bridge from the Japanese, Anderson was forced to make another decision he 'loathed to take'. He had to leave his wounded behind in order to facilitate the escape of the rest of the column through the swamps to the west. The wounded were subsequently tortured and killed by the Japanese; but the 2/19th 'was still an effective force to build on' and the remnants of the 2/29th Battalion and 65th Field Artillery Battery 'invaluable veterans' for the upcoming defence of Singapore.[137]

Galleghan has been compared unfavourably to Anderson as lacking the moral courage to make difficult decisions. The way he held his artillery during the Gemas ambush because he was unsure of the situation facing his forward company has already been discussed. The next day, as the Japanese advanced on the main battalion position, he again stopped the artillery from firing, despite the urging of the artillery officer attached to his headquarters, because elements of two of his companies were still forward of the main position.[138] The judgement that he lacked moral courage might be unfair, but he does seem to have lacked the physical reserves needed to bolster moral strength. The strain of battle undermined Galleghan's hard-man façade, contrary to his vociferous efforts to elevate his reputation post-war. He was seen crying during the retreat from Gemas; Maxwell recalled he 'rather cracked up' during the withdrawal from Namazie Estate; and by the time the 2/30th reached Singapore his assistant adjutant astutely noted: 'The old, aggressive spirit was subdued, and I have often wondered whether this was only because he was ill, or was it because he shrunk from the possible butchery of his battalion, when the rested and revived Japanese hit us?'[139]

Long after the end of hostilities, while serving as the senior Australian officer in Changi, Galleghan rebuked the officer who returned from the Burma–Thailand Railway with the emaciated survivors of the 2/30th Battalion: 'What have you done to my beautiful battalion?' Another of Galleghan's officers used this remark to reflect on his suitability for command: 'I cannot but compare his remarks [with] the thoughts of Charles Anderson at Parit Sulong Bridge, when his battalion was in a desperate situation and fighting for its life. If he had said, "What are you doing to my beautiful Battalion?", would he have been capable of extracting his unit in one piece.'[140] Galleghan most likely cosseted his battalion out of genuine concern for his men, although it is possible he also sought to protect it as an extension of his own ego. As Campbell has argued, however, the way Galleghan clung to the preservation of the 2/30th is evidence that he was not as hard as his legend suggests, nor perhaps did he possess the moral courage needed to command.

It is ironic to consider that Galleghan and Maxwell, two commanders with little regard for each other, were in fact very similar. Maxwell's performance as a brigadier was continually undermined by the strength of his regard for the men of the 2/19th, which was not even part of his brigade. During the 2/19th's retreat from Bakri, Maxwell made an unsuccessful request to Bennett for 500 men with which to break through to Parit Sulong and open the road for Anderson, despite having to command the 27th Brigade's own withdrawal. Later, after learning that Anderson had been ordered to get out 'as best he could' and head for Yong Peng, Maxwell was ordered to blow the bridge there. He went forward to supervise the demolition personally, the fate of his old battalion clearly on his mind. He delayed the demolition by four hours: 'I'd exceeded my orders by 4 h[ou]rs but we'd got in 189 men mainly of *my old battalion* [my emphasis].'[141]

Ultimately, Maxwell was confronted by the dilemma he had feared when he returned to combatant duty in 1938. His humanitarian ethos as a doctor overwhelmed the resolve needed of a soldier, proving that he would have been a poor CO in action. It is clear from post-war interviews that Maxwell considered the defence of Singapore a pointless waste of life.[142] This attitude caused him to disobey his orders, and order the units under his command to withdraw from the Causeway sector, unhinging the entire defence of northern Singapore, thus expediting the island's fall. Maxwell subsequently gave premature withdraw orders on a further occasion as well as absenting himself from his command to plead with his superiors to surrender, and in doing so he lost all control of his brigade.[143] Maxwell's actions might also throw further light on his decision to remove Boyes from command of the 2/26th and replace him with Oakes from the 2/19th. If his early withdrawal was premeditated, then perhaps he was seeking an officer he knew would comply with his orders without question. Maxwell's actions on Singapore resulted from another professional ethos and an alternative moral ideal, but as a commander they were acts of the greatest negligence and, ironically, left his brigade even more vulnerable.

Maxwell's actions affected his subordinates, who had to reconcile the obedience demanded of them as officers with their duty to their men and their own appreciation of their mission. As in the Middle East, the abiding loyalty of COs in Malaya was to their superiors. When the orders to withdraw from the Causeway sector were passed to the 2/30th Battalion, George Ramsay, now CO, was uneasy and his officers incredulous. They were well aware of the vital significance of the sector and considered their positions 'good and firmly held and able to withstand any attack'. Although Ramsay's doubts led him to consult with Oakes, to whom authority for the withdrawal had been devolved, he never sought to represent his doubts to Maxwell nor to go outside the chain of command and seek to contact Bennett.[144] We have already seen that personal loyalty to his superiors was a strong component of Ramsay's character, and it could be suggested that to question orders was anathema to him. Ramsay's commitment to adhering to the chain of command is further demonstrated by his reluctance on 10 February to send troops forward to the Mandai Road junction at the behest of the 11th Indian Division, without orders from Maxwell or at least official confirmation that the 27th Brigade had been attached to the Indians.[145] Ultimately, however, the situation overwhelmed Ramsay's commitment to the system. Ordered to counter-attack at Bukit Panjang the next morning with no artillery support and no information as to Japanese strength or dispositions, and finding no sign of the supporting battalion at the rendezvous, he abandoned the assault and withdrew.[146]

Ramsay's experience was not unique. Varley had been faced with a crisis of loyalty when Brigadier Taylor ordered him to withdraw while his D Company was fighting for its life at Nithsdale in January. Varley was more forthright in his response. He initially questioned his orders, then sought to avoid them by withdrawing his uncommitted companies and then sending one forward again – not to become involved in the action but to occupy a defensive position and wait for the engaged companies to fight their way out.[147] In the meantime, Taylor reiterated his orders. This time Varley gave in and complied, abandoning D Company to its fate.

It is instructive to step beyond Malaya to consider the experience of another of the 8th Division's COs, Len Roach on Ambon. On discovering that the defence of the island was simply beyond the capabilities of his battalion, he strongly and repeatedly represented his views on the situation to Sturdee at Army Headquarters in Melbourne and eventually to General Wavell at American, British, Dutch, Australian Command. Roach's protests led to his removal from command and the lasting ire of Sturdee: 'He was a squealer from the moment he got to Darwin and . . . From the time that he arrived at Ambon he never let up. His final message was demanding that ships be sent to Ambon to take the force out; that was before the Japs arrived . . . indicating to me that he had lost his punch.'[148] Roach's dismissal confirms that despite the lip-service paid to the individual initiative of officers, the AMF still demanded silent obedience from its COs. Theirs, indeed, was not to reason why.

With the fall of Singapore on 15 February 1942, the surviving COs of the 8th Division faced the entirely new challenge of commanding a demoralised and defeated army amidst the recriminations and deprivations of enemy captivity. The issue of command in captivity is beyond the scope of this study, and much work has already been completed on this subject by Joan Beaumont, Peter Henning and Hank Nelson.[149] In any case, as research by Rosalind Hearder has revealed, the defeat undermined the authority of many combatant officers. They were often supplanted in their leadership role by medical officers, who were invested with large stores of expert power owing to the fact that the principal objective of the 8th Division was now simple survival.[150]

Concluding the 8th Division's report on the operations in Malaya and Singapore, Colonel Jim Thyer referred to the role of inexperience in the division's performance, noting that Australian troops in Libya and Greece had behaved in 'a similar manner in parallel circumstances'.[151] These remarks highlight the 8th Division's major weakness: it was an immature military organisation that was denied the advantage of circumstance bestowed on its sister formations that served in the Middle East. In selecting the division's COs, Sturdee took a gamble. It was the same gamble taken by Blamey: that the battalions would have ample training time and that their first operations would be relatively easy, allowing younger command talent to develop. In the Middle East, this gamble paid off. But Blamey had the benefit of having skimmed the cream off the AMF's small pool of command talent and being able to expose more junior officers to modern tactical thinking at British Army schools. In taking his Malayan gamble, Sturdee lost. The operations there would have tested the best and brightest military minds; they were beyond the capabilities of many of the men Sturdee dispatched overseas.

The 8th Division never had the opportunity to mature as a military organisation. The appointment of the inexperienced Andrew Robertson to the command of the 2/19th Battalion over the head of the tried and respected Tom Vincent, merely because he was senior, is a simple but damning example of this. A command meritocracy, founded on success in battle, had not had a chance to develop. In Johore, when given their chance, and provided with appropriate leadership and direction, Australian troops fought with courage and determination, often to the very last. As the material of which armies are composed, the men of the 8th Division were no different from their countrymen who served with the other three AIF infantry divisions in the Middle East. The rise of such men as Vincent demonstrates that it was not the raw material that was lacking in the 8th Division, just the time, resources and good sense required to develop it to its full potential.

While commenting on drafts of Wigmore's official history, Oakes sought to defend himself and his fellow officers: 'In any case, in a withdrawing action such as we fought in Malaya, with a lack of sea and air support, nobody could possibly do the right thing as far as criticism from a war-winning point of view is concerned.'[152] Oakes makes a valid point that must be considered when assessing the performance of Australian battalion commanders in the Malayan campaign. They were part of a larger command system, which was as ill-prepared as they were, fighting with inadequate resources and in accordance with ill-founded strategy. Ultimately, it was this system that failed. Thyer concluded his remarks on Australian units in Malaya thus: 'Without in any way casting a slur on the individuals who constituted the force, the Japanese units outclassed the units of the AIF in Malaya.'[153]

CHAPTER 6

'NO PLACE FOR HALF-HEARTED MEASURES'
Australia and Papua 1940–42

In March 1942 a circular issued by Headquarters Northern Command warned that the AMF could soon be required to 'fight in defence of our hearths and homes, our own people and our own ideals' against a 'ruthless' enemy: there was 'no place for half-hearted measures in training or operations'.[1] Australia's year of crisis in the Second World War was 1942. The proximity of the Japanese threat exposed Australia's lack of preparedness to fight what was perceived as a war of national survival. The same year also saw the full mobilisation of the Australian nation and the full mobilisation of its army. For the first time, the two components of the AMF – the all-volunteer AIF and the part-volunteer, part-conscripted militia – would go into action together. The discrepancies in equipment, training and leadership that had prevailed since the start of the war at last had to be tackled.

Sapped of its best officers and men by the AIF, the militia had been allowed to stagnate during 1940 and 1941. Most critically, the measures implemented to improve the standard of battalion command before the war were not applied rigorously or universally, and hence no concerted effort was made to develop command talent to replace that lost to the AIF. The solution adopted when the threat of invasion revealed the militia to be far below operational standards was to transfer large numbers of AIF officers, recently returned from the Middle East, into militia battalions. This included a wholesale replacement of militia battalion commanders with no precedent or antecedent in the army's history. The introduction of AIF officers into the militia units, however, complicated command

165

relationships. They had to establish their authority while still outsiders as well as implementing significant change in command appointments and procedures, discipline and tactics.

In action, the COs of both militia and AIF battalions found that their experience in the Mediterranean theatres could not be readily applied to warfare in Papua. The terrain, vegetation and climate of the tropics all severely affected the basic elements of the conduct of operations: communications, tactics and the ability to manoeuvre. The challenges COs faced in Papua were similar to those confronted by their compatriots in Malaya, but they had the added complication of having to fight on a logistical shoestring with little fire support for most of the campaign. Additionally, the sense of desperation that prevailed in the higher echelons of the government and the AMF reinforced a rigid command relationship in which it was expected that orders would be followed without question and at all costs, thereby stifling COs' initiative, impeding their ability to adapt to the new conditions and forcing them to ignore many of the general principles that had underpinned their training.

Battalion command in Papua became a very personal affair. COs fought with their own limited resources close to their troops. In a campaign where soldiers of all ranks and experience had to delve into their deepest reserves of physical and mental endurance, leadership was paramount. The battles of Papua made the reputations of some of the war's most renowned commanding officers but utterly destroyed promising starts made by others.

BATTALION COMMAND IN THE MILITIA 1940-41

In early 1942, as the Japanese advanced rapidly through the islands of the Dutch East Indies and invasion of Australia seemed increasingly likely, Australia's defence remained in the hands of the militia: a shadow force. The formation and maintenance of four AIF divisions overseas had stripped it of the bulk of its command talent. Instead of seizing on this situation as an opportunity to develop a new generation of battalion commanders, the AMF hierarchy had opted to resurrect an older generation. The proceedings of the Promotions and Selection Board for 1940 show a succession of younger officers being passed over for command appointments in favour of veterans of the First World War.[2] Throughout 1940 and 1941 younger militia officers were denied the opportunity to gain command experience because they lacked command experience.³

Curiously, many of the older officers appointed were not serving at the time and, in some cases, had not been active in the militia for many years. Between October 1939 and December 1941, 12 militia battalion commanders were appointed from either the Reserve of Officers or the Unattached List. All had been absent from active duty for a year or more; one had not seen any military service for 21 years. In September 1939 only two officers who had commanded battalions in the First World War were militia COs, but during the period up to December 1941, a further seven officers with this experience were appointed.

The emphasis in command appointment policy at this stage would appear to have been on the ability to organise and train the battalions, rather than training commanders. Militia COs during this period faced a considerable organisational challenge. Their units assembled only periodically for several months of full-time training before standing down for a similar period. Each concentration period would involve the training and integration of large numbers of fresh personnel to replace those who had been lost to the AIF and the other services. This task was further complicated by the shortages in equipment and supplies that would continue to plague the militia until 1943.⁴

The net result was that by February 1942 the average age of the battalion commanders of the militia was 46, two years older than it had been in October 1939. The proportion of COs with First World War experience had similarly increased. In October 1939 there were two veteran COs for every non-veteran; by February 1942 there were almost three. By comparison, there were only five commanding officers with First World War experience remaining in the 27 operational battalions of the AIF, and the average age of its COs had fallen to 38; the youngest were 31.

Some efforts were made to further the training and experience of the militia COs during 1940 and 1941 but these were not universal, and there was no centralised coordination of syllabi or monitoring of standards. Senior command courses were run under the auspices of several of the regional headquarters. Southern Command conducted a month-long senior officers course in mid-1941 for all commanders and staff officers. Among the instructional staff was a British Army major seconded to the AMF to pass on the lessons learned from the operations in France.⁵ All of the COs in Western Command attended a similar course in late 1941. Lieutenant General Iven Mackay visited the school during his tour as GOC Home Forces and commented positively on its standards.[6] The remedial aspect of this course is highlighted by a final report, which noted that the tactical knowledge of Alfred Proud, CO of the 28th Battalion, was 'now good'. Previously, it had been observed that the atmosphere in the 28th was sluggish and there was little 'drive or snap' in the troops.[7]

Lavarack's Command and Staff School was also still operating, but its resources seem to have been partly wasted assessing the fitness for command of the old generation, instead of concentrating on developing the abilities of the new. The experience of three officers who attended the C&SS COs course in the second half of 1941 provide a snapshot of the potential offered by the veteran cohort of officers. Albert Stewart, whom we first met reraising the 37th Battalion, lost that command after attending C&SS. Stewart was found to lack sufficient knowledge to command a battalion but, although 'unimpressive', was still regarded as 'reliable'; it was recommended he command a garrison battalion – a unit of over-age and medically restricted soldiers formed for guard duties in Australia.[8] His classmate, the 'dogmatic' Richard Moss of the 57/60th Battalion, had 'difficulty in adjusting himself to modern conditions' and was also found unsuitable for command.[9] What is most telling is that both officers were only relatively recent appointments, Stewart in August 1939 and Moss in August 1940. Both had been recalled from inactive duty to take command and were described in glowing terms,[10] but both were subsequently found wanting when assessed by modern standards. Another veteran classmate, Rupert Sadler of the 7th Battalion, passed the course and was recommended for command, although he had seen no military service since 1919. Sadler's report noted, however, that he was 'slightly handicapped by deafness' raising the question of how fit he was for operational service.[11]

MILITIA BATTALIONS IN EARLY 1942

In *South-West Pacific Area – First Year: Kokoda to Wau*, Dudley McCarthy writes that the standards of the militia battalions and their strengths varied widely but that the calibre of some veteran officers was such that they would not allow either the training or the spirit of the battalions to be maintained at a 'second-line standard'.[12] This judgement is difficult to sustain. At the beginning of the war with Japan the militia battalions were ill-equipped and poorly trained, and lacked the depth of morale and internal cohesion that would be necessary to see them through the trials of battle. In some battalions this situation would persist into 1943. Government policy was, in part, to blame, but it was exacerbated by the appointment of COs who were past inspiring their young charges to the effort required.

In February 1942 Mackay reported to the Minister for the Army that the spirit of the militia was 'exceedingly high' and that both officers and men would 'fight with the greatest vigour and tenacity'.[13] It was a curious statement for Mackay to make as it had little grounding in reality. At the end of the previous year he had written despairingly of militia exercises, wondering when the 'real battle training' was going to commence.[14] The reports of experienced senior officers appointed to command militia formations following their return from the Middle East in early 1942 noted almost universally low standards of discipline and training in the militia.[15] John Lavarack, now commanding First Australian Army, was scathing in his criticisms of an unsophisticated exercise conducted by the 31st Battalion, which had made no attempt to reproduce battle conditions or to represent the employment of supporting arms or air cooperation.[16] In the Northern Territory, the newly promoted Brigadier Ivan Dougherty was so dissatisfied with the standard of training and discipline in the 23rd Brigade that he issued instructions for the training of sections and platoons and on occasion personally conducted lessons.[17]

The situation was most dire among the militia battalions deployed to Papua. A signal dispatched from the headquarters of New Guinea Force in June 1942 rated the 14th and 30th Brigades, respectively, as 'E: Units have completed training. A considerable amount of brigade and higher training is required' and 'F: Unit training not yet complete'.[18] Brigadier Porter, the new commander of the 30th Brigade, catalogued its ills in July 1942: 'bad discipline, sloppy office work, slipshod methods'.[19] Sturdee had previously branded the brigade's 49th Battalion the 'worst battalion in Australia'.[20] The first battalion deployed to Papua, the 49th was bedevilled for a long time by a running feud between its CO, William Oliver, and its 2iC, Major Harold Barker; the commander of New Guinea Force, Major General Basil Morris, confided in his diary that 'everything about [the] unit was wrong – efficiency, discipline, etc'.[21]

Among the likes of Porter, Mackay and Lavarack there was little doubt as to where the responsibility for the militia's poor standards lay. On taking command as GOC Home Forces, Mackay had addressed a circular to all officers reminding them that the 'standard of training and fighting' in their command depended on their 'influence and leadership'. Officers were directed to be decisive in their actions and clear and explicit in their orders, and to free themselves and their subordinates from 'inertia in all its forms'.[22] Lavarack observed that the poor performance of the 10th Brigade's battalions in mid-1942 was the result of a 'lack of drive on part of B[attalio]n and Co[mpan]y Com[man]d[er]s'.[23] In a similar vein, Brigadier John Crawford bluntly condemned the 11th Brigade's senior regimental officers as being 'far too old'.[24]

A lack of imagination in training was also perceived as a by-product of advancing years, although the humdrum training cycle of 1940–41 probably also contributed to wear the COs down. Tom Louch attributed the low standards of the 29th Brigade to 'unimaginative' training and noted that the practice of having battalions return to the same training area for successive periods of continuous training had resulted in 'lazy' programs, with the same exercises being repeated 'ad nauseam' because officers could not 'be bothered to think out something different'.[25] Even when there was an 'earnest desire to train solidly', a lack of up-to-date tactical knowledge undermined training.[26] After observing 4th Division exercises, Jack Stevens told his senior officers that they were 'criminally stupid' and liable to become 'first class murderers' of their troops should they continue to 'ignore sound tactical principles, create unreal situations' and convey 'wrong ideas of how to fight'.[27]

There are indications that a generation gap may have contributed to undermining command relationships, and therefore efficiency, in the militia battalions. There was a disjunction between the perception of the motivations of the young trainees and the reality. A generation removed from the First World War, they could not be expected to respect automatically the authority of that experience. Dougherty believed that the young trainees were actually suspicious of senior officers as a result of the popular mythology of the First World War. He recalled that during one lesson he conducted for militia troops they grew 'restless and suspicious'. Eventually one soldier asked, 'But has this ever been tried out?' Dougherty recalled that this was typical of the suspicion with which officers were regarded. He summarised this attitude: 'He *must* be a bungler and a "brass hat". Hadn't *Smith's Weekly* said so?'[28]

Unable to draw on such expert power, some COs resorted to founding their command on legitimate and coercive power and struggled as a result. In 1941 Frederick Hale, a long-serving militia officer and First World War veteran, commanded the 58th Battalion. Hale's confidential report for 1940 described him as 'zealous', 'good tempered' and of 'sound judgement'.[29] There are indications, however, that his young soldiers regarded him differently. Jeffery Pooler joined the battalion as an

18-year-old conscript and subsequently served with it in three campaigns, under three different COs. He described Hale as a 'pompous old goat' and 'a figure head who loved parades'. Under Hale, Pooler continued, the battalion was a 'pretty disorganised bunch' with low morale that 'couldn't have cared less'. He concluded that 'Hale was too old and didn't inspire anyone'.[30] The battalion history partly supports Pooler's observations, noting that morale in the 58th Battalion was low at the end of Hale's tenure.[31]

The perception of the Japanese threat also retarded battalion training programs. Throughout early 1942 many Australian-based battalions moved through a succession of invasion stations. Land had to be cleared, facilities developed and defences prepared, leaving little time for constructive and incremental training.[32] Major James Maitland, a former 2/11th Battalion company commander who was administering command of the 43rd Battalion, noted in July 1942 that weapon handling and field craft among his troops was weak and that several months of 'solid elementary tr[ainin]g' were required 'provided war conditions permitted'. The 43rd had undergone little field training before its arrival in the Northern Territory and for most of its time there had been 'scattered' 'on defence and labour duties, which did not improve tr[ainin]g or discipline'.[33] In Papua, the men of the 14th and 30th Brigades spent much of their time labouring on the docks or the construction of infrastructure and defences around Port Moresby.[34] Mackay criticised some COs for being overly preoccupied with preparing for their defensive role at the expense of the training of their troops. In March 1942 he urged all his commanders to balance defensive preparations with training. He also questioned the tendency to hoard ammunition for use by 'indifferent shots'. A small portion allocated to training, he directed, would have an exponential influence on standards of weapon proficiency that would easily offset the reduced quantities available in the case of an invasion.[35]

Several militia battalion histories discuss what one terms the 'scrappiness'[36] of training during this period, but curiously many absolve the COs from any responsibility for this state of affairs. This would seem to indicate an acceptance of the CO as part of the battalion group and that, at least on one level, some militia COs were functioning as effective leaders. Albert Stewart's shortcomings have already been discussed, and the 37th Battalion history describes the training he oversaw as 'listless'. Equipment shortages, and not Stewart, are blamed.[37] Demonstrating how one source of leadership power could compensate for another, Stewart was regarded highly owing to the sensitive manner in which he exercised his command.

His knowledge of modern tactics and training methods might have been lacking, but he had an insight into the motivations of his men. Before militia soldiers had access to 'wet' canteens, Stewart arranged, illegally, for his troops to be provided with beer via the battalion sergeants mess. When married soldiers were charged, any fine he awarded was always half of that meted out to single men. On Stewart's departure from the 37th many of his soldiers gathered spontaneously around the officers mess to farewell him with three cheers.[38]

In *Crisis of Command*, David Horner asserts that the 'elderly World War I veterans were still better equipped to train the newly formed units than the young inexperienced officers who had not joined the AIF'.[39] It is difficult to test this assertion because younger officers were not given widespread opportunities to develop as commanders during 1940 and 1941. Although they lacked active service experience, younger officers might have been more willing to engage with modern tactical concepts and display the imagination and drive that senior commanders believed was so sorely lacking in the veteran cohort. At the very least, a younger group of COs could not have done any worse. Battalions would probably have still not been battle-ready, but a new generation would have at least had the chance to grapple with the administrative challenges of battalion command. Ultimately, what the militia needed to fit it for battle were COs possessing a unique combination of high motivation and up-to-date tactical knowledge born of recent operational experience. Such experience was to be found only among the officers of the 2nd AIF.

THE REPLACEMENT OF MILITIA COs

When Major General Edmund Herring took command of Northern Territory Force in March 1942, he was so dismayed at the state of the two militia brigades under his command that he sacked their commanders and sought the promotion of two young and experienced lieutenant colonels, recently arrived from the Middle East, to reinvigorate training. Roy King took command of the 3rd Brigade and Ivan Dougherty the 23rd. Herring's action was ruthless and immediate, and driven by a perception that invasion was imminent. It was quickly repeated in the brigades. Dougherty was likewise convinced that he had to be 'ruthless' and avoid 'soft heartedness'; he 'got rid of all his colonels'.[40] A new wind was blowing through the militia: 'I realise the difficulty of getting sufficient efficient officers to fill all the officer appointments in our army which has so rapidly expanded, but not for that reason or any other, should inefficiency and lack of energy be tolerated.'[41] The 'Darwin Purge', as it became known, set a precedent for the wholesale replacement of militia commanders at all levels. Gavin Long seems to have had an insight into the thinking of the AMF's senior officers and mused in his diary in July 1942 that there had been such a 'widespread collapse of personal and regimental pride' that the only cure was to purge every 'ineffective or doubtful' officer.[42] Formation commanders had been directed in April to review their officers and provide a return detailing all who were over-age, unfit or inefficient. Commanders were warned that the retention of over-age or unfit personnel could have potentially disastrous consequences. Militia age limits were lowered to comply with those of the AIF, which 'had proved reliable on active service'.[43] The sackings began in the first half of 1942, then continued at a steady rate until mid-1943, across Australia and into the operational areas.

At Port Moresby, the purge had only just begun when operations intervened. When ordered forward to Kokoda in July 1942, the 39th Battalion had just received several 7th Division officers and a new CO, William Owen, who had served with the 2/22nd Battalion in New Britain. No changes were made in the 3rd,[44] 49th and 53rd Battalions before they were committed to action, although the latter two battalions had reasonably young and well-regarded militia COs, but AIF COs were appointed to both the 36th and 55th. In the 7th Brigade at Milne Bay, none of the militia COs were replaced, possibly out of a desire not to unsettle units as they readied defences; Brigadier Field's relative maturity – he was 43 at the time – might also have tempered his actions.

The extent of the purge at battalion level can be gauged by examining the experience of the February 1942 cohort of militia COs. In that month, there were 52 officers permanently appointed as the COs of militia battalions. Within six months, 27 had been posted from their battalions, without having seen action, never to command again; only three were promoted to command brigades.[45] By February 1943, another 11 had gone: within 12 months 75 per cent of the militia COs had been replaced. Only four of the February 1942 cohort would command their battalions on operations.[46] It is generally assumed that the purged battalion COs were replaced with veterans of the fighting in the Middle East, but they comprised only 56 per cent of the officers who followed the February 1942 cohort. Further replacements did occur, however. Ultimately, 65 per cent of militia battalions would begin their first operational deployments with COs who had experienced active service with the AIF earlier in the war.[47] If the battalions deployed to Papua before August 1942 are excluded, this percentage rises to more than 75 per cent.

The purge brought the standards for battalion command in the militia into line with that of the AIF. The replacement COs had an average age of 37, and at least 18 had completed a senior command course of some description before taking up their new appointments.[48] Echoing the experience of the AIF, the purge also saw the appointment of the first two militia COs who were younger than 30. The powerfully built William 'Bull' Caldwell, a former 2/2nd Battalion company commander, became the CO of initially the 5th, then the 14th Battalion, and Jack Amies, a veteran of the 2/31st Battalion, was appointed to the 15th Battalion. Both men were only 28.

During the same period, the British Army also initiated a purge of its older officers. The circumstances under which it occurred and its results were remarkably similar to the Australian experience and provide further evidence of a widespread belief that officers in their mid-forties lacked the physical and mental stamina to command battalions on operations. In mid-1941 the average age of infantry COs in the largely home-based British Army was close to 46.[49] Divisional commanders reported that many of the COs were showing signs of losing their 'drive and efficiency as a result of their age', and it was feared that the army would soon be committed to widespread operations with a number of COs who were likely to fail in battle.[50] As a result, a review of all officers over the age of 45 was initiated, and, as in Australia, 'ruthless' was one of the adjectives used to describe its conduct. Close to 18,000 officers faced scrutiny. Approximately 78 per cent were retained, 11 per cent were moved to less active employment and 10 per cent were retired.[51]

In Australia, old patterns of patronage continued to influence the appointment of many of the new COs. On his appointment to command the 3rd Division, Savige requested four officers from his old brigade for promotion to battalion commands in the division.[52] The appointments made within Northern Territory Force in early 1942 demonstrate the various relationships that guided promotion in the AMF. Rupert Sadler was replaced as CO of the 7th Battalion by John Wilmoth. A Melbourne lawyer, and Blamey's original ADC, Wilmoth had not served with a battalion on operations and seems an odd choice for the revitalisation of a unit on Australia's front line. Wilmoth, although not doubting his own abilities, attributed his appointment in large part to being known to both Blamey and Herring through the Victorian legal fraternity.[53] The 7th's sister battalion in the 23rd Brigade was the 8th. Its CO, Robert Wallis, was replaced by Keith Montgomery, formerly 2iC of the 2/4th Battalion. It is probable that Montgomery's close working relationship with Dougherty contributed to his new appointment.

The commanders of both the 27th and 43rd Battalions, Lindsay Farquhar and Albert Baldock, were replaced by officers from the equivalently numbered AIF battalions: Alexander Pope from the 2/27th, who had relinquished command of the 43rd Battalion in early 1940 to join the AIF as a major, and Clifford Frith, the former 2iC of the 2/43rd Battalion.[54] A quarter of the COs appointed to militia battalions during the purge took command of units from their home states, indicating that some senior officers might have been trying to preserve the regional identities on which the AMF was based. It was not an official policy, however, and Dougherty reflected that it was 'dangerous thinking, since it could result in a unit not being commanded by the best leader available'.[55]

Although the replacement of COs deemed to be over-age or inefficient was enacted ruthlessly, it was followed through in a style more in keeping with the gentlemanly ethos of the senior members of the AMF's officer corps. Most of the replaced officers were found roles in garrison battalions, with the Volunteer Defence Corps or in rear-area postings and in some cases were 'kicked upstairs'. Muir Purser of the 36th Battalion was dismissed when approximately 600 members of his battalion, due to embark in Sydney for Port Moresby, went AWL in protest against pre-embarkation leave that was deemed too short.[56] Purser, however, was subsequently appointed as CO of the 8th Infantry Training Battalion. Similarly, when Robert Sadler was removed from command of the 7th Battalion he was appointed to command the 14th Line of Communications Sub-Area. To some of the younger officers who replaced them this did not seem wise. They believed the lack of drive thought to be endemic to the group would be as detrimental to support units as it had been to battalions. With the training of his battalions suffering from a range of supply shortages, Dougherty wrote to Herring bemoaning the 'fallacious idea' that officers without 'sufficient "drive" to carry out their duties in a forward unit' were capable of efficiently filling a rear-area position at the same or higher rank.[57]

REVITALISING THE MILITIA BATTALIONS

'Drive' was not a characteristic lacking in the newly appointed COs. Major John Rowan had been awarded a Military Cross for his leadership in the Wadi Muatered attack at Bardia, and his first reaction on taking command of the 37th Battalion in September 1942 was: 'Just how are we going to turn this bloody rabble into soldiers?'[58] Motivated by their

The results of the purge: Lieutenant Colonel Rupert Sadler, CO of the 14th Line of Communications Sub-Area, Northern Territory Force (right), inspects the produce of the 1st Australian Farm Company with the Governor General, Lord Gowrie, Adelaide River, 3 August 1943. Sadler was a First World War veteran who had previously been CO of the 7th Battalion and, like many of his militia contemporaries, was moved into a logistics command when replaced by a younger 2nd AIF officer. (AWM Neg. 055940)

insights into the demands of modern warfare gained in the Middle East, a perception of imminent invasion,[59] and their superiors' expectations that they would bring about change in a hurry, the new COs generally acted with speed and resolution. Their commands exhibited the same strict discipline and intensive training that characterised those of 9th Division COs in the lead-up to Alamein. Officers and NCOs were replaced, disciplinary standards were set and rigorously enforced, and training was pursued with a vigour and sense of purpose not previously witnessed in the militia. One soldier recalled that the new CO of the 6th Battalion, Major Bill Egan, a veteran of fighting in Libya, Greece and Crete, had the battalion 'working like dogs': 'He turned night into day – for months we lived out of our packs, no tents, and reveille was at 6pm. He had us all buggered. He said, "You're fit, but not tough enough, so the battalion will march out tomorrow on another stunt."'[60]

John Wilmoth was typical of the new regimes in the militia battalions. A product of the Melbourne University Rifles and an embodiment of its elite ethos, Wilmoth described himself as 'very bloody strict'. He set the tone for his command within moments of his arrival; he found the battalion partaking of its afternoon siesta and immediately abolished the practice. On his first parade he told the battalion he would not waste time with rhetoric, but laid down the standards he would henceforth expect – including a complete cessation of 'filthy language' with a fine the equivalent of one week's pay to encourage compliance. Wilmoth believed a CO should not be overly friendly with either his officers or his men. He also adhered to the maxim that there were no bad soldiers, only bad officers, and his reforms began with them. On one occasion a junior officer was unable to tell Wilmoth what his platoon had eaten for breakfast: the officer's meal was immediately thrown out, and subsequently no officer was allowed to eat before ensuring that his men had been properly fed. Leading by example, Wilmoth often ate in the soldiers' mess. He expected perfection from his officers and competence from his NCOs, and if either failed to measure up the consequences were simple: 'I fired them.'[61]

The measures implemented by the new COs were often not popular.[62] They destroyed established routines and modes of behaviour and thus failed to conform to the existing expectations of a CO, leaving little foundation for trust. The 29/46th Battalion's history speaks of the 'shock' caused by the 'tougher training' and 'stricter discipline' implemented by Kenneth Cusworth, the former 2iC of the 2/14th Battalion; Wilmoth had no doubts that his troops were not initially impressed by his actions; and the 6th Battalion history records Egan's period of command as 'not necessarily popular'.[63] As the new regimes continued, however, and expectations and practice were brought into alignment, the new COs started to gain the trust of their men. As the 29/46th's history records, '... the men soon realized that the new routine was producing results.'[64] The expert power conferred on the new COs by their experience in the Middle East provided firm footings on which to base their leadership. One of Egan's men reflected: 'When we got Egan, we had someone who knew about the game.'[65] Another soldier echoed these sentiments: 'It was [Egan's] influence and tough disciplinary style that turned the unit around. The men accepted this because Egan had been through it all in the Middle East.'[66]

The success of these stern disciplinary regimes was aided by the insights into the needs and motivations of the troops provided by active service overseas. Wilmoth made his men feel that their opinions mattered by hosting an open forum after the weekly church parade. The troops were encouraged to vent their frustrations, as long as they were also prepared to offer solutions.[67] Among other measures, he also implemented a wide sporting program within the battalion to foster morale, had fish traps built on a nearby river to provide varied rations and made sure that praise was delivered when due: 'If you gave them a pat on the back they were happy to try harder next time.'[68] In the 6th Battalion, Egan introduced more liberal leave policies and had kegs of beer introduced to the soldiers' canteen to reduce the time spent waiting for refreshment at the end of a long day's training.

Divisional and brigade reports chart a steady improvement in the standards of the militia battalions throughout 1942 and into 1943. The full extent of the contribution of the newly appointed COs to this improvement, however, is difficult to measure. New divisional and brigade commanders were also appointed during the purge, and in late 1942 and early 1943 increasing quantities of stores, weapons and equipment began to reach the militia battalions. Additionally, the lot of the new COs was made easier by a degree of manning stability in their battalions that their predecessors had not enjoyed. What reports make clear is that neither the purge nor the improvements it brought about were uniform. By late 1942 command appointments in Northern Territory Force had largely been settled. The force's records for the period reveal a steady improvement in standards and increasing complexity in training activities.[69] The 4th Division's war dairy for the same period, however, reveals a GOC still dissatisfied with the performance of his battalions and members of the February 1942 cohort of COs still being replaced.[70] There are clear indications, however, that divisional commanders believed the changes in command were having the desired effect. The Allied Land Forces combat efficiency report for October 1942 notes a steady improvement in the standards of the battalions in the 11th and 29th Brigades following the appointment of several new COs.[71] By the following month, the improvements in the 29th Brigade attributable to the new COs, and now a new brigadier as well, were described as 'very pronounced'.[72]

COMMAND IN PAPUA

While the bulk of the militia battalions were made battleworthy in Australia, some were already fighting the Japanese in Papua, alongside AIF battalions of the 6th and 7th Divisions recently returned from the Middle East. The Papuan campaign essentially consisted of two phases. The first was the struggle to blunt and turn the Japanese offensive, and it included the withdrawal along the Kokoda Track from Awala to Imita Ridge between July and September 1942 and the defence of Milne Bay in a fortnight spanning late August and early September. The second phase was the pursuit and destruction of the Japanese once logistic overstretch and reverses on Guadalcanal brought about a retreat to the Papuan north coast and the establishment of several defensive strongholds. This phase featured a grinding advance back along the Kokoda Track between September and November, followed by costly and drawn-out battles at Gona, Buna and Sanananda that dragged into the new year.

The 28 infantry COs of the Papua campaign represented considerable military experience. Twenty-one had already seen service in the Middle East, the majority in Lebanon, and most had proven themselves capable company commanders or battalion 2iCs. The group included several decorated officers, but for all but one of the group[73] the Papuan campaign was their first experience of operational battalion command. Reflecting the experience of the Mediterranean campaigns on promotion within the AIF, the average age of the battalion commanders in six of the seven infantry brigades involved in Papua – four AIF and three militia – ranged between 33 and 38. The physical demands of the Papuan campaign, in particularly the Kokoda Track operations, are revealed by the fact that four of the five COs aged 40 or older in these brigades had been replaced before its end. The exception was the 7th Brigade at Milne Bay, the COs of which had an average age of 47 and remained in command long after the active operations had come to an end. The Milne Bay operations, however, were relatively short in duration and static in nature.

Experienced though they were, COs in Papua faced a challenging operational environment. Phil Rhoden, who as a 27-year-old captain took command of the 2/14th Battalion during its withdrawal from Isurava on the Kokoda Track, quipped, 'Give me open warfare any day. Jungle I wouldn't wish on anyone.'[74] The terrain of the Mediterranean theatres, particularly the desert, had offered several advantages. Generally, it provided wide fields of observation and fire, the ability to manoeuvre and, barring enemy interdiction, reliable lines of communication. Papua, however, was different. Jungle, swamp, kunai patch and highland ridge all limited fields of fire and observation, prevented the use of motorised transport, and severely restricted and channelled foot movement.

Although the campaign saw the steady development of resupply by air, logistics away from the coastal fringe were reduced to what could be man-packed. This limited the number of troops who could be deployed forward

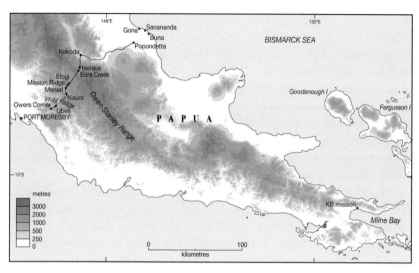

Map 6.1: Papua and the Kokoda Track

and initially led battalions to abandon much of their signals equipment and almost all of their heavy weapons. The battalions of the 21st and 30th Brigades fought during the retreat along the Kokoda Track with only one 3-inch mortar per battalion: the 21st Brigade battalions had yet to be issued Vickers guns, and the 30th's left theirs behind. This placed them at a particular disadvantage to the Japanese who, equipped with lightweight mountain guns and heavy machine guns, were able to outrange all of their weapons.[75] One report noted: 'It is very disheartening to watch an enemy 3000 yards [2.7 km] away inflicting casualties and not be able to get at him except by patrols taking 4/8 hrs [to] return.' The battalions of the 16th and 25th Brigades carried reduced numbers of Vickers guns and 3-inch mortars during the advance back to Kokoda and beyond, but their use was limited by both the terrain and the ammunition supply.[76] Hence, as the 2/10th Battalion's historian noted, what had begun in the Western Desert as a '"modern" war with motor transport and tanks and long-range weapons' had reverted to 'primitive' close-quarters fighting in which only 'bayonet and bullet and hand grenade could clench [sic] the argument'.[77] The campaign presented COs with infantry command in its purest form.

As an officer who believed that an 'eyeful was better than an earful', Phil Rhoden found that one of the greatest difficulties imposed by the jungle was the way it undermined his control. Other COs echoed his

'Only bayonet and bullet and hand grenade could clench [sic] the argument': Australian infantry armed with a Thompson sub-machine gun and an American M1 Garand rifle, close with a Japanese bunker, Buna, Papua, 28 December 1942. The difficulties presented by the terrain of Papua meant that for much of the campaign Australian troops fought with very limited fire support and only the weapons they could carry. (AWM Neg. 013930)

sentiments. Frank Sublet of the 2/16th recalled that he could have a platoon 50 metres away and 'not know they were there', and Paul Cullen of the 2/1st similarly complained that 'you could never see anything' and that his subunits got 'lost all the time'.[78] With few field telephones, wireless equipment that in the extreme humidity was 'subject to more temperament than a prima donna',[79] and a 'ten to one chance runners wouldn't come back',[80] there was little chance of a CO being able to communicate with subordinate commanders if they were separated, by either design or circumstance, from the main battalion position.

These conditions generally resulted in narrow frontages in the advance and attack, and tight perimeters in defence. When moving, COs sought to position themselves much further forward than had become practice in the Mediterranean operations.[81] During the 2/31st Battalion's advance along the Kokoda Track, Colin Dunbar's command group moved immediately to the rear of the forward company, 'much closer than would have been the case in Syria – but so fierce was the enemy's fire and so dense was the jungle that they could not gain an accurate picture of the battle' otherwise.[82] Similarly, during the 2/27th's attacks at Gona, Geoff Cooper positioned himself between the battalion's left and right forward companies: 'It was no good being 50 yards ahead or 50 yards behind in that sort of country.'[83] Being at the very forefront of battle, however, was a dangerous place for a CO. Any 'telltale command or gesture' was a magnet for Japanese snipers; Cooper was wounded by a Japanese bullet at Gona.[84]

In defensive positions, Cooper recalled having little difficulty exercising his command because his companies were close at hand and he could utilise runners to maintain control. He compared this situation to the 2/10th's experience at Tobruk where companies, platoons and sections were spread out over hundreds of metres.[85] He did, however, concede that the exercise of command at Tobruk was made easier by the largely static nature of the operations there, whereas during the fluid fighting on the Kokoda Track it was easy to lose control of scattered detachments. Tight defensive positions, however, had a number of tactical weaknesses. They presented concentrated targets for indirect fire weapons, were easy to outflank and encircle, and limited a CO's ability to manoeuvre his subunits for tactical advantage. The defensive position occupied by the 2/10th Battalion at KB Mission during the battle for Milne Bay was not much more than 300 metres across at its widest point. After visiting the site, Peter Brune commented, '. . . it seems such a small battleground for a force of this [battalion] size.'[86] With no wireless or line communications in the battalion, the 2/10th's CO, James 'Jimmy' Dobbs, was dependent on his voice and runners to exercise command. Even within this tight perimeter, however, Dobbs was unable to maintain control once Japanese attacks began.[87]

As the fighting in Papua progressed it became apparent that COs would regularly have to accept diminished control, or even a complete loss of control, over their subunits.[88] In addition to the control difficulties imposed by dense tropical vegetation, the cover it provided also meant that contact could occur with little warning, and the enemy could withdraw just as quickly. As the 21st Brigade report on the operations commented, it resulted in a situation in which subordinate commanders had to have a clear understanding of their CO's intent and be 'trusted implicitly'.[89] Wherever possible, Rhoden sought to communicate with his subordinate commanders in person to ensure that they understood what was expected of them, and it became his practice for the rest of the war.[90] Likewise, Cullen learnt to trust in the abilities of his officers and soldiers: 'We prevailed because the quality of our soldiers was marvellous.'[91]

The observation and control difficulties imposed by the jungle, combined with the fight and withdraw delaying tactics employed by the Japanese during their withdrawal from Ioribaiwa, resulted in the development of the first battle drills in this theatre. Essentially these involved the first subunit that contacted a Japanese position maintaining pressure to the front, while successive subunits in the line of advance automatically deployed to locate and exploit the Japanese flanks.[92] Although simply a means to overcome the inertia of sudden contact, these drills seem to have become a tactical end in themselves. The 25th Brigade report on operations in Papua advised that these battle drills be used to replace 'normal battle procedure', which was prevented by time and the lack of observation. This statement illustrates a dangerous conclusion reached by both the 25th and 16th Brigades: that detailed reconnaissance by senior officers was 'impractical', that waiting for patrols 'caused unnecessary delay' and that it was almost impossible to gain any accurate information on enemy dispositions or strengths by these means.[93] These conclusions are contradicted by the 39th Battalion's report on its operations at Gona. It details successful means by which small patrols could be used to locate Japanese posts and their arcs of fire.[94] Flanking manoeuvres were the key to success in many engagements in Papua, but the danger inherent in launching them in an instinctive, blind fashion is well illustrated by the experience of C Company of the 2/14th Battalion at Gona on 28 November 1942. Having unexpectedly encountered a Japanese position, the 2/14th's CO, Hugh Challen, committed C Company to a flanking move. The unreconnoitred enemy front was too wide, and C Company suffered heavily, losing all three platoon commanders as well as its OC.

The fate of C Company on 28 November highlights the high casualties suffered in the fighting in Papua and the strain it imposed on the chain of command, and particularly a CO's control, as the numbers of experienced personnel dwindled. After fighting at Buna, the 2/9th Battalion received

15 reinforcement officers and was committed to battle at Sanananda with company commanders who had never commanded a platoon in action, let alone a company. Major William Parry-Okeden, the 2/9th's 2iC, was in command at Sanananda after the wounding of the CO at Buna. He recalled having to go forward to encourage a company, commanded by a reinforcement lieutenant, to keep moving: 'Lloyd had completely lost control of his coy. They had come under heavy fire . . . and Lloyd wasn't doing anything about it. The poor bugger really didn't know what to do.' Parry-Okeden had to explain the principle of fire and movement and arrange some mortar support. Lloyd was killed not long after.[95]

A dearth of experienced officers could also have a critical effect on the maintenance of morale in a battalion. Unlike most of the other militia battalions, the 53rd Battalion went into action in Papua without experienced AIF officers. Poorly trained, its companies floundered during the fighting around Alola and Abuari in August, particularly after the death of its CO, Kenneth Ward. The battalion subsequently played no effective part in the Kokoda Track fighting and was eventually merged with the 55th Battalion. The 30th Brigade report on the campaign noted that after Ward's death 'there was no one to undertake a deliberate policy of marshalling and repairing' the battalion's 'impaired morale'.[96]

DESPERATION AND COMMAND

The proximity of the Papuan operations to the Australian mainland, and the unbroken run of success enjoyed by the Japanese to this point, resulted in their being subjected to considerable scrutiny by both the Australian Government and General Headquarters South-West Pacific Area (SWPA). Both organisations lacked experience, and the attitude that prevailed could be best described as one of desperation, and at worst, panic – encapsulated by Prime Minister John Curtin's remark following the fall of Singapore that the 'battle for Australia' had begun. Both General Douglas MacArthur, Supreme Commander SWPA, and Blamey, now Allied Land Forces Commander, feared for their jobs should the Japanese not be quickly turned back and defeated in Papua. Operational commanders were continually badgered about the progress of their operations, with little regard for their logistical difficulties or the conditions being endured by the fighting units. Withdrawal was seen not as a rational tactical option, but rather as indicating a lack of resolve in the face of the enemy; offensive operations had to be pursued with all speed and at all costs.[97] Formation commanders operated under extreme pressure, which was summed up in a memo from Vasey, as GOC 7th Division, to Dougherty, commanding the 21st Brigade at Gona: 'Canberra must have news of a clean up and have it quick or we will both go by the boot.'[98]

On the battlefield this sense of desperation resulted in battalion commanders often being bustled into action without the opportunity to rest exhausted troops, conduct reconnaissance and proper battle procedure, concentrate their force or wait for the accumulation of adequate supplies of artillery ammunition. In addition brigadiers attached and detached companies at will, undermining both control and unit cohesion.[99] COs routinely protested against the folly of such orders and requested more time to prepare, but to no avail.[100] COs were expected to act 'without equivocation or mental reservation of any kind', despite the dubious nature of the operations they were ordered to conduct. Any hesitation in coming to grips with the enemy was viewed as a lack of aggression, indicating an unfitness to command. Cooper was widely criticised in senior quarters for 'avoiding making contact' with the Japanese at Menari and Nauro during his withdrawal from Mission Ridge.[101] This was despite the 2/27th having been cut off behind enemy lines with little food or ammunition and encumbered with more than ten stretcher cases.[102] To have attacked at either Menari or Nauro to attempt to regain the main track or harry the rear of the advancing Japanese would have resulted only in the 'piecemeal and fruitless annihilation' of the battalion.[103] Instead, Cooper bypassed both villages and, although his command during the withdrawal was not faultless,[104] he eventually regained Australian lines with approximately 250 of his men; another large party of 2/27th troops, separated from Cooper early in the withdrawal, also returned safely.

The result of precipitate action forced on battalion commanders was badly conceived and executed operations and horrendous casualties. Examples of such operations occurred on all of the major Papuan battlefronts. At Milne Bay, Dobbs and the 2/10th Battalion were rushed forward on the afternoon of 27 August 1942 to meet an enemy of undetermined strength and intentions, but definitely known to be equipped with tanks. The men of the 2/10th had had little sleep for the past 48 hours, and the battalion would have no support once it left the main defensive positions around the airstrips. Its foray had no clear objective: it was partly reconnaissance in force, partly an embryonic counter-attack, and would ultimately become an ill-prepared blocking force at KB Mission.[105] The 2/10th was lost to Milne Force for no result; the eventual defeat of the Japanese as they collided with well-sited defences with clear fields of fire at

Number Three Strip, and their destruction by pursuing forces thereafter, confirms the premature folly of sending the 2/10th forward.

At Eora Creek, as they followed up the Japanese withdrawal, the 2/1st and 2/3rd Battalions confronted one of the most formidable natural defensive positions along the length of the Kokoda Track. The Japanese were ensconced on high ground that dominated the two bridges over the creek and the track immediately beyond.[106] 'Well aware of the need for haste', Brigadier Lloyd, commanding the 16th Brigade, ordered that a frontal attack be made straight across the Japanese killing ground at dawn on 23 October. In doing so, he ignored the protests of the COs of both battalions.[107] Cullen, who had already dispatched a company to outflank the Japanese position to the west, urged Lloyd to develop this advance into the brigade's main effort but was told, 'You have your orders.' He was well aware of the pressure being placed on Lloyd to 'get on quickly' but has also reflected that Lloyd was exhausted and his tactical imagination slipping: 'He didn't know what I was talking about . . . he had a Somme mentality.'[108] After the 2/3rd had secured the two bridges under cover of darkness, the 2/1st would make the assault on the main Japanese positions, despite lacking two of its rifle companies: one was already committed to the outflanking move, and Lloyd had earlier detached another to reconnoitre an alternate route to Aola. The 2/1st had no hope of securing its objectives. Meanwhile, the 2/2nd remained uncommitted to the rear.

Ultimately, the 2/3rd did not secure the bridges, but Cullen was able to infiltrate his troops across after a risky personal reconnaissance and with the assistance of a great degree of luck. With dawn, however, they were pinned down beneath the main Japanese position and made little progress for the ensuing five days. The battle for Eora Creek was brought to a successful conclusion only once a concerted assault was mounted on the Japanese west flank between 27 and 28 October. Characteristic of the confused and piecemeal fashion in which force was concentrated during the Papuan battles, this attack was mounted by three companies of the 2/3rd, with a company of the 2/2nd attached.[109]

At Buna, Brigadier Wootten ordered the 2/9th Battalion into action against the extensive Japanese bunker system around Cape Endaiadere the morning after it had completed an approach march of 30 kilometres through swamps and tidal creeks. Its CO, Clement Cummings, had wanted to rest his 'buggered' battalion for a day so that his men could have 'a good sleep and plenty of tucker' before the attack.[110] An operational pause would also have allowed a reconnaissance of the well-sited

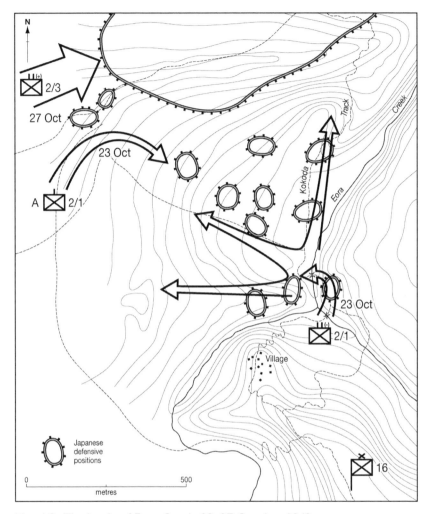

Map 6.2: The battle of Eora Creek, 23–27 October 1942

and heavily camouflaged Japanese positions and liaison with the supporting armoured unit. Instead, the 2/9th entered battle on 18 December tired and blind; the closest Cummings came to a reconnaissance was surveying the battlefield from the stump of a coconut palm, and he saw no sign of the enemy.[111] The ensuing fight lasted for six days and cost the 2/9th 374 casualties – more than 50 per cent of its strength.

For several COs, these battles were a trial of their training and the values that underpinned their commands. All of the Papua COs interviewed for this book reflected that their worst experience in command occurred there. Cullen remembered that he stood 'trembling with emotion . . . with fury, with irritation' when Lloyd reiterated his orders at Eora Creek, and similarly Cooper recalled 'terrible moments of horror' watching his troops being cut down by Japanese machine guns at Gona.[112] Cooper also felt slighted that his professional judgement was construed as a lack of personal courage. Expressing doubts about orders, he maintained, was 'not being cowardly; that's just insisting on time to make the right preparations and the right analysis of the situation'.[113] Yet, in spite of both the moral anguish and professional reservations many COs felt, there is no documented incidence of any of them directly refusing to act as ordered. As had been the case in earlier operations in the Mediterranean and Malaya, loyalty to the chain of command retained its primacy. Asked if he had considered refusing Lloyd's orders, Cullen replied: 'You have no alternative, you either do it or you tell the brigadier to go and get to hell . . . That would have done no good.'[114] Cooper expressed similar sentiments about the terminal authority of orders. He questioned his at Gona, but the orders for the attack in dispute were confirmed and it took place: 'You can't make a song and dance . . . it's an order and it has to be carried out.'[115]

The reflections of COs, however, point to a more complex code of behaviour than simple obedience to the legitimate authority of a superior commander. A CO's perspective was limited; he was not always aware of the situation on other parts of the front and how his actions would affect other friendly troops. Although he was greatly concerned by his orders for the 39th Battalion's first attack at Gona, Ralph Honner believed that 'they must have fitted into some plan that was beyond my conception'.[116] Orders were ultimately carried out without question out of loyalty to the officers and men of other units. Cooper reflected that, as one part of a larger organisation, he couldn't 'say "Well, I don't like this" and sit down and boil the billy' while other fellows went ahead and got 'shot to bits'.[117]

The strength of the code of loyalty described by Cooper is manifest in bitter and persistent rumours about the actions of Alan Cameron, the CO of the 3rd Battalion, at Gona. Cameron was ordered to manoeuvre his battalion to support the southern flank of an attack being mounted by the 2/27th on 1 December. He objected on the basis that this required the 3rd to realign its north-facing positions to the west and that the orders were conveyed only at 21.30 hours for a dawn attack the next morning.[118] Nevertheless, the 2/27th attacked as planned. Despite the vicious Japanese machine-gun fire that was a characteristic of all of the

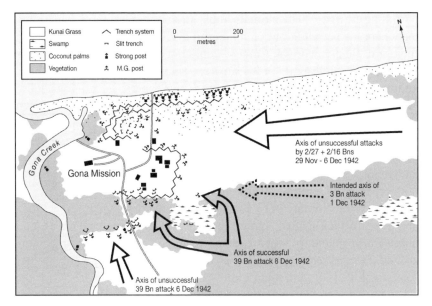

Map 6.3: Australian attacks at Gona, 29 November – 8 December 1942

ill-fated attacks on Gona, an attached company from the 2/16th managed to secure a precarious foothold in the heart of the defences around the village, but was ultimately repelled with crippling casualties. Cameron's battalion never took part.

Although his orders had specified both a time and a direction for his advance, Cameron held the 3rd Battalion back because Cooper's southernmost company was not sighted and, it was therefore assumed, had not advanced. Dougherty later wrote there was 'no excuse' for the 3rd not having moved forward, which should have advanced 'on the timing given' because in 'country like that physical "linking" is not practical'.[119] The 3rd was subsequently found much further south than it had been reporting, possibly as a result of both poor maps and a failure by the 25th Brigade's headquarters to determine accurately the position of its units on the ground.[120]

Honner and the 39th Battalion relieved the 3rd, and Honner believed it had not even been in contact with the enemy, although Cameron had been reporting grenade duels to brigade headquarters. Writing in 1955, Honner noted that the covered approaches to the Gona defences through the 3rd Battalion had been overlooked 'while bloody losses were sustained in reckless attacks elsewhere', implying a criticism of Cameron's patrolling efforts.[121] The ultimate implication is that Cameron had knowingly kept his battalion out of contact to avoid casualties and thereby let the rest of the force down. Cameron does not seem to have been popular with either his colleagues or his superiors, being regarded as a 'bombast' and 'unreliable'.[122] The extent to which the experience of Gona influenced these judgements can never be fully determined. Honner's own code of behaviour is clearly evident in his actions at Isurava on 28 August when he requested that the 39th Battalion be allowed to remain in place, despite having been relieved by the 2/14th, because he expected the coming Japanese attacks would be too strong for one battalion alone. He was not prepared to leave the 2/14th 'in the lurch'.[123]

There were alternatives to directly disobeying orders. After the 39th's first frontal attack against Japanese bunkers at Gona failed, Honner found himself in a quandary when ordered to mount another over the same ground. He was reluctant to disobey his orders, but 'didn't want another company killed off' and 'didn't know how to get out of it'.[124] When the air support for the intended attack landed behind the 39th's own positions, Honner was presented with a pretext to ask for its cancellation.[125] Dougherty agreed. Honner thus had the opportunity to conduct a personal reconnaissance of a covered approach to the flank of the Japanese defences, then recast the attack plans to exploit it. Launched on the morning of 8 December, Honner's second attack led to the fall of Gona and demonstrates what could be accomplished when time was allowed for proper battle procedure and there were sufficient reserves to reinforce success. Closing on the Japanese defences from the covered southern flank, one of Honner's companies breached the outer perimeter, and another two were sent through behind it to attack the main positions from the rear. The ensuing fighting was not easy, but the Australians were not attacking across cleared fire lanes, and by nightfall they had cleaved the Japanese position in half, prompting its abandonment after dark.[126]

Honner's plan included several subtle but critical details. He had the supporting artillery fuse their shells for delay so they would penetrate trees and bunker roofs before exploding and transmit their shock wave below ground even if they missed their target. He also adjusted the timings of the attack so that the assaulting troops would arrive at the first line of bunkers and trenches while they were still under artillery fire and their occupants were still sheltering below ground. These were arrangements that seem not to have been made during previous attacks at Gona. Whether it was owing to an inherent lack of tactical imagination on the part of other COs or an oversight caused by exhaustion and stress we can only surmise.

'Kill the bloody enemy': Frank Sublet (right) was one of several hardened and pragmatic commanders who emerged from the trials of the Papuan campaign. His predecessor as CO of the 2/16th Battalion, Albert Caro (left), struggled with the burden of continual pressure for aggressive action from higher command and the appalling casualties that resulted. (AWM Neg. 026752)

The Papuan battles also produced some COs of a rare mindset who were ruthless in the execution of their orders and prepared to push on at all costs literally without question. Sublet, who took command of the remnants of the 2/16th Battalion during the fight for Gona and would command it for the rest of the war, insisted that his main objective as an officer was to 'kill the bloody enemy' and that he was 'never one to query the orders' nor exercise his own judgement unless ordered to, 'and never was'. He observed that Gona broke the 2/16th's previous CO,

Albert Caro, because he was 'too gentle' and felt the loss of his troops too deeply.[127] Sublet's command ethos was pragmatic in the extreme: 'It's no use feeling too bad about your friend who's dead; you can't help him when he's gone . . . You have to think of the bigger picture . . . The show must go on . . . That's the attitude that commanders of all ranks have to take.'[128] It must be noted, however, that Sublet commanded the battalion only for its last attack at Gona. Major Ian Hutchison, who took command of the 2/3rd Battalion at Eora Creek and would also command it until its disbandment, was of a similar mould. McCarthy wrote that his 'deliberate manner suggested the possession of a brain which, having decided on a plan of action, would follow it through to the bitter end with bulldog fixity of purpose'.[129] In 1945, he was still regarded as 'a solid plugging type who would never give up'.[130] Cullen observed that armies need 'people like that', but the dogged resolve of officers like Hutchison and Sublet also appears to have been viewed with suspicion by their fellow COs as signifying a lack of imagination or compassion.[131] There are indications, however, that this demeanour did not necessarily come naturally to these men. Speaking about ordering his men into action at Gona, Sublet still conceded, 'It was pretty hard.'[132]

BATTLEFIELD LEADERSHIP IN PAPUA

When the 'battered, bruised and tired' survivors of the 2/9th Battalion 'limped' off the battlefield at Buna following their relief, they were 'feeling bloody sad and miserable', their acting CO recorded. The next afternoon the battalion was warned to prepare for operations at Sanananda. Upon receiving his orders the first thought that entered William Parry-Okeden's head was: 'Oh Jesus! What am I going to tell my men?' His reaction highlights the tremendous challenge of leadership faced by COs in Papua, particularly during the beachhead operations, of having to persuade sick and tired men to carry out their orders in the face of mounting casualties. Despite the series of rushed and badly conceived attacks mounted at the beachheads, there were surprisingly few instances of men refusing to go into battle, a phenomenon that, to modern sensibilities, is difficult to understand. Several veterans have attributed this to a 'Koitaki Factor', a line of interpretation subsequently followed by Peter Brune. During a parade at Koitaki on 9 November 1942, Blamey uttered the infamous remark that only the rabbit that runs gets shot. Perceiving this as a slur on their conduct during the withdrawal along the Kokoda Track, the men of the 21st Brigade are said to have thrown themselves into action at the beachheads to prove Blamey wrong.[133]

The Koitaki Factor, however, cannot be applied to battalions of the 18th Brigade nor the 25th. In his reflections on the beachhead battles, Parry-Okeden offers another explanation. He believed that what kept the battalion going forward was a sense of mutual pride in which even the CO was embroiled: 'We all hate showing cowardice in the eyes of our mates. I have seen chaps moving forward under heavy fire and not one of them would go to ground in fear that their mates may think they were afraid.' Privately, Parry-Okeden did not deny his own fear, noting that any soldier who maintained he had never been afraid had 'not been in a decent scrap or is a bloody liar'.[134] In battle, just as in barracks, the CO had to set the standard for his men to follow. Honner's adjutant recalled 'looking around to see what Ralph was going to do' when a Japanese mountain gun opened fire on the 39th Battalion perimeter at Isurava on the Kokoda Track: 'We were waiting for a move from him, because we didn't want to let him think we were frightened.'[135] The ability of COs in Papua to stand tall, as a beacon of inspiration, while battle swirled around them has been well remembered. We have already seen how Cullen led his troops across the first bridge at Eora Creek expecting machine guns to open fire at any second, and how Cooper was wounded among his forward troops at Gona. Both Cummings and Dobbs moved about the battlefield at Buna 'almost regardless of danger', the former 'standing up with his felt hat on with enemy snipers all round' 'swearing like a trooper'.[136]

Battlefield leadership in Papua was about providing a stable, predictable presence in an environment in which little else could be controlled. COs were expected to rise above the privations of hunger, disease and exhaustion, the wild confusion of jungle contacts, the desperation of harried retreat and the apparent madness of repeated attacks across enemy killing zones. Aside from their courage, one of the most remarked qualities of successful COs in Papua was their ability to project an air of calm confidence and restrained emotion – as one officer noted favourably of Cummings at Milne Bay, their 'coolness in the face of difficulties'.[137] A 2/16th sergeant attached to the 2/27th for its third attack on Gona remembered the calming effect of a few quiet words from Cooper on the eve of the battle: 'I've never met Colonel Geoff Cooper and I wouldn't know him if I ran into him, but I'd probably recognise his voice even now. Surprisingly I did get to sleep.'[138]

Veterans of the 39th Battalion remembered Honner as being unshakeable, not prone to angry or excited outbursts, and never visibly afraid.[139] His demeanour as a commander is well illustrated by an exchange with Captain William Merritt, one of his company commanders, while both were shaving to the rear of the battalion position at Isurava. Honner received a message that the enemy had attacked Merritt's company. Merritt remembered Honner as the 'coolest man' he had ever seen: 'The CO . . . grinned and said: "Captain Merritt, would you go up to your company when you've finished your shave? The Japs have broken through your perimeter."'[140]

Honner's fellow CO at Isurava, Arthur Key of the 2/14th, is remembered in much the same way. Key was also a quiet man and had been described by Vasey in the Middle East as having 'too much of the manners of the manager of the boys dept at Myer's to be an effective CO'.[141] One of his NCOs commented that he 'did not project an "I would follow you through hell or high water" air, but rather an "I need your assistance" air'.[142] But he was an officer who inspired his troops in action with his 'sureness of touch' and sound tactical ability; the battalion historian noted that under Key's command a spirit of 'complete mutual confidence existed in the battalion'.[143] The manner in which such a spirit was developed is well illustrated by Honner's approach to B Company of the 39th. After being ousted from Kokoda on 28 July, the company had been branded cowards by Cameron, then the 30th Brigade's BM and acting CO of the 39th, and he decided to break it up. When Honner took command he did not follow through with Cameron's dramatic knee-jerk reaction but instead retained it intact and posted it to the most dangerous sector of the battalion perimeter at Isurava – a display of quiet confidence that it could accomplish what needed to be done.[144]

As in the Middle East senior commanders regarded a calm, decisive manner as the mark of a competent CO. Seven battalion commanders were sacked as a result of their performance during the Papuan campaign. Brune, in his various works on Papua, makes the simple argument that those who questioned their orders and made 'sensible protests aimed at reducing the terrible losses' were removed from command.[145] On first examination this appears a sound premise. Caro, Cooper and Dobbs all protested against the manner in which their battalions were employed during the beachhead battles, as had Colin Dunbar of the 2/31st during the advance along the Kokoda Track,[146] and all were sacked. Cameron, Cummings and Honner, however, similarly protested but retained their commands. The reasons for the dismissals were more complex than simply questioning orders. The demeanour of the officers concerned must be considered. As the battle for Buna progressed, Dobbs became 'depressed, malaria-stricken and utterly exhausted';[147] Caro has been described as a man of 'pronounced' compassion and, given an angry and emotional protest to Dougherty, it is hard to believe he could have concealed his feelings from his troops;[148] and the 2/31st Battalion history notes that before Dunbar's dismissal it had become 'obvious' that his 'strength and powers of endurance were failing'.[149] If it is accepted that a calm and confident demeanour was critical to battlefield leadership in Papua, then obviously these officers no longer had the requisite attributes to continue in command.

Similarly, Dougherty harboured significant doubts about Cooper's decisiveness in action that extended beyond the baseless perception that he lacked aggression. Cooper halted the 2/27th's first advance at Gona while searching for an artillery forward observer who had missed his rendezvous with the battalion. This delay contributed, in part, to the battalion arriving on its start line an hour after its assault was due to commence, which had been timed immediately to follow an air strike on the Japanese positions.[150] Cooper's actions at Mission Ridge, when he resited the battalion's defensive positions, laboriously dug with helmets, bayonets and bully beef cans after Brigadier Potts had already resited them once, could also be seen as indecisive. Ultimately, whether Cooper was or was not decisive in his actions was irrelevant. He had lost Dougherty's confidence as well as that of some of his battalion. After the Gona fighting, Brigadier Murray Moten, the 2/27th's original CO, was approached by a deputation from the battalion to intervene on their behalf and see that Cooper did not return after he recovered from his wounds; Dougherty had already lodged an adverse report.[151] As Honner reflected, 'war is a confidence business'; without confidence, command relationships could not successfully function.[152]

The last Japanese stronghold in Papua – Sanananda – fell to Australian and American troops on 22 January 1943, thereby liberating Australia's largest overseas territory and generally quelling the sense of desperation that had pervaded the Australian Government and the Allied armies. The advancing Japanese had been stopped, and planning for a counter-offensive had begun. This strategic progress, however, had come at a heavy cost. The AMF suffered approximately 2000 fatal casualties in Papua, a total that surely would have been much less had the conduct of operations been governed by patience and imagination, as opposed to haste and dogged but predictable resolve. As Geoff Cooper reflected: 'There's a funny old belief . . . that determination can win any battle. Unfortunately the chap who thought it out hadn't seen machine gun bullets by the millions that lay men out.'[153]

For some COs, Papua was a moral as well as a physical ordeal. They found themselves trapped between a code of behaviour that demanded they carry out their orders without hesitation and their operational experience, which told them that to do so without thorough battle procedure was pure folly. As we have seen, that dilemma took its toll on some. Unable to continue to act in a calm, decisive fashion they lost the confidence of both their men and their superiors. This is not to demean their previous good records, nor their personal integrity and bravery, but simply to acknowledge the extreme demands made of COs by the unique nature of the Papua campaign. The experience would continue to haunt some for years afterwards. Jimmy Dobbs was so traumatised by his experience at Buna that he was unable to attend a 2/10th Battalion reunion until 1963.[154] For others, simply enduring brought its own reward. Phil Rhoden recalled the sense of achievement he felt after his period of command on the Kokoda Track: 'Everything was at stake then – and we got out of it . . . If we had failed there was no second chance.'[155]

In Papua's unique operational environment, battalion command was characterised by prominent personal leadership and a command system based on the direct passage of verbal orders. Tactically, the campaign was uninspired. The Australian withdrawal on the Kokoda Track is remarkable for the quality of the battlefield leadership and the courageous timing of the withdrawals and not for its tactical sophistication. Australian offensive operations in Papua were generally mounted without the benefit of intelligence and with minimal fire support; frontal assaults were the norm. The experience of COs in Papua validated the great purge that was still going on in Australia. Simple survival in jungle warfare, let alone victory, demanded COs who were fit and agile, in both body and mind, and able to inspire their officers and men with their physical and mental endurance. It is telling that the most successful and least costly battalion operations of the campaign were those conducted on the basis of sound intelligence that demonstrated a degree of imagination in the employment of manoeuvre and fire support. As we shall see in the next chapter, it was a point that was not lost on battalion officers or the AMF at large.

CHAPTER 7

'THERE IS NO MYSTERY IN JUNGLE FIGHTING'
The New Guinea offensives

The campaigns of 1943 and early 1944 marked a period of consolidation for the AMF. With the perceived threat to the Australian mainland defeated, the AMF was able to take time to reflect on how it conducted operations. The desperate days of 1942 had passed, and the AMF was imbued with a sense that it had wrested the initiative from an enemy who could be defeated.[1] The New Guinea offensives, as these operations were dubbed by the Australian official history,[2] demonstrate a more considered approach to warfare. There was a growing realisation that the jungle was not an alien environment, requiring unique doctrine and the desperately light organisations that were a feature of the Papuan operations, but rather one in which an adherence to established tactical and organisational principles remained critical to success on the battlefield. These needed to be adapted and applied, with the judgement born of knowledge and experience, not abandoned in the face of the unexpected or unfamiliar.

Reflecting this shift in attitude, there was an increasing emphasis on training COs and honing their judgement, not solely as jungle masters but as well-rounded infantry commanders. This emphasis was most manifest in the Senior Officers Wing at the Land Headquarters Tactical School. The experience of the New Guinea campaigns also highlighted the strain of long-term operational service on COs and the need to provide them with opportunities to reinvigorate mind and body. It confirmed the notion, established in the Middle East, that credible battalion command was the domain of the young and fit. In the field, the provision of adequate supporting elements, combined with an operational environment in which Australian forces generally held the initiative, meant COs were able to take a step back. Casualty figures readily demonstrate that there had been a change in the nature of the war that the AMF was fighting. Whereas ten COs became casualties at El Alamein and six suffered this fate in Papua, only four COs became battle casualties in New Guinea during 1943–44.[3]

Although the maintenance of their commands was not as heavily dependent on personal example as it had been in Papua, the COs nonetheless still continued to face challenges in controlling their troops that required them to make decisions about their position in battle and the latitude they accorded their subordinates. The tactics and procedures that evolved in New Guinea were built on the hard-won experience of the Papua campaign. The conduct of operations remained the ultimate test for COs, and the experience of the New Guinea offensives continued to validate the rationale that had underpinned the great purge. In both the AIF and the militia, the New Guinea campaigns were a period when new COs were appointed and established, heralding a period of command stability that would last until the end of the war.

The principal objective of the New Guinea offensives was the capture the Japanese bases along the New Guinea coast, thereby contributing to Allied control of the Vitiaz and Dampier Straits and the Bismarck Sea beyond. These operations presented COs with new challenges operating in mountains, grassland and river estuaries, integrating combined arms and cooperating with American ground forces as well as Allied air and naval forces. Salamaua and Lae were the first objectives and were captured in early September 1943. Salamaua's fall came only after the Japanese had been cleared from the hill country east of Wau, where Australian forces had first confronted them in January. Lae was captured by a pincer action involving the 7th Division air-landing at Nadzab and moving down the Markham Valley, and the 9th conducting an amphibious landing to the east.

With a strong foothold at Lae on the base of the Huon Peninsula, the AMF then began a long advance towards Madang. The 9th Division landed at Finschhafen on 22 September to make its way along the coast, an advance that would later be joined by 4th and 8th Brigades. Simultaneously, the 7th Division once again boarded Dakotas to be flown into the Ramu Valley to climb into the Finisterre Range via a precipitous feature that would become known as Shaggy Ridge. With the Finisterre heights secured, the 15th Brigade took the lead and eventually met up with the coastal advance at Madang on 24 April 1944, thereby ending

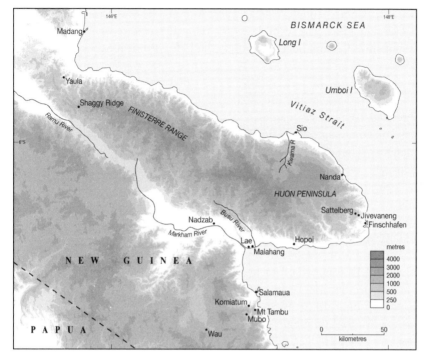

Map 7.1: Australian area of operations in New Guinea, 1943–44

a campaign that had confirmed Australian tactical dominance over the Japanese.

LHQ Tactical School

By 1943 it had been realised that COs needed to be specifically trained for their role on the battlefield just like their men: it was simply not enough just to rely on the accumulated experience of their military career. The central component of the AMF's system of training for COs was the Land Headquarters Tactical School at Beenleigh, near Brisbane.[4] During the desperate days of 1942, scant attention could be paid to the development of the AMF's training organisations. The Directorate of Military Training (DMT) did, however, recognise two critical deficiencies: first, that there was no establishment for the training of personnel for tropical warfare, and second, there was no sound organisation for the training of unit commanders.[5] In late 1942 measures were put in hand to address both problems. The first instance resulted in the establishment of the Australian

Training Centre (Jungle Warfare) at Canungra in southern Queensland in November 1942. The second saw the relocation and reorganisation of the Regimental Commanders School as the LHQ Tactical School in January 1943.[6]

The role of the new school was to disseminate a common tactical doctrine and teach 'the art of command'.[7] Its 'A', or Senior, Wing was intended to prepare majors and lieutenant colonels for battalion and brigade command, while the 'B', or Junior, Wing, trained company commanders. The school aimed to produce adaptable, well-rounded commanders who were not bound to any one operational environment. Hence, it began by teaching 'war as such and its principles in ordinary circumstances' before progressing to examine 'the differences of method which were necessary for fighting in tropical country'.[8] In accordance with this policy, the school's six-week senior syllabus was broad and included conventional operations in open country, the employment of combined arms and air support, obstacle crossings, defence against airborne troops, and the peculiarities of operations in a variety of different environments including built-up areas, mountains and the jungle.[9] It sought to equip officers for the command of both units and brigades and, as such, instruction was delivered at both levels. The course was also intended as a refresher for incumbent COs and was completed by several officers with extensive operational experience who had already attended other command and staff courses.[10] In many ways the school seems to have been modelled on METS, which had instructed students in general tactics and procedures before progressing to an examination of the specific requirements of desert warfare.

The inclusion of experienced commanders also highlights the educational culture of the school. It was as much about an open exchange of ideas between the students and their instructors, based on experience, as about the inculcation of an official line. An introduction to the school made clear that 'discussion was the basis of instruction'. Students were urged to take an active part in discussions 'irrespective of arm or f[iel]d experience'. Discussions dispensed with military formalities, and there was to be 'no occasion for diffidence or hesitation' as 'sound views based on experience or study help[ed] other students'. Less sound views were considered equally useful as they helped to clear students' minds of misconceptions and assisted instructors to illustrate more practical principles.[11]

There are indications, however, that instruction at the school did not always match its ideals and that, on occasion, experienced officers with strong ideas clashed with their instructors. On his return from the senior officers course, George Smith of the 24th Battalion wrote to his brigade commander about what he considered unsound teaching regarding defensive tactics in close country. He reported being 'bashed down' by an instructor and accused of unduly influencing fellow students, after conducting reconnaissance of an alternative site to that chosen by the instructor.[12] Some of the school's instructional practices contradicted its guiding philosophy. Instruction emphasised quick decision-making, verbal orders and a lack of rigidity in tactics, but it also taught almost immutable 'lessons'.[13] A guide to students, however, did warn that 'the "lessons" listed [were] NOT the whole of any subject'.[14]

The course focused on the practical aspects of command, and students worked hard during field exercises. They were required to issue orders and instructions, complete reports and returns, and use the appropriate means to seek additional information, just as they would on operations.[15] There were no exams and, in November 1943, Blamey ordered that the grading of students cease.[16] Although students could be withdrawn from the course, they could not be failed. Instructors, however, reported on each student discussing their tactical knowledge, their character and their ability to cooperate with their peers.[17]

The school also fostered the personal relationships that facilitated the smooth functioning of command on the battlefield. It brought infantry commanders together with the artillery and armoured corps officers on whose support they would depend.[18] Students were well provided with messing facilities and encouraged to participate in team sport. Phil Rhoden of the 2/14th Battalion described the school as 'more like a club'. Rhoden subsequently reflected that the school served a secondary purpose, which, as we shall see, was critical to the longevity of COs in the field. Attendance removed COs and company commanders from the 'rigours of life in a unit' and exposed them to a completely 'different atmosphere'.[19] Free from the responsibility for the lives and conduct of their men, often sent to the school while operations were in progress, an officer was given six weeks during which he could develop his mind and reinvigorate his body. A rest day was programmed into each week of the senior officers course, and students were expected to maintain their physical fitness while at the school.[20]

The school was used to identify and test potential command talent.[21] Its importance can be gauged by the fact that although established by the Directorate of Military Training, which also guided its policy, it was commanded directly by Advanced Land Headquarters. The increasing significance attached to attendance by COs and potential COs can be discerned by examining their service records. In the 25th Brigade, for instance, all three battalion COs attended the school before being deployed for the Lae operation.[22] In the 8th Brigade, the two COs[23] who had yet to see active service in the war attended the school as part of the brigade's preparation for its deployment to New Guinea. By early 1944 attendance at the school had become a prerequisite for appointment to battalion command. Of the 29 battalion COs appointed for the first time from January 1944 until the end of the war, 22 had completed the senior course before taking up their command[24] and one completed it soon after.[25] Of the seven remaining, four had attended the Regimental Commanders School in 1942,[26] and three were graduates of staff schools in the Middle East and Australia.[27] Additionally, at least 17 incumbent COs, many with extensive operational experience, completed the course during 1944 and 1945.[28]

Ultimately, the LHQ Tactical School provided COs and potential COs with a reservoir of knowledge and a common base of understanding. Like all doctrine, that taught at the school had to be applied with judgement. The school sought to create an intellectual environment in which officers could engage with tactical principles in the light of their own experience as well as benefiting vicariously from the increasingly diverse operational experience of the AMF as a whole. A senior British liaison officer reported that the school was 'most efficiently conducted'.[29] Summing up the utility of his time at the LHQ Tactical School, Phil Rhoden commented, 'You learned lots of things that were useful, and lots of things that you would never use.'[30] In many ways, the LHQ Tactical School embodied the principles that underpinned the establishment of the Command and Staff School in 1938. It was a centralised organisation directed by the AMF's highest level of command, committed to promoting a common standard of command proficiency, adapted to Australian conditions and organisations. The school was just one component of an 'extensive and well-defined training and educational organisation' that developed throughout 1943 and 1944 and contributed to increasing standards of operational proficiency in the AMF.[31]

THE TROPICAL SCALES BATTALION

In early 1943 the experience of the fighting in Papua brought about the most extensive revision of the AMF's structures and organisation that would occur during the war. Six infantry divisions, three AIF and three militia, were restructured as 'jungle' divisions. Recognising the impediments to wheeled transport presented by the jungle, the new divisions had greatly reduced allocations of artillery and motor vehicles. Changes were also effected to the composition of most constituent units, including the infantry battalion, guided by the principle that all 'units, sub-units, transport and equipment . . . not essential for general operations in jungle conditions' be eliminated from the division to lighten the logistical burden.[32]

With only a single regiment of artillery in the new jungle division, a CO's principal fire support would have to be provided from within his battalion.[33] Hence, the number of 3-inch mortars in the mortar platoon was increased to eight, and an anti-tank platoon, armed with four 2-pounders, and a medium machine gun platoon with four Vickers guns were added permanently to the battalion.[34] The anti-tank platoon, in keeping with AMF's aggressive ethos, was soon renamed the tank-attack platoon because it was believed that the former name engendered a passive and fearful attitude towards tanks. Even this title was a misnomer because, in addition to providing a CO with an organic anti-armour capability, the platoon was also intended to engage landing craft and other light vessels in the defence of a beach or to destroy Japanese bunkers.[35] With a reduced air threat and the jungle canopy supposedly providing protection against air attack, the anti-aircraft platoon was removed from the battalion.

Conventional transport had proved of little use during much of the fighting in Papua so both the battalion's Bren gun carriers and its wheeled vehicles were absorbed into divisional units. If the terrain permitted, appropriate transport would be provided from the divisional pool. A need for some organic transport was later recognised with the addition of eight jeeps and trailers to the battalion establishment in early 1944.[36] In some conditions, this lack of transport was likely to reduce the amount of fire support available to a CO. For example, although he had eight 3-inch mortars at his disposal, if there were insufficient local carriers available he lacked the resources to move all eight and the ammunition with which to fire them; the personnel of the mortar platoon could carry only two mortars, or four at a stretch, with adequate supplies of ammunition. In one respect the battalion had not changed: its mobility was still 'dependent on a soldier marching on his feet', and without transport he would be carrying a 'very heavy load'.[37] The light scales of the new battalion, however, meant that it could be readily transported and sustained by air or by small boats in coastal areas – a characteristic that in the Papuan operations had proved extremely valuable.

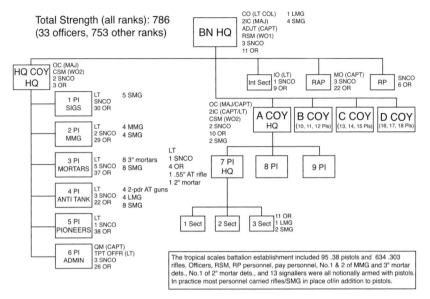

Figure 7.1: The Australian tropical scales infantry battalion, 1943

Communication difficulties in Papua had proved that effective wireless communications were essential to mastering jungle warfare. The tropical scales battalion was issued with sufficient wireless equipment to maintain communications with both higher and lower headquarters, as well as having spare sets to maintain redundancy and equip isolated outposts and patrols. The introduction of the American-designed SCR-536 set, the original 'walkie-talkie', in 1943, reduced the burden of battalion signallers and allowed wireless communications to be routinely extended down to the level of section patrols. Despite the steady improvement of the quality of wireless equipment, the terrain and climate of the South West Pacific continued to interfere with its reliable operation, and field telephones would remain a critical component of a battalion's communications system.

The tropical scales battalion was a reflection of an organisation that had learned from its trials. It exemplified the doctrinal approach that had been a key motivation behind the establishment of the Command and Staff School in the 1930s; that is, taking the doctrine and structures of the British Army, with which Australia wished to remain interoperable, and modifying them for the particular conditions in which Australian troops were to fight. In the tropical scales battalion, a CO finally had a flexible

Carrying a very heavy load: a soldier of the 2/16th Battalion carrying mortar ammunition forward for an attack on the Pimple feature on Shaggy Ridge, New Guinea, 27 December 1943. Although the tropical scales battalion was organised to maximise the firepower available to a CO and optimised for transport by air or coastal barge, its operational effectiveness still depended heavily on the carrying capacity of the individual soldier. (AWM Neg. 062323)

organisation, with adequate supporting weapons under his own control. These could be employed as single-fire units to produce a concentrated effect or as detachments to support the operations of individual companies. The CO also had adequate quantities of the most modern signal equipment available, which allowed him to readily pass and receive the information that was the very basis of his command in the field.

Jungle tactics

Reflecting on his advance from Red Beach to the Busu River heading for Lae, Hugh Norman of the 2/28th Battalion noted, 'On all occasions it has been found that our textbooks may need adaptation but never alteration.'[38] Norman's remarks reflect the growing tactical confidence of the AMF during the New Guinea offensives. Knee-jerk measures implemented as a result of the experience of fighting in Papua were examined in the light of continuing experience, and the jungle was found not to be as alien an environment as first thought. The 9th Division report on the Lae and Finschhafen operations of September to November 1943 commented that operations in the jungle demanded the 'correct application of normal principles of war'. It continued: 'Specialised tactics or "drills" should not be allowed to take the place of quick appreciations and the tactics and dispositions proper and appropriate for any particular situation.'[39] Although the means of providing it had often to be adapted, the experience of the New Guinea offensives also proved that it was possible to 'give inf[antry] the full aid of supporting arms' and, indeed, some of the earlier operations around Bobdubi Ridge during the drive towards Salamaua demonstrated the consequences of attacking without it.[40]

The nature of the terrain largely dictated the AMF's offensive tactics. As in the Papuan operations, tracks and ridges remained the primary axes of advance. The Japanese naturally sought to control these. They used the ground to their advantage and wherever possible attempted to refuse their flanks by using supposedly impassable terrain, such as swamps and steep forested ridge faces. Japanese defensive positions in New Guinea, however, had little depth, and few resources were devoted to securing their lines of communication.[41] Learning from their adversaries, AMF tactics for the attack became heavily based on infiltration and outflanking manoeuvres. In the advance it became standard practice for units or sub-units, when held up, to fix the front of the enemy position immediately and start feeling for its flanks. In the attack these flanks were exploited by 'bush bashing' and, as the battalions and their officers gained more confidence, subunits would infiltrate enemy positions, often to threaten Japanese lines of communications and withdrawal routes. After the extraordinary efforts of his battalions to capture the heights of the Finisterre Range, Brigadier Fred Chilton was well placed to advise in his post-operational report: 'Seldom is country found which is impassable to determined troops. No ground must be regarded as "too difficult" for operations. In offensive operations this fact should be considered as a means of achieving surprise.'[42] If the operations at the Papua beachheads and the early attacks on heavily fortified positions in the Salamaua hinterland had proved anything, it was that 'to slug away at [the enemy] frontally gained nothing but heavy casualties'.[43]

Hugh Norman successfully employed infiltration tactics on 14 September 1943 during operations to capture Malahang Anchorage east of Lae. He infiltrated three companies of the 2/28th Battalion through a coastal swamp, which was so deep and glutinous that some of his troops became so bogged they could not get out without assistance. Nevertheless, by nightfall he had three companies astride the road out of the anchorage, behind the main Japanese position. He later wrote: 'I am positive . . . the Japanese appreciated that such an approach as we made was impossible.'[44] Another example of the successful employment of supposedly impassable terrain to overcome fortified positions was the attack on the Prothero feature in the Finisterres on 21 January 1944 by the 2/12th Battalion, under the command of Charles Bourne.[45] The battalion attacked from Canning's Saddle after an approach march of more than seven hours, in the final leg of which ladders had to be used to reach the start line.[46] Being hurried up the slope by his company commander, one soldier replied, 'A monkey with extra claws in his backside wouldn't and couldn't move any quicker.'[47] The operation resulted in the Japanese being 'manoeuvred out of very strong positions at comparatively small cost to the attackers'.[48]

The razorback ridge along which the 2/12th attacked the Prothero feature allowed the use of only one company forward, which highlights that, although terrain could be exploited to achieve surprise, it also limited a commander's options regarding the deployment of his troops. The battalion-scale attacks mounted by the 18th Brigade along Shaggy Ridge or the 26th Brigade at Sattelberg were the exception rather than the rule in New Guinea. As Lieutenant General Morshead pointed out to Gavin Long, the war in New Guinea was on a completely different scale from that in the Middle East: 'M[orshead] picked up the day's intelligence summary and read something like "The gun at 965476 is now identified as a

Map 7.2: 2/12th Battalion attack on Prothero, 20–21 January 1944

light AA gun. A Jap was killed by a booby trap at 543267," and gestured as much to say what kind of a war is this. He has come back from a war in which divisions fought as divisions . . .'[49] Owing to terrain or the extent of enemy resistance, most attacks were mounted at the company level, and even the battalion attacks referred to previously would be better described as mutually supporting company attacks.[50] In the advances to contact that were a defining feature of the New Guinea offensives, companies moved in self-contained groups, generally widely spaced. At night, they would bivouac on the line of march and retain their spacing, thereby establishing a succession of company-defended localities that provided defensive depth should the Japanese try to attack or outflank.[51]

Terrain was also the defining factor in Australian defensive tactics in New Guinea. Although it could be traversed, some terrain was simply impossible to occupy in strength while features that would have been of critical importance in open warfare, such as high ground, were tactically useless because vegetation precluded any degree of observation.[52] As in Papua, the capacity of the lines of communication meant that only relatively small forces could be maintained in forward areas and, given their often large areas of responsibility, commanders could not hope to maintain a continuous front.[53] Hence, just like the Japanese, the Australians sought to occupy only key features, usually those in a position to control tracks, which often meant the wide dispersion of subunits. The possibility of infiltration had to be accepted but was countered by mutually supporting positions and the maintenance of patrols and listening and observation posts in between the defended localities.[54] Discussing its experience of defence around Mubo, a 17th Brigade report noted that the 'unit of defence is the co[mpan]y'.[55] A 9th Division report counselled against dispersing platoons: 'A few strong positions from which defensive patrols operate are better than many small det[achment]s which are liable to be defeated in detail.'[56]

Some COs were still unnerved by the potential for Japanese infiltration between their subunits. During his advance towards the Busu River, Norman pulled his companies into a battalion perimeter at night and later devised a drill to achieve this concentration rapidly.[57] Several post-operations reports, however, warned of the dangers of such close perimeters, which had been apparent since the fighting in Papua.[58] The counter to Japanese efforts to surround a defended locality was simply to hold ground, a tactic validated by the 17th Brigade's experience west of Salamaua: 'Every time a Jap force attacked and cut our L of C, it was evident that he expected us to get out and was bewildered and defeated when we held our ground.'[59] The 9th Division's report on defensive operations around Jivevenang and Pabu reflected its experience at Tobruk. When defensive localities in the jungle were 'apparently "cut off"' the maintenance of offensive patrols proved vital. Thus the flanks of the attacking force were determined, 'roundabout routes' to other friendly positions located and the 'siege' brought to an end.[60] These tactics were exemplified during the 2/17th Battalion's stand at Jivevenang between 7 October and 5 November. The battalion was assailed from three sides and the track to the Finschhafen beachhead cut. Patrolling found a route through to the 2/15th Battalion's position at Kumawa and henceforth, 'although not an easy trip or entirely without hazard',[61] the 2/17th was kept supplied by daily carrying parties.

In both offensive and defensive operations, Australian troops were reluctant to operate at night. Apart from a few approach marches and administrative moves, which were often hailed for their audacity,[62] Australian operations were largely confined to daylight hours. Movement in the jungle by night was considered difficult and, owing to the cover offered by the vegetation, was not thought to be worth the trouble.[63] Australian attacks were usually called off at nightfall, often allowing the Japanese to break contact and escape.[64] This led to the 9th Division report on the Huon Peninsula operations advising commanders to 'test' Japanese positions after an 'overnight pause' to ensure the enemy was still in occupation.[65] There were also very practical reasons to avoid attacking at night or late in the afternoon. The inevitable casualties would have to be carried out by night, which meant a slow and torturous journey, if indeed it was possible at all.[66]

Command in the Jungle

If there was a single defining characteristic of Australian tactics in New Guinea in 1943 and 1944 it was dispersion. As a précis from the First Australian Army School of Infantry counselled, however, dispersion increased the difficulty of control.[67] Because observation and communications were severely limited by the jungle, the 'speedy movement of a commander . . . to meet a tactical situation [was] impossible'.[68] Earlier, in the Salamaua hinterland, the companies of 24th Battalion were so dispersed that a visit to every post entailed an estimated journey of more than 30 days' continuous walking.[69]

The solution to such situations built on the experience of Papua. The 15th Brigade report on the Finisterres operations counselled that it was essential for a commander to anticipate where his presence was 'likely to be needed and move to that spot'. In offensive operations, this was immediately to the rear of the forward company, where the CO 'could gain early information and so be in a position to influence the battle'. He could appreciate the 'exact moment' when following troops had to take over from the forward elements to maintain the momentum of the advance and 'make a better decision regarding the employment of his reserve'.[70] The 18th Brigade report of the same campaign reveals a similar experience: commanders of all levels needed to be well forward to take the best advantage of all opportunities.[71]

Often, despite his best efforts, a CO was not able to place himself in the type of position that the 15th and 18th Brigades found to be optimal for command in jungle operations. The 3rd Division's report on the Wau–Salamaua operations commented that higher commanders were often forced to rely 'to a great extent upon the judgement of the junior commander'. Although it was determined by the operational environment, such an approach to command embraced the principle of empowering subordinates – a key component of effective leadership. The 3rd Division report emphasised the significance of mutual trust in such an arrangement, which was said to be achieved through the 'junior knowing

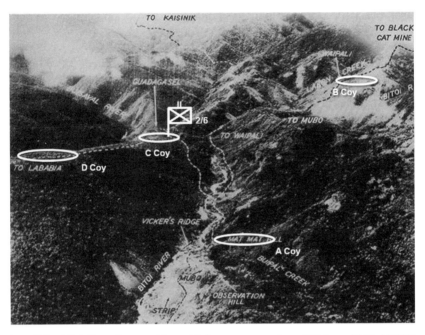

Map 7.3: 2/6th Battalion positions, Mubo area, morning of 21 June 1943. (AWM 101113)

that the superior has confidence in his ability to carry out his task; and also by the superior seeking his opinion on matters upon which, from his local knowledge, the junior is better fitted to give an opinion'. It counselled commanders not to interfere with a subordinate's 'normal function of command' as to do so could create an 'inferiority complex' and stifle the initiative critical to success, particularly if he was required to act semi-independently.[72]

The fighting for Lababia Ridge between 20 and 22 June 1943 illustrates the effective employment of devolved command in the 2/6th Battalion. The action was primarily fought by D Company under the command of Major Walter 'Bill' Dexter, the brother of the official historian. Bill Dexter was an original officer of the 2/6th Battalion and, wounded leading a patrol at Bardia, had already won himself a reputation as a nonsense man of action before the battalion deployed to New Guinea.[73] Twenty-nine-year-old Dexter was tough and alert, and his company was regarded as the best in the 17th Brigade.[74] Lababia Ridge was the key to the Australian position south of Japanese-held Mubo and, illustrating the principle of dispersion, was held by a company-strength defended locality on its western end. Battalion headquarters was located approximately four kilometres, as the crow flies, to the south, and communications were provided by field telephone.

When the Japanese attack developed on 21 June the 2/6th's CO, 'Freddie' Wood, was absent from his headquarters visiting another company position at Mat Mat. Dexter directed the company's defence throughout the day, employing his tiny reserve when needed and absorbing reinforcements in the form of C Company when they arrived in the early evening. Around the same time, Wood returned to his headquarters after being advised by radio of the Japanese attack.[75] Dexter's defence continued throughout the next day, and Savige subsequently described the manner in which Wood exercised his command: 'Nearby was his B[attalio]n Commander, who had already proved his great qualities of leadership. Wood watched every move and gave instant support to Dexter's needs. Above all he did not interfere with Dexter's management of the fight.'[76] For his actions at Lababia Ridge Dexter was awarded the Distinguished Service Order, a rare award for an officer of his rank, recognising his 'personal courage and leadership' and the firm grip he retained on the defending force, which eventually amounted to almost half of the battalion.[77]

The competence of company commanders was an important factor in the success of Australian battalions in New Guinea. In *Bravery Above Blunder*, John Coates reflects on their contribution on several occasions, including during the successful attack on Sattelberg – an imposing feature north-east of Finschhafen that dominated the coastal plain. One of the strengths of the 2/48th Battalion, Coates observes, was that its CO, Robert Ainslie, 'was wise enough to recognise that he had some quite irrepressible talent beneath him and used it. His four rifle company commanders, Brocksopp, Hill, Isaksson, and Morphett, were sound and experienced leaders.'[78] All four had embarked with the battalion in November 1940, and both Olof Isaksson and Hurtle Morphett were recipients of the Military Cross.[79] The 2/12th Battalion's historian has similarly remarked that the battalion 'was indeed fortunate in the general competence of its company commanders'. He recounts the story of the officer commanding D Company, Captain Kevan Thomas, who, left without orders following the wounding of Charles Bourne during the attack on Prothero, took the initiative and ordered his troops forward when he saw that the forward company was not making 'much progress'.[80]

Ultimately, the success or failure of an operation could rest with the judgement of a company commander. The success of the advance by

C Company of the 2/9th Battalion to capture Crater Hill on 31 January 1944 was attributed to the 'speed with which the Company Commander exploited the disorganisation of the enemy after the failure of his counter-attack'.[81] Conversely, the 57/60th Battalion's drive to capture Yaula on the night of 31 March 1944 floundered owing to the inability of the commander of the leading company, Major John Connell, to retain control. Connell was moving with the rear platoon and thus lost touch with the developing situation as his two forward platoons became isolated and communications were cut. The two forward platoon commanders subsequently decided to withdraw, but the battalion's report on the operations reflected: 'Had the Company Commander been forward, the favourable ground and the effect of the fight on the enemy would no doubt have influenced a decision to remain, whilst B[attalio]n HQ restored communications and made provision for rations and ammunition.'[82] Major Connell's experience highlights the trap that the tactics of dispersion and devolved command could also pose to unwary COs.

During the operations of the 2/5th Battalion on Mount Tambu, south of Salamaua, in late July 1943 Thomas Conroy was criticised for maintaining his headquarters too far away from the site of the action. According to Brigadier Moten, the efforts of Major Mick Walters and A Company to seize a foothold on Mount Tambu went unsupported because 'Conroy was trying to run the battle from too far back'.[83] A Company established itself on the southern spur of Mount Tambu around 1800 on the evening of 16 May, were heavily counter-attacked throughout the night and the next day but, although resupplied, were not reinforced until the morning of the 18th.[84] Later attacks by the 2/5th Battalion on Mount Tambu were also hampered by a lack of appreciation of the difficulties presented by terrain and the Japanese positions, which would seem to stem from a lack of personal reconnaissance by Conroy. Neither of the attacking company commanders was given sufficient time for thorough reconnaissance, committing them to costly and ill-fated frontal attacks.[85] On Mount Tambu Conroy lost the confidence of both his company commanders and his superiors; he never commanded in action again.[86]

Owing to the inability of a commander to observe events directly and rapidly manoeuvre his forces, the fight for information in New Guinea was of much greater significance than in open warfare. The 15th Brigade's report on the operations around Bobdubi Ridge recognised intelligence as the basis of mobility in the jungle: 'As movement of troops is restricted by long and exhausting marches, the only means of gaining any mobility of force to counteract the enemy's plan was by having early, accurate information. If the information is passed back quickly and accurately, commanders can then anticipate and act. By this anticipation the loss of time entailed by the slow movement of troops can, to some extent, be overcome.'[87] Extensive use was made of aerial photography as a source of tactical intelligence in New Guinea, and COs were advised to make themselves proficient in the interpretation of air photos.[88] Patrols, however, remained a CO's main source of information, particularly regarding the selection of routes and objectives for attacks.[89] Hugh Norman wrote that the 2/28th Battalion did not carry out any operation that was not preceded by patrolling.[90] In the 15th Brigade, COs were directed to brief and debrief patrols personally wherever possible as they were the 'eyes' through which they could 'visualise the ground hidden by thick timber'.[91]

The poor initial performance of the 58/59th Battalion around Bobdubi Ridge, and in particular Old Vickers position, in July 1943 was in part attributed to its 'inability and inexperience to patrol'.[92] The 58/59th's CO, Daniel Starr, had already been removed from command of the 2/5th Battalion following their withdrawal from Wau in February and March, for failing to keep in close touch with the Japanese and failing to maintain the flow of information along the chain of command.[93] The increasingly impatient messages directed to Starr by Brigadier Moten reveal that higher headquarters were just as dependent on patrols for the intelligence on which to base their operations as were COs. On 28 February Moten wrote: 'Essential your sitreps give outline of patrol activity . . . you must not lose contact . . . planning of air ops dependent on prompt receipt of sitreps.'[94] Another handwritten message was blunt in its directions: 'Instns our 0291 are emphasised. FIND THE ENEMY.'[95] As CO of the 58/59th Starr was also prone to establishing his headquarters too far back.[96] This was only part of a wider malaise affecting Starr's commands, which will be discussed presently.

Given this dependence on information at all levels of command, it is not surprising that the 15th Brigade's COs were instructed to become 'communications conscious' and plan to ensure communications at all times.[97] In addition to ensuring a smooth flow of information, reliable communications also facilitated the effective command of dispersed sub-units allowing a commander to be 'put into the picture . . . and so exert his influence on the operation'.[98] We have already seen how Wood was able to be summoned back from a remote outpost when the attack at Lababia Ridge developed. The challenge presented by the maintenance of communications is demonstrated by the fact that during its operations in the Finisterres the 15th Brigade was employing 482 kilometres of field telephone cable; its war establishment was only 70 kilometres.[99]

Continuing reliability problems with wireless equipment meant that the field telephone remained an important means of communication. Shortages of line, operational contingencies and terrain impediments often meant that a single line provided communications from the forward companies all the way back to brigade headquarters. On occasion this served to complicate command relationships because it gave brigade headquarters the opportunity to monitor battalion communications and talk directly with company commanders, without reference to battalion headquarters.

After-action reports reveal that confidence in the use of wireless communications grew as equipment improved and operating procedures were developed to enure its maximum effectiveness.[100] The flexibility provided by a truly portable radio set, like the SCR 536, was greatly appreciated within battalions, and a 7th Division report on the Lae operations reported, 'COs are enthusiastic in their praise of the sets.'[101] They proved 'invaluable' in controlling dispersed platoons and companies during the Lae operations and were critical to effective infantry–armour cooperation during the fighting for Sattelberg.[102] Like all elements of a command system, it was essential that redundancy be maintained in the communications system.[103] Cutting field telephone lines was a standard Japanese tactic, and atmospheric conditions often prevented the use of even the most reliable wireless equipment. An Australian observer attached to an American unit on New Britain reported scathingly on their communications practices, which provided wireless equipment only for communication within the battalion. When this failed owing to prolonged exposure to humid conditions, they were forced to rely on runners.[104]

THE AUTHORITY OF EXPERIENCE

By the middle of 1943 most of the battalions of the AIF were commanded by officers with extensive operational experience. Many of the officers of the 7th Division had endured the desperate fighting of the Papuan campaigns, and those of the 9th Division had fought in the cauldron of El Alamein. Their average age – 36 – was a reflection of the physical demands of that experience. They had been tested, tempered and selected by the unforgiving experience of battle: they were a group of survivors. For the majority, however, the New Guinea offensives were their first experience of battalion command, and a third had held their commands for less than six months before embarking on operations. The actions of many were characterised by what could be termed the authority of experience. They exhibited a determination to do things their own way, not to be bustled from afar and to make full use of the support available to them. In doing so, they earned the trust of their subordinates, grounding their command firmly on the exercise of expert power.

The COs of the AIF battalions in New Guinea were men of diverse personality, and their actions were usually a product of their character. The range of personality to be found, demonstrating that there was no single archetype of an effective CO, is illustrated by the example of the 20th Brigade. In New Guinea, its three battalions – the 2/13th, 2/15th and 2/17th – were commanded respectively by George Colvin, Colin Grace and Noel Simpson. All three had seen action at Tobruk or El Alamein, and by 1945 all three had been awarded the DSO, Colvin and Simpson twice.

George Colvin had taken command of the 2/13th Battalion on the first day of the battle of El Alamein when the CO, Robert Turner, was mortally wounded. He commanded the battalion with distinction in several attacks, earning his first DSO in the process, and was eventually evacuated wounded himself.[105] Colvin was a flamboyant character known to his peers and subordinates alike as 'Gorgeous' or 'Flash' George, but was also described a 'first class CO'.[106] His brigadier, Victor Windeyer, himself an accomplished battalion commander with a refined insight into human character, remembered Colvin as 'eager, vigorous, personally demonstrative . . . ready to seek glory in the cannon's mouth, proud of his battalion in which he had served since it was raised'.[107] During the Lae operations Colvin's enterprising spirit manifested itself in the occupation of Hopoi airfield. The 2/13th had been ordered to secure the right flank of the Lae beachhead by advancing to the west bank of the Bulhem River, and no further. Colvin was not content to stop there and pushed two companies across the river and onwards to Hopoi.[108] Windeyer recalled that Colvin wanted to venture even further west on 'undertakings of his own'. Windeyer reined him in because the 'broad plan' involved the 2/13th simply 'holding the right flank securely', and he did not want the battalion to be too far away, dispersed or committed 'when the next phase arrived'.[109]

Colin Grace, 'dour' and 'bespectacled',[110] was a vivid contrast to Colvin. Windeyer described him as 'careful and cautious', a 'studious and discerning officer' who was 'scrupulous, unselfish and loyal'.[111] His colleague Simpson found him quiet and logical, and observed there was something of the 'academic' in him.[112] Gavin Long, on meeting Grace in late 1943, gained the impression he was too much of a thinker, noting in his diary that Grace was 'argumentative' and 'inclined to see the difficulties more clearly than he sees the objective'.[113] Grace's cautious attitude towards command is best illustrated by the detailed reconnaissance and planning he personally undertook before committing the 2/15th to battle. His DSO for the Huon Peninsula operations was principally awarded for a 'marked disregard of danger in getting, by close personal observation of the ground, the information he wanted for the formation of his plans'.[114] He carried the responsibility of his command heavily and was not as inclined as some of his peers to devolve it to subordinate officers. Coates has observed that Grace was 'meticulous in ensuring that his troops had a maximum of fire support, which he usually coordinated himself'.[115]

One of Grace's company commanders has reflected that 'he achieved the ultimate objective in command – success in battle with a minimum of casualties. I don't recall any incident when the Battalion or a company was ordered by him to do something which resulted in a breach of these two aims.'[116] Grace's actions in New Guinea were consistent with earlier behaviour at El Alamein. As acting CO during Operation Bulimba in September 1942, Grace had ordered the withdrawal of the 2/15th's scattered and hard-pressed companies contrary to Windeyer's expectations that they should hang on to their vulnerable salient, an action that probably condemned his chance of being permanently appointed to command at that time.[117] Illustrating that a battalion can have a character, the 2/15th reflected Grace's persona. Assessing its performance on the Huon Peninsula, Coates wrote that it was 'a study in imperturbability and self-reliance; it scarcely put a foot wrong from Scarlet Beach to Sio'.[118]

The 20th Brigade's third CO in New Guinea, Noel Simpson, was one of the AIF's most capable COs in the Second World War but has not achieved the acclaim that has been accorded to a select few. Windeyer regarded the 2/17th Battalion's defence of Jivevenang, under Simpson's command, as the best single battalion exploit of the Pacific war.[119] Slightly built, outwardly unemotional – quite shy, Windeyer thought – Simpson was known for his coolness and composure in battle.[120] At Alamein he had moved up to the start tapes on the night of 23 October wearing only his soft cap until a brave liaison officer tossed it into the darkness and replaced it with a helmet.[121] Later in the battle, armed with only a fly whisk, he was heard to remark, 'This is really quite exciting.'[122] Simpson's command was founded on the maintenance of the highest of standards that 'produced a uniformity of dress, equipment and procedures'.[123] He led by the example of his own conduct. His orders and instructions were

The 'Red Fox': Lieutenant Colonel Noel Simpson, CO of the 2/17th Battalion and later the 2/243rd. Cool and composed, Simpson was one of the most capable of Australia's infantry COs during the Second World War. (AWM Neg. 087669)

clear and precise, and he never bullied nor badgered; the standards he expected were clear, and any divergence from them was met with an incisive but calm rebuke and punishment if appropriate.[124] Simpson's strict approach had not initially endeared him to the 2/17th, but by the time of Jivevenang it is clear that it not only trusted him but also identified itself with him.[125] For reasons that remain obscure, Simpson's nickname was the 'Red Fox'.[126] At Alamein the OC of the 2/17th's carrier platoon adorned Simpson's carrier with a 'rampant red fox' and on the night of 23 October 'Tally Ho! the Red Fox!' was adopted as the battalion's battle cry.[127] It was later used at Jivevenang as a means for the troops of the 2/13th and 2/17th Battalions to identify each other.[128]

Equipped with the intellectual authority of their experience, COs in New Guinea increasingly questioned the direction provided by higher headquarters when it urged action they considered ill-conceived, premature or badly prepared. The fact that they were able to do so, have decisions changed and still remain in command indicates the respect accorded to their judgement by superior officers and of the more cautious and considered operational approach being adopted in New Guinea. It also indicates a changing attitude in the command relationship: loyalty to superiors was no longer unconditional. During its advance from the Busu River towards Lae, the 2/28th Battalion requested more 9mm ammunition for its Owen guns, a critical component of infantry minor tactics in the jungle. At that point they had only 20–30 rounds per gun. The DAQMG at divisional headquarters offered only half the battalion's demand, stating that they were in a better place to judge its requirements. Hugh Norman personally responded: 'No ammunition, no move.' He gained the support of his brigadier, Bernard Evans, and the ammunition was in due course provided. The continuing inability of the 9th Division to support the coastal advance towards Lae with anything but minimal waterborne logistic support meant that the ammunition was indeed later required, vindicating Norman's stand.[129] Advancing on Sattelberg, Robert Ainslie of the 2/48th also refused to be hustled from afar. Coates describes him as a 'sound, imperturbable commander' who resisted pressure from Wootten to move ahead more rapidly. He persisted with the use of tanks and all available fire support to capture well-sited positions with a 'minimum cost in casualties'.[130] It was an approach for which the men of the 2/48th were thankful. One young digger commented: 'Thank God it's keeping dry . . . If it rains it'll stop the tanks.'[131]

Ainslie's approach exemplifies the style of operations increasingly conducted by the AMF in New Guinea, particularly during the pursuit phase of the Huon Peninsula campaign. When Japanese rearguard positions were encountered the first action would often be to withdraw the forward troops and subject the position to bombardment by artillery or mortars, and on many occasions this would be sufficient to cause the Japanese to withdraw.[132] Following the horrendous losses of the Kokoda Track and the Papua beachheads, there seems to have been an increased sensitivity to casualties among the AMF's senior leadership. Windeyer recalled a directive that 'no Australian lives were to be lost in this coastal pursuit merely to avoid delay' and that 'full advantage was to be taken of artillery support'.[133] Approaching Nanda late on the afternoon of 31 December 1943, Grace opted not to commit the 2/15th Battalion to an attack that day owing to the imminent approach of nightfall and his orders to avoid unnecessary casualties. Instead he employed an harassing artillery

'Thank God it's keeping dry . . . If it rains it'll stop the tanks': tanks of the 1st Army Tank Battalion link up with troops of the 2/48th Battalion ready to begin the attack on Sattelberg, New Guinea, 17 November 1943. The New Guinea offensives of late 1943 heralded a new style of fighting for the Australians that dispensed with the desperately light scales and panic of Papua and replaced it with carefully prepared combined arms action. (AWM Neg. 060593)

program on Nanda throughout the night, after having to argue for the use of guns that had been forbidden to fire at night owing to their proximity to divisional headquarters. Nanda was occupied without opposition the next day.[134]

This increasing reliance on fire support did not mean that COs became dependent on it, nor did it mean that their superiors tolerated undue hesitation. Countering claims that he had not attacked at Nanda because he was waiting for tanks, Grace replied, 'Lack of tanks had nothing to do with the decision not to attack that night . . . it was nice to have them . . . but [I] didn't waste time waiting for them.'[135] Grace's remarks are supported by a report on the use of tanks in the Huon operations, which commented that there were many obstacles for them to surmount in jungle terrain and that infantry should always be prepared to go forward without them. To hold the infantry back waiting for tanks would delay an advance 'unwarrantably'.[136] The tanks supporting the 2/48th at Sattelberg were twice held up, but Ainslie did not hesitate to order his companies forward, and they 'responded immediately'.[137] After commenting on Grace's efforts to keep casualties to a minimum, the company commander quoted earlier continued: 'Yet we were never coy when it came to carrying out our responsibilities in the overall scheme of things.'[138]

Nor were the 2/28th. Before halting for want of ammunition Norman had pushed his battalion across the flooded Busu River in a courageous operation on 8 September 1943. No bridging equipment or assault boats were available, and the river, 700 metres wide and close to two metres deep in places, was flowing at 10–12 knots and 'increasing in fury' by the hour.[139] To halt at the river would sap the momentum of the 9th Division's advance on Lae, allowing the Japanese time to strengthen their defences on the Busu's west bank, then concentrate against the 7th Division's advance from the west. Mindful that during the attack on Ruin Ridge at Alamein the 2/28th had crossed a minefield on foot and was subsequently cut off from support on the other side, leading to the capture of most of the battalion, Norman made the most difficult decision of his command.[140] He ordered '700 men, many of whom could not swim, to cross a raging torrent under enemy fire'.[141] Many were swept away, thirteen were ultimately drowned and the battalion lost 25 per cent of its automatic weapons and 80 rifles, but it established a bridgehead on the west bank and the advance continued the next day.[142]

The dithering of Percy Crosky of the 4th Battalion at the crossing of the Kwama River between 25 and 26 January 1944 compared unfavourably to Norman's bold action at the Busu. On the first day of its advance towards Madang, the 4th encountered the Kwama in flood, neck-deep and running at approximately ten knots.[143] Nevertheless, a platoon was put across the river on a makeshift bridge. That bridge was washed away, and additional attempts to replace it were unsuccessful. The river subsided during the night, but when he went forward at around 1200 the next day Brigadier Claude Cameron found the battalion had yet to cross and the situation was 'most unsatisfactory'.[144] Crosky, despite the unmolested presence of his platoon on the far bank, was worried by reports from Papuan Infantry Battalion patrols that there were 'hundreds of Japs' on the other side and was unwilling to take the risk. Cameron ordered him to get more troops across immediately, and the battalion eventually crossed the Kwama and proceeded without incident.[145] The brigade advance, predicated on the maintenance of speed to overtake the retreating Japanese, had been delayed a day.[146] The benefits of experience are obvious on this occasion. Crosky had never seen active service, let alone commanded a battalion on operations. The exhortations to keep casualties to a minimum probably weighed heavily on his mind. Crosky never recovered from this initial setback. Cameron later told David Dexter that he never lived it down, and he was replaced at the end of the campaign.[147]

THE STRAIN OF COMMAND

Command in battle imposes a significant mental strain even in the best of climatic and geographical conditions. The environmental conditions in New Guinea during protracted operations there compounded this strain, as did the fact that most COs had carried the responsibility of command at varying levels for several years. The 15th Brigade report on operations around Bobdubi Ridge clearly expressed the arduous nature of command in New Guinea: 'The time required for reconnaissance and planning together when added to the time required for normal administration means long hours for a commander. Further, this administration has to be done usually after exhausting walks. Under these considerations it is difficult for commanders to maintain a high standard for long periods. Good organisation within a unit is necessary to give the commanders an opportunity to rest and to save fatigue.'[148] The fighting in New Guinea exhausted COs and ultimately broke some. We have already met Daniel Starr of the 2/5th Battalion. In March 1943 he was removed from command as a result of the general air of sluggishness about the battalion and its operations. Writing to recommend Starr's dismissal, Moten noted his failure to maintain close contact with the enemy, the requirement to 'continually' urge him on and a 'general air of slackness' about the 2/5th. There are clear indications that Starr was worn out: 'He lacks the ability to make a decision and the necessary driving force to carry out an order when it is given to him.' He had not always been so. In the Middle East he had been a capable company commander, and his initial period as CO of the 2/5th there had been well regarded. Even when his drive and control over his battalion were being questioned his static positions remained tactically sound.[149]

There may also have been a personal dimension to Starr's first dismissal, highlighting the continuing impact of personal feelings on command relationships. Moten was quiet, reserved and thoughtful.[150] Savige recalled him as being a 'glum sort of fellow [who] doesn't talk much'.[151] Starr, however, was much more outgoing, and his character seems to have rankled with Moten, who described him as 'easygoing' and a 'playboy'.[152]

Some members of the battalion shared similar suspicions about Starr's removal: 'Danny Starr, God bless him, was a magnificent CO but a man of independent ideas – too independent for the hierarchy.'[153] Subsequent events, however, justify Moten's decisions.

Starr took command of the 58/59th Battalion on 28 April 1943 at the instigation of Lieutenant General Iven Mackay, then GOC of New Guinea Force, who was concerned that three COs in the 3rd Division were only substantive majors.[154] Mackay, as previous GOC of the 6th Division, was aware of Starr's previous performance in the Middle East. Taking command of the inexperienced battalion in its first campaign, Starr was faced with the extremely difficult task of capturing the Japanese strongpoint of Bobdubi Ridge, and doing so with minimal fire support and with one of his four companies detached to protect the brigade line of communications.[155] Successive attacks on the position regarded as the key to the ridge – Old Vickers – resulted in failure owing to 'lack of effective support, insufficient preparation, and the use of too small an attacking force to crack a stronghold capable of resisting a force three or fours times as strong'.[156] Morale in the battalion plummeted and remained low even after a well-planned and supported attack, largely the brainchild of Lieutenant Leslie Franklin, the OC of C Company, captured Old Vickers on 28 July.

Reports of the poor condition of the 58/59th, including 13 self-inflicted wounds from the beginning of June until the middle of July, are legion.[157] The 15th Brigade's medical officer, William Refshauge, later commented that the battalion was the worst he had ever seen: 'They had no idea of how to live along the track, admittedly they were terrible tracks. The offrs were hopeless, morale non-existent. If a m.g. opened up ahead of them they just came back.'[158] In a candid interview, the brigade commander, 'Tack' Hammer, was much more succinct: the 58/59th were 'unshaven' and 'poor as piss'.[159] Starr defended himself by making reference to the 'criminal lack of fire support'. The battalion only had artillery support for the last attack on Old Vickers: 'These lads were not fools. They could size up a situation . . . they seemed under the impression they were being sacrificed to the military ambition of the Bde Comd. My attempts to combat these ideas was not the least of my responsibilities.'[160] Administration in the battalion, however, was severely lacking. Near the end of Starr's command two companies at Erskine Creek had not been supplied with a hot meal for a week.[161] As a 15th Brigade report would later comment, beating the jungle itself was critical to the success of warfare therein. A well-organised battalion could overcome the 'most uncomfortable living and fighting conditions' with 'a consequent rise in morale'.[162] A member of the battalion recalled, '"Danny' Starr never got the gist of what we were at."'[163]

Starr was removed from command on 13 August 1943. Hammer informed Savige that Starr was 'mentally and physically unfit for ops for some time due heavy strain prolonged ops'.[164] Even the battle-hardened Hammer displayed some of the old-world ethos discussed in previous chapters. He had doubts about Starr's suitability from their first meeting but felt he 'had to give him an opportunity to prove himself'. He reflected, however, that he had procrastinated too long and that being kind to Starr was being unkind to the 58/59th.[165] Once he decided to act, it was with characteristic ruthlessness. He asked Savige to have Starr returned to Australia for immediate dismissal. Savige, however, demurred and arranged to have him evacuated through medical channels. This would become Savige's *modus operandi* in such cases for the rest of the war. He told David Dexter it was a 'racket without a blush'. He recognised that the officers concerned were unfit to command but also 'remembered . . . their past splendid service' and acted in a way he thought 'proper'.[166] Hammer wondered about the original decision to appoint Starr, noting that it was not sound to appoint a 'mediocre', 'tired' or 'disillusioned' commander to an inexperienced battalion.[167]

Henry Guinn of the 2/7th Battalion was another experienced Middle East veteran who never commanded in action again after his experience in New Guinea. His battalion was one of the first deployed to defend Wau in early 1942, and it later occupied the key position on Lababia Ridge. Relieving the 2/7th there in May 1943, Bill Dexter observed that all was not quite right with the 2/7th. Forward troops were reluctant to show his 'fellows the country and normal security patrol routes', and the battalion was able to please itself 'regarding dress and general appearance which, wherever you are, is a bad thing and a definite moral [sic] reducer'. Dexter found Guinn himself to be 'a bundle of nerves'.[168] The 2/7th rested for approximately a month in the Wau area following its relief, but Guinn's rest was interrupted by a period acting as commander of the 15th Brigade. During the march forward to resume operations in late July Guinn's health deteriorated, and he was ordered out to rest. He returned briefly to duty before being evacuated to hospital on 5 August, then eventually to Australia, suffering from pneumonia and an 'exhaustion state'.[169] Hammer, under whose command Guinn was operating, wrote to Savige: 'Lt-Col Guinn's present condition, I am certain, has been brought about by his devotion to duty and the demands that have been made on him for constant movement from place to place over terrain which

'A bundle of nerves': Lieutenant Colonel Henry Guinn, CO of the 2/7th Battalion during the New Guinea offensives. An experienced Middle East veteran, Guinn's health was worn down by the demands of jungle warfare. (AWM Neg. 122339)

can only be appreciated when it is personally experienced.'[170] Moten, somewhat unkindly, described Guinn as 'a terrific walker with no great mental capacity', but his further remarks – 'When the former went there was nothing there' – make a pertinent point. Guinn's DSO citation for the defence of Wau aerodrome reveals a CO whose command was heavily based on positioning himself forward and 'encouraging and setting an example to his men with complete disregard of personal danger'. Moten concluded that when Guinn became sick the '2/7th slipped'.[171]

Guinn returned to the battalion on 1 December but struggled to maintain standards. Between June and November 1944 there were a number of disciplinary incidents involving senior officers including the 2iC, several company commanders and the adjutant.[172] Moten noted that Guinn was lacking the 'enthusiasm, grip and drive which he had previously displayed'.[173] In early October the situation came to the notice of the divisional commander, who directed that a change in command occur if there was not an immediate improvement in standards. Guinn was sacked on 3 November 1944. Highlighting the AMF's old-world ethos, and perhaps a sensitivity to combat exhaustion among its commanders, no report was made against Guinn that would 'adversely affect' his future employment.[174] Ultimately, his experience was validated by appointment as Commandant of the LHQ Tactical School.

The Jungle Breed

Recommending Henry Guinn's removal from command of the 2/7th Battalion, Moten wrote that the battalion needed an officer 'possessing a strong forceful personality' who could restore it 'to its former plane'.[175] Battalion COs in New Guinea exhibited a wide variety of personality types. There was, however, a remarkable congruence in the nature of many of the officers appointed to replace those removed from command. While they possessed varying character traits, what they shared in common was a strong, aggressive personality and a command style founded on the vigorous exercise of referent power. This suggests an organisational ideal of the type of CO needed to succeed in the jungle environment. The numerous references in official correspondence to officers of 'strong forceful personality'[176] also support the view that the young, inexperienced troops of the militia battalions responded best to this vigorous leadership. Writing of the 15th Brigade, Refshauge commented: 'The troops were young and they would follow a leader . . . If a leader was keen and enthusiastic the troops would follow him.'[177]

The epitome of the forceful, aggressive CO was George Warfe, who was appointed to replace Starr. Aggressive, impulsive but exacting in his expectations, Warfe 'displayed no hesitation in killing the King's enemies'.[178] He had won a reputation for personal bravery commanding the 2/6th Battalion's carrier platoon at Bardia, which remained with him throughout his service.[179] His first command in New Guinea was the 2/3rd Independent Company, and Warfe was not daunted by either the terrain or the enemy. The historian of the 24th Battalion has recalled the attachment of one of its platoons to Warfe's company to gain jungle experience. Observing the tendency of inexperienced troops to 'generally pussy foot around' and not 'advertise their presence' while close to the enemy, he described Warfe striding in among the cowed gathering, raising two mess tins aloft and crashing them together while yelling, 'Righto, you yellow bastards, we're here waiting for you – come out and fight!'[180]

On taking command of the 58/59th Battalion, Warfe set in place an immediate and vigorous program of change the character of which was exemplified by an instruction to one junior officer: 'Find that b— so-and-so . . . and when you reach him wrap your rifle around his head and tell him that's from me.'[181] Hammer recalled that Warfe changed the outlook of the battalion 'overnight'.[182] He established kitchens to get hot food forward to the companies on a regular basis and built a rest camp, sheltered from enemy fire and observation, to give tired troops a chance to rest. His 'blitz' on the battalion's cooks has gone down in its annals: '"Cooks," he said, "exist only to cook and cook they bloody well will." He threatened "to banish to the utmost weapon pits" any cook who failed to get at least two hot meals per day to forward troops.'[183] The Warfe way was initially a shock to many in the 58/59th, but his 'dynamic personality and ruthless drive . . . quickly welded the unit into an aggressive and confident band of jungle fighters'.[184] One of Warfe's men has reflected: 'The by word of the troops was "We will show this son of a bitch how good we are" and we did . . . Wharfie was laughing all the way to the bank for his strategy worked, kick a few bums, yell at 'em, threaten 'em and see what happens . . . we became real soldiers.'[185] Identifying with its leader, the battalion 'fed on stories of his latest exploits',[186] including several mad rushes at Japanese patrols.[187]

Although Warfe's success with the 58/59th Battalion cannot be challenged, perhaps his own command maturity should be. Leading section-level attacks was not the most sensible thing for a CO to do, but was perhaps the corollary of the youthful vigour – Warfe was only 31 – needed to overcome the operational challenges of New Guinea. Further, it is unlikely Warfe's demands on the battalion would have been tolerated had he not practised what he preached.

Originally Warfe's company commander in the 2/6th Battalion, George Smith was of similar character. A confidential report submitted in October 1942 commented: 'He is a strong type physically and mentally, slightly impulsive, and he has energy and drive and can get the job done. He is capable of producing and maintaining a very high standard of training from men and can extract the last ounce of effort out of them.'[188] Smith also had an eccentric innovative streak that earned him the nickname 'the Mad Major' in the 2/6th and did not always 'endear him to his superiors'.[189] The stern Hugh Wrigley described Smith as 'capable but not consistent [he] has some brilliant ideas but these are seldom completed. Is inclined to try to do everyone's job as well as his own.'[190] After a period in command of a training battalion, Smith was appointed CO of the 24th Battalion on 9 July 1943, replacing the 'temperamental and difficult' Alexander Falconer.[191]

Smith brought rapid changes to the 24th Battalion, making his presence felt by his 'boundless energy, his enthusiasm, and his unorthodox solutions to all problems'.[192] Refshauge has recalled that the 24th was the first battalion in the 15th Brigade to have a real sense of discipline about it.[193] Smith visited all of its scattered outposts on foot and, where possible, instituted training programs and firing practices to remedy some of the battalion's training deficiencies. In addition to his physical toughness, tactical innovation and down-to-earth language, Smith has been remembered for his 'legendary kindness'. He kept a record of every man's birthday and continued to send birthday greetings after the war. Combining this respect for the individual with an unerring commitment to the fight against tropical disease, when he left the battalion in late 1944 he sent a letter to every man of the 24th to say farewell and remind them to continue taking their Atebrin on leave to ward off malaria.[194] (Smith had been forced to leave the battalion owing to a troublesome knee injury sustained in the course of operations at Markham Point near Lae.)

There is little doubt about the success of Smith's command. He enjoyed the respect of Hammer, who remembered him for having 'a crack at anything'.[195] In May 1944 he wrote to Smith that the 24th 'had won the admiration of all for its determined fighting and sound administration. The extraordinarily good sickness figures of the unit in New Guinea have set an example to the whole of the Australian Military Forces.' Smith was awarded a DSO for his efforts during the Salamaua and Lae operations, and his citation referred to 'splendid organisational ability', 'aggressive spirit and drive' and a clear appreciation of the situation on the 7th and 9th Division's flanks owing to well-disposed OPs and patrols.[196] The cohesion he engendered is clearly evident in the 24th Battalion's return to Melbourne in late 1944. In the Middle East, Smith had been nicknamed 'Larry the Bat' after a contemporary cartoon character owing to his enthusiasm for night operations and an uncanny ability to find his way in the dark. When the train carrying the 24th pulled into Spencer Street station it carried a placard proclaiming 'the Bat's Boys' were home.

COs from the warrior mould were bestowed on several other battalions during this period. The 22nd, for example, got John 'Burn-em-up' O'Connor, a regular soldier and athlete, who was 'full of fight and breeziness'.[197] In the 7th Brigade, Brigadier Field sought a 'young AIF CO' to replace the over-age Edward Miles when he was posted in January 1944.[198] John McKinna, whom Field described as a *beau sabreur* – a fine dashing soldier[199] – was appointed in his place. For another of his battalions, Field requested the appointment of Bill Dexter.

Dexter's appointment provides an example of the manner in which COs were appointed at this stage in the war and how personal relationships still played a role. Dexter was originally suggested to Field as a potential CO for the 61st Battalion by Savige, who was aware not only of

'Larry the Bat': Lieutenant Colonel George Smith, CO of the 24th Battalion. Energetic, enthusiastic and innovative, Smith epitomises the Middle East veterans who assumed command of militia battalions during 1943. The success of Smith's command is borne out by the fact that his battalion branded themselves 'the Bat's Boys'. (AWM Neg. 100624)

Dexter's exploits at Lababia but also, as former commander of the 17th Brigade, of his performance in the Middle East.[200] Field then spoke to 2/6th's CO, Fred Wood, on 29 February 1944. He interviewed Dexter on 2 March, requested his posting via Savige's headquarters on 6 March, and received notification that Land Headquarters, meaning Blamey, had approved the appointment on 9 April. Dexter assumed command on the 15th.[201]

By this stage in the war, officers had reputations that extended beyond their battalions and brigades. After sacking Charles Geard from the 2/10th

Battalion in August 1944, Brigadier Fred Chilton actively sought Tom Daly as a replacement after hearing of his reputation from both inside and outside the battalion.[202] Officers aspiring to commands also sought patrons to help them along their way. While attending a senior officers artillery course in April 1943, Brigadier Field was approached by three officers, two from motor regiments and another from a militia battalion, seeking active commands.[203]

In Dexter's appointment we also see the hand of Blamey still active in the selection and appointment of COs. Early in 1943 Brigadier Dougherty had sought the appointment of Major John Bishop to command the 2/27th Battalion. Bishop, a graduate of the Middle East Staff School, had spent much of the war in staff postings and aspired to a field command. Blamey, however, prevented the appointment because Bishop had been identified for a senior staff position. Blamey eventually relented, but Dougherty was told that Bishop could have only one campaign, a proviso of which Bishop was not informed. Bishop was never quite content as a staff officer 'and thought he was made' when given command of the 2/27th.[204] When it 'transpired that [he] was to return to the gilded staff', despite a short but successful tenure as CO, he was confused – 'why, I don't know, as my pants are shiny enough already' – and filled with gloom.[205]

Further demonstrating both the high regard in which a posting as a CO was held and Blamey's power is the experience of Major Don Jackson. A PMF officer, he too had spent much of the war in staff postings. Back in Australia he had been posted to Land Headquarters, and Blamey had taken to calling him by his first name, leading him to believe he was 'amongst the favoured ones'.[206] Impatient to return to active service, Jackson arranged himself a posting through official channels to the 2/28th Battalion. It proved unfortunate that he did so when Blamey, unbeknown to all but the most senior figures in the AMF, was in London. Also unknown to Jackson was that he was earmarked to be the postwar Director of Military Intelligence. The next time he saw Blamey, the conversation took the following course: 'You have annoyed me, Jackson. You chose to go against my wishes in your employment and did so when I was out of the country . . . I know what you have been after, a battalion command . . . Well, you might do so but it will be in the rank of major.' Jackson was appointed to command only in the last days of the war following the death of his CO and was not promoted: 'It seemed then that General Blamey's curse was in operation.' Jackson, however, commented, 'It was a joyful moment to command an Australian battalion in war and I cared little for anything else.'[207] Gaps in the documentary record prevent a thorough analysis of Blamey's influence on postings in the latter years of war, particularly in misguided appointments such as those of Conroy and Starr. Herring, however, wrote of being amazed by Blamey's 'sureness of touch' regarding the selection of brigade commanders and COs.[208]

COs AND BATTALION CULTURE

Implementing change in any organisation is usually difficult, particularly when the agent of that change is an outsider. By late 1943 many battalions in the AMF had distinct characters as a result of their previous experiences and the influence of former COs. Establishing themselves in a new battalion proved a challenging experience for some COs during this period. We have already seen how the dramatic change embodied by Warfe caused an initial shock in the 58/59th Battalion, although it did not last long once Warfe began improving conditions, partly because, as a battalion new to active service, it had few ingrained traditions or modes of behaviour. When Warfe later took command of the veteran 2/24th Battalion in 1945, his methods were regarded as an affront to their experience.[209]

The experience of Robert Ainslie on taking command of the 2/48th Battalion similarly demonstrated the mutual respect that veteran soldiers demanded from a CO. Quiet and methodical, Ainslie was a marked contrast to the battalion's previous two COs: Hammer and Windeyer, both men of enormous charisma and visibly driven. A 6th Division veteran who had already spent a year in command of a militia battalion, Ainslie joined the battalion at Milne Bay as it prepared for Lae. Camp facilities were poor, and the troops were 'a bit scruffy'. On his first parade Ainslie, perhaps, like much of the AMF, being suspicious of the 9th Division's self-assurance, told the battalion they were not as good as his previous command. Bob Lewin, a platoon commander at the time, told Coates of their reaction: 'Since our boys had seen a lot of action while he and his battalion had not, we were a bit incensed. I'm glad I was near the rear of the parade because I thought the men would surge on top of him.'[210] In a single address, Ainslie lost any hope he had of quickly gaining the trust or respect of the battalion. Even as the battalion edged its way up Sattelberg, with Ainslie doing his utmost to ensure its operations were well conceived and supported, the troops still displayed a marked lack of trust. Long recorded these doubts and the men's pride in the battalion when he visited the 2/48th on 20 November 1943:

'We'd be on Sattelberg now if we had our old CO.'
'Or Windeyer, the first CO.'
'And the company commanders we lost at Tel el Eisa.'
'Hammer was the bloke.'
'He's a brigadier now.'
'Two brigadiers from this battalion.'[211]

It was only in a manner akin to his operational style – slow and steady, building on success – that Ainslie was able to gain their respect; by the Tarakan operations in 1945 he enjoyed their trust, but it seems to have been based on his professional aptitude rather than personality.[212]

An example of an even less successful new CO colliding with the ingrained culture of a battalion is provided by Robert Joshua and the 2/43rd. Joshua, awarded the Military Cross at Tobruk and wounded on three occasions, had proved himself a brave and capable company commander.[213] Brigadier Bernard Evans, who personally instigated Joshua's appointment to the 2/43rd, described him as a 'good fighter'.[214] Joshua replaced Bill Wain, considered too old for command, on 26 July 1943. Wain was a popular CO, revered since his time as a company commander in the 2/16th for being prepared to stand up for his troops against the demands of higher authority. Few questioned his courage, but several of his peers believed that he was not up to the job and devolved too much of his command to his company commanders.[215] Under first Crellin, then Wain, a highly consultative command culture had taken hold in the 2/43rd. Wain told Joshua on his arrival, 'The battalion runs itself.'[216]

This became the root of the trouble. Joshua was reluctant to delegate, the battalion was used to being run on a committee basis and, as a result, an internecine feud flared between the CO and his senior subordinates.[217] He lost the confidence of battalion headquarters staff and, as he tried to recoup the authority that had previously been yielded to company commanders, a whispering campaign of rumour and innuendo further undermined him.[218] Company commanders chafed at what they saw as undue interference in their work and believed Joshua always wanted to be the forward scout and not the CO.[219] Contrasting dramatically with Wain, as well as other 9th Division COs such as Ainslie, Joshua's loyalty was perceived as being always with his superiors and never with his troops.[220] Conversely, it was also observed that Joshua's decisions were continually scrutinised and questioned by a highly critical clique of original battalion officers.[221]

This situation could not be allowed to continue because, as the senior officers squabbled, standards among the platoons and sections began to slip.[222] Brigadier Selwyn Porter, now commanding the brigade, reluctantly had Joshua removed from command in mid-December 1943. Porter felt that Joshua, robust and tactically able, would have been fine had he been given a good battalion.[223] As Coates points out, this was a 'back-to-front' argument because a strong, insightful CO has the ability to make a good battalion. Noel Simpson was transferred from the 2/17th to address the problems in the 2/43rd Battalion and thereafter it never looked back.[224] In keeping with the unofficial personnel policies of the AMF and AIF–militia hierarchy, Joshua was given a month's leave before being appointed CO of an Australia-based militia battalion for the rest of the war.

The experiences of Ainslie and Joshua illustrate that, particularly in the case of experienced and cynical battle-worn battalions, new COs could not hope to simply rely on the legitimate power bestowed by the army. They needed to foster trust and acceptance, founded on the exercise of referent or expert power, and for outsiders without a prior reputation this was a difficult task. Of the seven COs appointed to AIF battalions in which they had not previously served during 1943–44, three were sacked and another two encountered significant difficulties in gaining the respect and trust of their men. It could, however, be just as hard for an officer to gain acceptance when promoted to command from within his own battalion. One example will suffice as an illustration. When Fred Wood, who was reputed to be highly strung, became CO of the 2/6th Battalion he was viewed suspiciously by many of his troops and was not regarded as being particularly popular by outside observers.[225] It was not until he was wounded, while visiting a forward post, that he began to win over the battalion. After being hit in the head by a bullet he evacuated himself on foot to the RAP and in doing so earned himself the nickname 'Fearless Freddie'; from that day he was made.[226] A reputation for bravery established, Wood was then able to consolidate his authority with a demonstration of sound tactical and administrative prowess as he gained confidence in his role.

The New Guinea offensives of 1943 and 1944 were a period of consolidation for the AMF. During this period the operational procedures and tactics with which it would see out the war crystallised. Although the two theatres of war could not have been more different, the evolution of command practice in New Guinea in 1943–44 echoed that which had taken place in the Middle East in 1941–42. A series of earlier defeats and close-run victories had demonstrated weaknesses in the battalion's command system and organisation that were remedied with improvements in communications technology, more capable weapons, the reinterpretation of doctrine in the light of experience and the centralised training of officers.

The experience of New Guinea further demonstrates the situational nature of battlefield leadership in the AMF. With its burgeoning qualitative and numerical superiority over its enemy, the AMF was not pervaded by the same sense of desperation as it had been in 1942. It was on the offensive, with all the confidence inherent in such action, and therefore there was little need for COs to inspire their troops with acts of personal courage. In the experienced and battle-hardened battalions of the AIF, what troops expected from a CO was competence to match their own and the provision of the supplies and the support they needed to do their job. The actions of many AIF COs in New Guinea clearly demonstrate the exercise of prudent judgement to minimise risk, which is the hallmark of military competence. Many of the militia battalions new to combat, however, lacked the competence and therefore the confidence of the AIF battalions. They needed a different style of leadership – forceful, aggressive and enthusiastic – to instil the determination needed to defeat the privations of the jungle environment and ultimately the enemy.

CHAPTER 8

'EXPERIENCED, TOUGHENED, COMPETENT'
1945

By the beginning of 1945 the AMF had been at war for more than five years and committed to active operations for four. In its ethos, structure and proficiency it bore all the hallmarks of a long-established professional army. The men commanding its battalions were more experienced and more thoroughly trained than they had been at any time since 1939. As Peter Stanley has pointed out, this was a force that was among the best in the world – and it knew it.[1] Stanley was referring specifically to the original units of the 2nd AIF, but his remarks are equally applicable to the operationally experienced battalions of the militia, particularly those of the 4th, 8th and 15th Brigades.[2] Although popularly lauded for its amateur virtues, veterans of the AMF of 1945 have in fact spoken of its professionalism.[3] In his foreword to *Nothing Over Us*, the history of the 2/6th Battalion, Frederick Wood recalled the words of a senior Australian officer: 'You are not professional soldiers but you are professional men of war.'[4] Asked to expand on his description of the 2/1st Battalion in 1945 as a 'professional' force, Paul Cullen replied: 'We were a very experienced, toughened, competent army, with a wonderful team of company commanders and platoon commanders . . . It was a marvellous team to be the captain of.'[5]

Cullen's words highlight one of the great strengths of the AMF by this point in the war: the insight, trust and mutual respect that underlay many command relationships, both between battalion and brigade and within the battalions themselves. These relationships were the product of shared training, ethos and experience and, in many cases, the type of firm and genuine friendships that are the product of a common ordeal. Demonstrating the main weakness of any human organisation, however, some command relationships in the AMF continued to be undermined by personal animosities, suspicions and idiosyncrasies as they had been since 1939.

In a material sense, the AMF was supplied better than it had ever been before. It was now over-supplied with ammunition and most major items of equipment and ordnance. This, however, did not necessarily translate into a bounty of destruction on the battlefield. The Dickensian aphorism 'It was the best of times, it was the worst of times' is apt. In the Borneo campaign the battalion commanders of I Australian Corps had access to the full range of supporting arms, enabling them to conduct true combined arms operations in the modern sense and thereby limit their cost in sweat and blood. Elsewhere, battalion commanders who had been trained to employ such support without hesitation longed for its provision. These operations were conducted on a logistical shoestring, forcing COs and formation headquarters to argue and quibble over the use of single landing craft and meagre allocations of artillery shells.

Logistical difficulties were not the only frustrations with which battalion commanders struggled in 1945. In a political and strategic sense the operations of that year were the most challenging fought by the AMF during the war. The operations essentially fell into two categories. First, there were the campaigns conducted to clear the Japanese from Australian-mandated territory in Bougainville, New Britain and the Aitape–Wewak region of New Guinea. These operations had no real strategic significance apart from being a statement of Australian sovereignty as the forefront of the war against the Japanese had long since moved to the Philippines and the islands of the central Pacific. Second, there was the Oboe series of operations in Borneo, which involved landings on Tarakan island, in British North Borneo and at Balikpapan in Dutch Borneo. The Oboe operations were conceived as part of a dubious strategy to retake Java, but in reality had little more strategic relevance than those conducted in the mandated territories.

Despite being far removed from the decision-makers in Canberra, Brisbane and Hollandia, the political machinations surrounding these operations often significantly affected the freedom of action of battalion commanders. In several instances their tactics and decisions were subjected to unprecedented scrutiny from higher command, and the resulting direction often completely contradicted established doctrinal principles. In addition, it was readily apparent that the operations were contributing

'Experienced, toughened, competent': the tactical headquarters of the 2/23rd Battalion atop Tank Hill on Tarakan, 1 May 1945. The CO, Lieutenant Colonel Frederick 'Fag' Tucker (third from left), had served with the 2/48th at Tobruk and Alamein before taking command of the 2/23rd during the Huon Peninsula campaign. (AWM Neg. 090930)

little to the ultimate defeat of Japan. Many COs struggled with ordering operations that had the potential to kill men who in some cases had survived five years at war, and were angered by having to conduct them in a fashion that they believed increased the likelihood of casualties.

Like the AIF in 1918, the AMF in 1945 was at the height of its development. Its personnel were well trained, its doctrine was sound,[6] its equipment, although not always provided in sufficient quantity, was appropriate to the task and developed to the limits of the technology of the day. Like the AIF of 1918, however, the very experience that had resulted in its proficiency was also its most significant weakness. Within the battalions experienced personnel were tired and stretched increasingly thin. Government decisions to reduce the size of the AMF and discharge those with five years service threatened to break up close-knit command teams. Newly appointed COs were younger than ever before – partly a sign of the triumph of merit-based promotion, but also an indication of the strain imposed by prolonged command.

Youth

After meeting Bernard Callinan in northern Bougainville in February 1945, Gavin Long remarked in his notebook how young the CO of the 26th Battalion looked. He was quiet and calm, but at the same time 'cheery'. Long also commented that he was 'lean', appeared very fit, and surprisingly for the times, did not smoke – 'like a considerable number of young officers'.[7] When promoted to lieutenant colonel, Callinan was 32. He replaced John Abbott, a 38-year-old veteran of the campaigns in Greece and Libya. Callinan had commanded the 2/2nd Independent Company with distinction on Timor in 1942; had completed a junior staff course and served a staff posting on the headquarters of First Australian Army; and, before taking command of the 26th, had been 2iC of the 31/51st Battalion and attended the LHQ Tactical School. He epitomised the new generation of COs who were being appointed in 1945. They, more than any other group of COs, were 'professional men of war' and remarkably alike. Most had begun the war as platoon commanders and thus been exposed to the operation of all of a battalion's subunits. They possessed a common doctrinal understanding and were fit, determined and earnest in their application to their commands.

In 1945 the AMF's COs were younger than they had ever been before. There had been another concerted campaign in the latter half of 1944 to reduce the age of the AMF's senior officers. Brigadier John Field, commander of the 7th Brigade, noted in August of that year that 35 was now regarded as the maximum age for lieutenant colonels and therefore the maximum age for battalion officers; his counterpart in the 4th Brigade, Brigadier 'Boss' Edgar, had already 'got rid of' all of his officers over that age.[8] In February 1945 the average age of AMF COs was 35; the average age of the initial cohort of AIF COs in 1939 was 43, and climbed to 45 – retirement age – in the 8th Division. Most telling is the average age of the COs appointed during 1945: 31. The declining age of COs, reflecting the physical and mental demands of war in the Pacific theatre, was in keeping with a broader trend for younger senior officers in the battalions. A representative sample of AIF and militia battalions – the 2/6th, 2/14th, 2/17th and 58/59th – reveals that the youngest rifle company commanders were 26 and that none were older than 32. Only three senior officers – two HQ company commanders and a battalion 2iC – were older.[9]

The youngest Australian CO of the Second World War was Charles 'Charlie' Green, who took command of the 2/11th Battalion in March

1945, aged just 25. When Green was appointed there were already two battalion commanders younger than 30,[10] and another two would follow him in succeeding months.[11] Despite his youth, Green had considerable military experience. He had first enlisted in the militia as a 16-year-old private in 1936, had been commissioned in March 1939, and was selected in October as one of the original platoon commanders of the 2/2nd Battalion. He served with the 2/2nd in Libya and Greece, and by the end of 1944 he had commanded a company in action, been seconded as an instructor to the First Australian Army Tactical School and the Junior Wing of the LHQ Tactical School and had completed the senior course at the latter. He was described as an 'outstanding student'.[12] His second CO, Fred Chilton, has reflected: 'Although quite young at the time he was very mature; a quiet, calm man, obviously with exceptional reserves of and force of character.'[13]

During his meeting with Bernard Callinan, Long also observed that the 26th Battalion's 2iC was older than 40 and reflected that this was a common occurrence in II Corps.[14] Long's observations raise the issue of age and its influence on authority. In most human societies age has its own inherent authority. As Long's remarks make clear, in 1945 many officers and soldiers in the AMF were being commanded by men much younger than themselves. Such young COs were often viewed with suspicion on assuming command. In the 2/10th Battalion, Tom Daly, who took command at the age of 30 in October 1944, was initially known by some as the 'Boy Bastard'.[15] Thirty-year-old Peter Webster was appointed to command the 57/60th Battalion in March 1945. Webster was never popular in the battalion, and his 'confident, superior air' was a striking contrast to the kindly paternal attitude of the previous CO, 'Happy Bob' Marston. The battalion history makes it clear that his age and quick rise to command were at issue: 'On 13 March when Webster was announced as Lieut Col in the Routine Orders he had been in the army for not quite 5½ years, and had risen from private to CO. To attain the same position Marston had been in the army 24 years, under two years of which was in action. It is no wonder that they were such different men . . . no wonder the men who they commanded had such contrasting views of the two . . .'[16]

Similarly, Green was not initially welcomed into the 2/11th with open arms. Several officers considered him too young for the appointment, and his quiet, serious demeanour earned him the ironic nickname 'Chuckles'. David Butler has astutely suggested that '"Chuckles" may have been an earnest young man in the loneliness of first command'.[17] It must also

'Chuckles': Lieutenant Colonel Charles Green, CO of the 2/11th Battalion. Aged just 25 at the time of his appointment, 'Charlie' Green was the youngest Australian infantry battalion commander in the Second World War. (AWM Neg. 097970)

be considered that Green was the first outsider to be appointed to command the 2/11th, a particularly parochial Western Australian battalion.[18] The battalion's preferred candidate at the time of Green's appointment was its 2iC, Major David Jackson.[19] It is instructive to note that Jackson was 28 at the time of Green's appointment, highlighting that age was not the only factor in the latter's slow acceptance. Many in the 2/2nd Battalion lamented Green's loss, where his reputation as a commander of determination, skill and integrity was firmly established.[20] Chilton observed that even as a platoon commander most of Green's men were a lot older than he was but 'there was never any question of his authority or of the respect in which he was held and the confidence he inspired'.[21]

It seems Green was conscious of his age on taking command in the 2/11th. He wrote to his wife: 'They are a very good lot and particularly the senior officers, who are all old chaps . . . and have all the experience necessary.'[22] These comments throw light on the reasons for Green's ultimate success. He respected and used the talent in his command team, and it in turn came to respect his experience and judgement, as did the battalion at large.[23] His divisional commander commented: 'He proved himself a gallant soldier, possessed of sound common-sense, initiative and administrative ability. He exercised a firm and wise control over his men and had their respect. His personal character and conduct are beyond reproach and I found him completely trustworthy.'[24] Tom Daly's earnestness and strictly regimental manner did not endear him to all in the 2/10th, although it was probably as much a product of his PMF background and experience with the British Army in India as it was of his relative youth. Les Peterson has recalled that he did not like Daly much 'as a chap' because he was 'devoid of a sense of humour'. Peterson's recollections, however, prove that it was not a sense of humour that mattered. He observed that Daly was a 'soldier and a half' and a 'superb administrator' and that on Balikpapan he had the battalion's operations 'nutted out down to the last round'.[25] Roger McElwain has identified a similar source for the authority of younger COs of the New Zealand Expeditionary Force (NZEF). It was based on 'shared adversity . . . on their competence as commanders and on their willingness to look after their men'.[26]

The retention of 40-year-old 2iCs, passed over for command of their battalions, might be an indication that the AMF hierarchy was still in the process of accepting such young COs. Steady counsel to guard against youthful exuberance, perhaps? Given that the age range of the other ranks in a battalion could still stretch from 18 to 45, retaining a senior officer with an additional ten years of life experience also seems a sound man management policy. The retention of such officers could also be seen as another sign of the high value placed on operational experience. Previous campaigning had clearly demonstrated that COs needed to be young and fit to endure the physical demands of New Guinea. The role of the 2iC, however, was an increasingly static one, principally concerned with the administration and supply of widely dispersed forward companies from a static and generally well-established B Echelon. As 2iCs, with their wealth of experience and no doubt a few tricks up their sleeves, older officers still had a useful role to play.

The Australian experience was mirrored by that of other Allied armies. In July 1944 Long observed that most American battalion commanders were aged between 30 and 36.[27] By the war's end the average age of British

Army infantry battalion commanders was 32, and 30 was considered the optimum; in the New Zealand Army's 2nd Division it was 37, and in the Canadian Army in 1945 'a battalion commander in his late 20s was not uncommon'.[28] In 1946 the British Army's Military Secretary reported that the experience of both world wars had 'conclusively proved' that the best age for arms corps unit commanders was in the range 28 to 36.[29]

EXPERIENCE

In 1945 the greatest contributing factor to the success of AMF battalions on operations was the experience of their senior officers and NCOs. They were the thread of continuity that linked the battalions of 1939 with those of 1945 and provided the conduit for the collective experience of the intervening years. By 1945 casualties, promotions and postings to staff and training appointments had reduced the numbers of experienced personnel in the battalions to small minorities. In the 2/6th Battalion, for example, at the end of 1944 there were only 78 officers and ORs remaining who had first sailed with it for overseas service in April 1940; more than 55 per cent of the battalion were reinforcements who had not seen active service.[30] This situation was common in most of the battalions of the original AIF and many of the militia units. In the 2/48th Battalion only about half of its strength had seen action before 1945, and in the 2/24th less than half of the battalion had served in New Guinea, less than a fifth in the Middle East, and only 70 members were 'originals'.[31]

In terms of personal experience, the battalions of the AMF were more inexperienced than they had been at any time since the first battles of 1941 yet, collectively, most proved more proficient than they had ever been. As David Hay has pointed out, it was the 'solidity of the leadership group'[32] that held the battalions together. COs formulated hard, focused training programs, implemented by a solid core of experienced company commanders and senior NCOs, which inculcated fresh reinforcements and junior officers with the skills, doctrine and ethos that had been proven in battle. As the 2/10th Battalion's report on the Balikpapan operations observed, quick decisions by a commander were of no use unless the battalion was 'capable of equally rapid action'.[33] Well-rehearsed battle drills, combined with the collective experience of the battalion command teams and the mutual trust it engendered, brought about a situation where, as Tom Daly reflected, battalions could be put 'into action with about three paragraphs over the telephone'.[34] Similarly, Brigadier Hammer praised the drills and intuitive action of his units, which could establish themselves with 'amazing speed and efficiency', thus ensuring that the 'commander never lost the grip of his command throughout the battle'.[35]

In 1945 AMF infantry COs were more experienced than ever before. Of the 55 COs of the AMF's operational battalions in February 1945, there was not one who had not seen active service in at least one theatre of war. The overwhelming majority – more than two-thirds – had served both in the Middle East and in Papua or New Guinea.[36] Lesser proportions had served in the Middle East or New Guinea alone and in other theatres of operation, such as Malaya and New Britain. Twenty-four (a little less than half) had already officially commanded a battalion in action, and there is little doubt that a large proportion had also experienced periods of acting command.

The experience of the COs was backed up by that of their senior officers. The extent of the shared experience and depth of talent to be found among the command teams of the battalions – in this instance defined as the 2iC, the five company commanders and the adjutant – is readily apparent on a brief investigation of the service histories of the officers involved. During the Aitape–Wewak campaign the 2/6th Battalion was commanded by Fred Wood, one of its original company commanders. Its 2iC was its original signals officer, its adjutant had joined it as a reinforcement officer in 1940, and four company commanders had been with the battalion since 1940, including two who had originally enlisted in the ranks.[37] An insight into the *esprit de corps* and mutual trust this common background and shared experience created is provided by an entry from Captain Ken Brougham's diary on 9 May 1945:

> I have realised another of my ambitions – to command my original company. In spite of the fact that the Company Sergeant Major (Wally Reddick) and I are the only two surviving members, it is still the best. We are all very happy in spite of the bad conditions and lack of equipment . . . B Company has now rejoined us so the old firm are operating together – Ernie Price with A Company, David Hay with B, Bevan French with C, myself with D, and Mick Stewart with HQ . . .[38]

The complexion of the command teams across the battle-seasoned battalions of the AMF was remarkably similar. In February 1945, 14 of the 27 original AIF battalions were commanded by original members.[39] In the 2/14th Battalion only one member of the command team, a rifle company commander, had joined the battalion since 1940, and all but one had served in at least three of the battalion's campaigns.[40] In the

9th Division, the 2/17th Battalion was commanded by John Broadbent, the original 2iC of D Company. Similarly, all but one of his command team were battalion 'originals' – his 2iC and three company commanders had been subalterns in 1940, and his adjutant and another company commander had been ORs.[41] Even in the 2/1st Battalion, the bulk of which had been captured at Retimo in May 1941, five members of Cullen's command team had joined the battalion in 1939.[42]

Owing to the more limited operational experience of the militia battalions, their senior officers were not necessarily as close knit as were their peers in the AIF battalions. By the end of February 1945 no militia battalions were commanded by an original member. As Long points out, the senior officers in militia battalions in 1945 were a mix of those commissioned in 1940–41 and officers transferred from the AIF battalions.[43] This situation, however, still represented a sound base of command experience on which COs could rely. When the 58/59th Battalion began its advance down the Buin Road on Bougainville in March 1945, it was commanded by William Mayberry, a 29-year-old veteran of the fighting in Libya and Greece, who had considerable staff experience. His adjutant and two company commanders had been with the battalion when it first deployed to New Guinea in 1943, while his 2iC and one rifle company commander had served with 6th Division battalions in the Middle East, Papua and New Guinea.[44]

Representative of the militia battalions that saw active service for the first time in 1945 was the 7th Battalion, commanded by Harry Dunkley. Thirty-three years old, Dunkley had originally enlisted in the 2/6th Battalion as a private in 1939. He earned a Military Cross at Bardia and subsequently served in Greece and New Guinea, where he was severely wounded. In April 1945 none of Dunkley's officers were older than 30, and only one, Major Francis Rowell, his 2iC, had seen active service.[45] Providing another example, the 31/51st, commanded by Joe Kelly, was a mix of the old and new militia. Its adjutant and three company commanders were aged 35 or older, and none had seen active service. The remaining three company commanders, aged 26, 26 and 30, were AIF veterans of the fighting in the Middle East and New Guinea.

The influence on operations of the personal insights and relationships developed over five years of war are exemplified by Tom Daly's experience with the 2/10th Battalion at Balikpapan. Although he had been away from the battalion for four years, Daly knew the company commanders and many of the senior NCOs from the time when they were platoon commanders and soldiers and Daly was adjutant.[46] His subsequent posting as brigade major of the 18th Brigade had, to a degree, allowed him to maintain his relationship with many of the officers. Daly commented that it was essential for a CO to understand the characteristics of his company commanders and know what 'made them tick'.[47]

These insights subsequently guided the tasks they were given in battle. Major Frank Cook, OC C Company, was an 'up and go, get at 'em' type, and hence his company was held in reserve to conduct the attack on the key feature of Parramatta.[48] In an example of the mutual trust that existed between the two men, Daly recalled simply telling Cook, 'Right, Frank, off you go' when the time for that attack came.[49] Captain Alan 'Donkey' Bray, OC B Company, Daly continued, was a 'bulldog type': 'You wouldn't give him anything that required too much initiative because you knew very well if you asked him to sit on a hill nothing in the world would tip him off.'[50] Bray's was one of the first companies ashore at Balikpapan; he 'was told to ignore all opposition on the way to his objective, which he did, covering the 800-odd yards in the planned time of 15 minutes. The great haste was in order to secure some sort of depth to the beachhead before the enemy recovered from the bombardment.'[51] Similarly, care had to be taken to ensure that Captain Roger Sanderson, OC A Company, 'wasn't going to lose himself'.[52]

The best example of Daly's extraordinary sensitivity to the motivations of his subordinates is provided by Captain William 'Bill' Brocksopp. Daly explained that Brocksopp had had a 'mixed career', including evacuation from Tobruk with anxiety neurosis, and was 'anxious to redeem himself'. He made sure that Brocksopp was given a task that he could 'get his teeth into' so he could go home thinking 'I wasn't that bad after all'. At Balikpapan, Brocksopp's company was ordered to secure the left of the landing beach on 1 July 1945 and was subsequently given a role on the battalion's left flank during the advance on the town itself, which Daly believed suited Brocksopp's temperament. Daly's deft handling of his subordinates was lauded in a confidential report by Brigadier Chilton: 'Possesses in a high degree the ability to obtain the utmost from his officers and men both in training and in action.'[53]

Many of the command relationships between the COs and their brigadiers were similarly strong and insightful. If not based on a long period of shared experience, they were grounded at least on the knowledge of demonstrated competence. In February 1945 all but two of the brigadiers in the AMF's expeditionary formations had commanded infantry battalions in action. There were five brigade commanders for whom the operations of 1945 were their second in command, and the brigadiers of the 20th, 21st and 25th Brigades were directing their third campaigns. In the 25th Brigade all three of Brigadier Ken Eather's battalion commanders – Thomas Cotton, 2/33rd, Richard Marson, 2/25th, and Ewan Robson, 2/31st – had already served under him in the Lae and Finisterres operations. Marson had also commanded his battalion under Eather on the Kokoda Track, while at the same time Cotton and Robson were both serving in more junior positions in their own battalions. For George Colvin and Colin Grace, the British North Borneo campaign was their second under the leadership of Brigadier Windeyer.

Although neither was formally appointed to command at the time, Phil Rhoden and Frank Sublet had also been serving under Brigadier Iven Dougherty since the operations in Papua. A small measure of the mutual trust that existed between Dougherty and Rhoden is provided by the latter's description of him as one of the finest commanders he ever served under. Rhoden recalled Dougherty questioning his caution during the operations to secure Manggar Airfield at Balikpapan in July 1945, but he came forward to allow Rhoden to explain the situation and thereafter left him to pursue his operations as he saw fit.[54] A similar relationship is exhibited by Chilton's willingness to let Daly push on up Parramatta without fire support: 'I advised the Brigade Commander of the problem and asked if he had any instructions. He said, "It's up to you, you know the situation." So I said, 'Let's go.'"[55]

In 1945, however, the AMF was facing a potential command crisis. As we have seen, command experience within the battalions resided principally with a thin layer of highly experienced company commanders that rested on a much larger strata of junior officers with limited experience. This layer was the source of new COs, and it was being stretched increasingly thin. This was particularly the case in the 6th and 7th Divisions, which had provided large numbers of experienced officers for the militia battalions in 1942–43. On 30 October 1944 the 6th Division's war diarist lamented the appointment of Major Arthur Anderson, 2iC of the 2/3rd Battalion, to command the 24th: 'This constant drain on our senior Majors for command of units outside the Div is placing the Div itself in a serious position for potential Comds.'[56] Major General Stevens was similarly concerned, telling Gavin Long that it was easier to identify potential brigade commanders in his division than it was to find battalion commanders.[57]

The decision in June 1945 to release all members of the AMF with more than five years service also had to the potential rapidly to drain the pool of command talent. More than half of the COs were immediately

Shared experience and demonstrated competence: Brigadier Ivan Dougherty, commander of the 21st Brigade (second from left), with his battalion commanders: Lieutenant Colonels Frank Sublet, 2/16th Battalion (far left), Keith Picken, 2/27th Battalion (second from right), and Phil Rhoden, 2/14th Battalion (far right). All had been serving since the beginning of the war, and Rhoden and Sublet had worked under Dougherty's command since the Papuan campaign. (AWM Neg. 113192)

eligible for discharge, and all but a few would be discharged by the end of 1945. In August Keith Picken of the 2/27th Battalion told Long that he would have lost 20 of his officers if the five-year men had gone out during the action at Balikpapan.[58] In the 2/10th Battalion, all of Tom Daly's company commanders and 2iCs were eligible for discharge under the scheme, but some had chosen to stay. Daly did not consider that such discharges would cause the battalion any loss of efficiency as there were plenty of lieutenants who could replace the company commanders,[59] but it is hard to conceive of the battalion being able to compensate immediately for the loss of the collective experience that these officers embodied. In a letter to his parents, Brigadier Eather was more pessimistic. He feared that the repatriation of the 'five year service chaps' would deprive the brigade of 'many battle trained and experienced leaders', making a 'tremendous difference in the fighting efficiency' of the battalions. He was convinced that had this occurred before Balikpapan his brigade 'would not have been capable of doing the job at such slight cost'.[60]

PERSONALITY AND COMMAND RELATIONS

Despite the professional ethos prevailing in the AMF, personality clashes still undermined command relationships, emphasising the value of the command cohesion discussed above. Headstrong and confident young COs often resulted in turbulent command relationships. In the 15th Brigade on Bougainville both George Warfe, 32, and William Mayberry, 29, clashed with Brigadier Hammer and sought reposting. Warfe was successful and was posted to command the 2/24th Battalion. Mayberry, probably because he lacked Warfe's previous command experience, was not, and he continued under Hammer for the rest of the war. Geoff Matthews of the 9th Battalion was described by Brigadier Field as 'impetuous' and in 1943 had quarrelled with his older colleagues in the 7th Brigade over who should administer command in Field's absence. He wrote petulantly in his diary: 'I am senior CO here but apparently only a small boy still.'[61] On Bougainville Matthews clashed with Peter Webster of the 57/60th. Reading Matthew's account of a meeting with Webster on 29 April 1945, one would be forgiven for doubting that these two were among the AMF's most experienced officers:

> His plans for the relief do not agree with mine, he is adamant that my rear coys will be relieved before the fwd ones . . . He wishes to redispose his coys and wanted to know if I was going to shift any of mine for him to make an easy handover. I refused. Wanted us to make a new dropping ground for him near C Coy. I refused. Wanted my Coys to . . . corduroy the road. I refused. Said our method of corduroying . . . was not in accordance with his method. That is his worry . . . A patrol from A Coy crossed the Hongorai and saw 12 nips on the other side . . . They returned and arty fire brought down. The leading scout['s] . . . name was forwarded tonight as the name of the crossing. Webster stated it was wrong to name a place until it was captured . . . it should be left until his battalion had made a bridgehead. He is obnoxious! I am afraid I was rude to him.[62]

The worst instance of dysfunctional command relationships on Bougainville was the 29th Brigade under the command of Brigadier 'Mad Mick' Monaghan. A difficult and erratic character, Monaghan quarrelled with both his divisional commander and his staff, many of whom he sacked or placed under open arrest.[63] There was no stability or cohesion in the brigade, and Savige wrote of Monaghan's relationship with his COs: 'Poor men were "tops", or good men were on the outer and forced to obtain transfers or seek their discharge.'[64] Monaghan sacked Norman Goble of the 47th Battalion, although not without reason, and also sought the dismissal of Herbert McDonald of the 15th Battalion on spurious medical grounds. In the early operations in southern Bougainville Savige graded the 29th Brigade 'in a low order of efficiency', and Monaghan was dismissed. Brigadier Field, whose brigade relieved the 29th, noted its disunity in his diary and observed of Monaghan: 'His actions are generally not indicative of team work in the Div.'[65] A further indication of the emphasis placed on harmonious command teams within the brigades is provided by the example of Eugene Egan of the 35th Battalion. Although variously described as a 'keen' and 'efficient' officer, as a CO he proved 'uncooperative' and lacking in 'team spirit', which led to his sacking in May 1945.[66]

The relationship between Brigadier Selwyn Porter and Hugh Norman in the 24th Brigade provides an example of how personal feelings could become entangled with operational matters and cloud our subsequent understanding of them. In 1945 relations between Porter and Norman were frosty. Don Jackson, Norman's 2iC, believed this was because both men were of 'considerable presence... and not a little vanity'. Porter gave the impression that he considered the 2/28th 'too big for its boots'. These attitudes poisoned relations between the respective headquarters. Jackson reflected: 'It was sad that the relationship with our superior headquarters grew that way, with the [brigade] staff seeking to lower us a peg and our officers dubbing the Brigadier "The Sawdust Caesar".'[67]

What is clear from Porter's writings was that he felt Norman was 'run down' and not up to his job. He was prepared to have Norman removed from command but relented because Norman, a 'sentimental, paternal commander who loved his unit', had 'begged' to remain. Jackson's reminiscences reveal that Porter, in addition to believing that Norman would not last long, was actively working to undermine his position in the battalion. On his arrival Jackson was told that Norman was not robust and that he was positioned in the battalion to take over. Jackson was 'not pleased to carry this confidence' and told his new CO, ironically reinforcing the confidence between the two while contributing to the animosity between Norman and Porter. Ultimately, Major-General Wootten removed Norman from command for losing control of his companies during operations on Labuan island in British North Borneo.

Correspondence with Long reveals Porter and Norman to be advancing competing narratives of their experience on Labuan until the 1950s: Norman emphasising the difficulties he faced as a result of the actions of higher command, Porter seeking to downplay the role of the 2/28th and advance that of the 2/43rd, under the command of Mervyn Jeanes.[68]

COMBINED ARMS WARFARE

Phil Rhoden recalled that the Balikpapan operations of July 1945 were the first in which he was able to employ the full range of supporting arms, provided with copious quantities of ammunition and 'finally send in the infantry without disaster'.[69] By 1945 the employment of combined arms was at the heart of the AMF's operational doctrine, and this approach was given added impetus by the perception that the campaigns were strategically irrelevant and not worth risking men's lives unduly. This style of warfare, combined with the nature of the enemy being faced – usually small groups in independent defended localities – had a considerable influence on the role of the battalion CO. Even more so than in New Guinea in 1943–44 the company was the primary tactical unit, and generally the CO became more akin to a tactical resource manager, ensuring that his company commanders had sufficient support and supplies to carry out their orders.

The variety and extent of fire support available to many COs in 1945 is illustrated by the inventory of destruction contained in the 7th Division's report on the Balikpapan operations:

> Support calls by infantry could normally be made on:
> - air support (bombing, strafing, NAPALM)
> - naval gunfire (cruisers, destroyers and, in some cases LCI (G) and LCI (R))
> - field artillery (25-pounder)
> - tank-attack artillery (6-pounder as sniping gun)[70]
> - 4.2-inch mortars
> - Matilda tanks
> - Frog flame-throwing tanks
> - manpack M2 flamethrowers
> - subunits of divisional machine gun battalion
> - engineer assault and demolition parties.
>
> In addition to the normal battalion weapons including:
> - 3-inch mortars (which can be brigaded)

- machine gun platoon
- tank-attack 2-pounder guns
- WP [white phosphorus] grenades
- PITA [projector, infantry, tank-attack].[71]

As the report went on to comment, there were few targets that a battalion could not adequately destroy or neutralise. It noted, with a curious degree of surprise, the 'small number of infantry troops required to capture even heavily defended positions when sufficient and co-ordinated support is given by supporting arms and naval and aerial bombardment'.[72] Discussing the operations to secure the 'the Pocket' on Labuan, a 24th Brigade report similarly reflected on the lack of resistance offered by a 'completely dazed' enemy after artillery bombardments were called down before each attack. Any remaining 'offensive spirit' was quickly quashed by 'Frog' flame-throwing tanks.[73]

The 7th Division at Balikpapan was probably the best-supported Australian division of the war, and other Australian forces in 1945 were not as well endowed. Nevertheless, the employment of the maximum available fire support to save infantry casualties became the fundamental principle of Australian tactical doctrine. Much of the tactical manoeuvring of infantry was designed to force the enemy into positions where Australian fire support could be employed to its greatest effect.[74] Patrols were used to determine the extent and strength of an enemy position as well as to drive in outposts to provide concentrated targets for artillery and mortar barrages and air strikes. Once the objectives were thus prepared, the infantry would be committed to the assault, wherever possible with the support of tanks in addition to continued indirect fire support. This style of warfare is commonly associated with American forces, but it was the product of a distinct Australian operational experience. Reflecting on the Balikpapan campaign, Phil Rhoden lamented the haste with which the operations at Gona in 1942 were conducted.

Rhoden's conduct of the 2/14th Battalion's operations around Manggar airfield, north-east of Balikpapan, in July 1945 exemplify the AMF's approach to combined arms warfare. At Manggar the Japanese were entrenched in the hills overlooking the airfield in considerable strength. Immediately to the north were two 6-inch coastal defence guns on a hill called Waite's Knoll. It was dominated by another higher feature known as Frost. Two spur lines, Brown and Green, descended south-east from Frost towards the Vasey Highway – the 2/14th's axis of advance – and on these were sited several Japanese machine guns and artillery pieces. Once

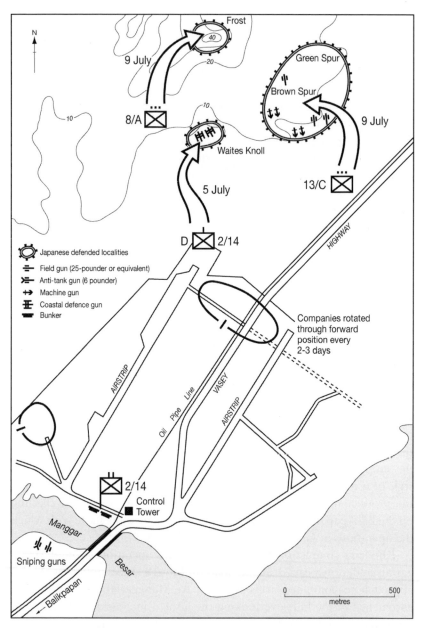

Map 8.1: 2/14th Battalion operations at Manggar airfield, 4–9 July 1945

the Japanese opened fire on the morning of 4 July, the operations proceeded with 'unhurried, calculated and deadly precision to pave the way for a final infantry assault with minimum loss'.[75] Rhoden recalled: 'There was no reason to do it quickly, it was a battle that perhaps need not have taken place. The whole Balik landing, I mean . . . I just wasn't going to waste men's lives needlessly.'[76]

For the next two days the Japanese positions were pounded with mortars, artillery, naval gunfire and air strikes, and two 'sniping guns' were moved into position to duel with individual Japanese emplacements. In the meantime, a series of observation posts pinpointed each Japanese position. On the afternoon of 6 July, demonstrating the devastating efficiency of the Australians' fire support, a 13-man fighting patrol from D Company was able to attack and capture the coastal defence batteries on Waite's Knoll, driving off around 50 Japanese in the process. The rest of the company was moved in to occupy the knoll and resisted enemy counter-attacks for two days; their fire support again telling. Rhoden had already initiated phase 2 of his operation, pushing patrols out to locate the positions on the spurs overlooking Vasey Highway. Another two and a half days of bombardment followed. Finally, on the afternoon of 9 July, a slackening of Japanese fire indicating their positions to be 'almost neutralised', Rhoden ordered D Company to dispatch fighting patrols to occupy Frost and Brown Spur. This was accomplished with little resistance. Characteristic of much of the fighting of 1945, Rhoden never had any more than two of his companies heavily engaged at the one time, although all four were in almost constant contact with the enemy through patrolling and infiltration attempts.[77]

By the time Brown Spur was occupied, 3458 3-inch mortar rounds, more than 5000 rounds of naval gunfire and more than 12,000 25-pounder rounds had been expended.[78] This was in common with other areas of operation. The preparatory fire for the attack on Wewak by the 2/4th Battalion on 10 May included 4500 25-pounder rounds,[79] and the 15th Brigade, during its advance down the Buin Road in southern Bougainville, employed 68,000 artillery shells, 38,000 mortar bombs and 768 tons of aerial bombs.[80] In his report on the Balikpapan operations Brigadier Dougherty noted the unfavourable comparison with the complete lack of fire support during the retreat from Kokoda or the mere 250 rounds of 25-pounder ammunition employed in the final attack on Gona.[81]

It should not be thought that all of the Australian infantry actions of 1945 were complete walkovers. Much arduous patrolling was required to locate and fix the enemy, and any attack still required the infantry to clear Japanese emplacements at close range. The Japanese remained particularly skilful in the siting of defensive positions, and these were often located to negate the advantages offered by Australian firepower. For instance, positions on the steep, narrow ridgelines of Tarakan were notoriously difficult to bombard effectively with artillery, and at Wewak in New Guinea the Japanese occupied deep caves that were impervious to any bombardment and could be dealt with only at close range by explosives or flamethrowers. Particularly during the early operations on Tarakan and Bougainville, logistical difficulties severely limited supplies of ammunition and, in the opinion of many involved in these actions, these shortages contributed to greater Australian casualties.[82] Whereas attacks referred to previously occurred after barrages of thousands of rounds, the daily battery ammunition allocation for the 2/7th Field Regiment during the third week of the Tarakan campaign was just a hundred rounds.[83]

For COs trained to use fire support to reduce casualties, such situations were frustrating. Geoff Matthews regularly chafed against the ammunition restrictions imposed on him in Bougainville. After requesting a thousand rounds of artillery fire for the 9th Battalion's attack on Artillery Hill, Matthews was told by the brigade major that 'higher authority would scream': 'I said "Let 'em" but it appears bodies are cheaper than shells and I shall have to go slow with arty.'[84] Matthews argued with the brigade major and Brigadier Field for two days before finally being allocated 600 rounds of preparatory fire, which he considered the 'barest minimum'. The battalion attacked on 18 December and captured Artillery Hill at a cost of five killed and eight wounded. Matthews reflected, 'We are very lucky!' Arguments over ammunition expenditure continued when the 9th was committed to the advance down the south-west coast in January. Having had to argue for 60 rounds of artillery fire to bombard a position encountered by one of his patrols, Matthew's frustration is palpable: 'War on the cheap again!!!'[85]

There was a significant morale imperative driving Matthew's insistence on artillery, but a contrary point relating to its employment needs to be considered. During the Aitape–Wewak campaign it was often found that the employment of artillery against small targets encountered by patrols resulted in more casualties rather than fewer. The Japanese often withdrew to a new position when shelled, and further casualties were subsequently incurred locating them. It was considered less costly to 'close in for the kill' and destroy small enemy parties with the infantry's own weapons rather than withdraw and lose contact while artillery fire was brought onto the objective.[86]

The effective employment of the diverse range of fire support available to COs required a thorough knowledge of the capabilities and tactical employment of each weapon, and 'great care' had to be exercised in the preparation of a coordinated fire plan.[87] Although battalion COs received training in cooperation with armour and artillery at the LHQ Tactical School, they accumulated a steadily growing group of attached advisers within their headquarters to provide specialist advice. During the 2/14th's operations at Manggar, Rhoden's headquarters included an air liaison party, a shore fire control party, an artillery battery commanders party, an artillery forward observation party, and armoured, engineer and machine-gun officers. As the 2/10th's report on Balikpapan noted, the management of all of these specialist officers within a headquarters was a considerable challenge, although it was greatly assisted by close personal relationships developed during the course of joint training.[88]

Daly managed this situation by forming a small 'command post' that essentially consisted of himself and the commanders of the supporting arms needed for immediate employment. It was usually forward, mobile and hence quite vulnerable. The rest of Daly's tactical headquarters was located to the rear but within voice range and sited to take advantage of the terrain to protect critical personnel, such as the supporting arms commanders not immediately required. In the less-mobile operations at Manggar, Rhoden commanded from a unified, static command post at the south-eastern edge of the airfield. This was the most common practice throughout the campaigns of 1945, both during lengthy patrol operations and the assaults in which they culminated.[89] Rhoden visited his forward companies only once during the engagement at Manggar, but his command post was the site of nightly conferences with the commanders of his supporting units to plan the following day's operations.

Rhoden's role is not dissimilar to that played by Walter Howden during the 2/8th Battalion's attack on the summit of Mount Shiburangu south of Wewak. The actual assault was entrusted to C Company, commanded by Captain Mick Dwyer. Dwyer and Howden initially discussed the problem, then Dwyer was left to make his preparations. After several days of patrolling and an aerial reconnaissance, he drafted his plan, then had a meeting with Howden and the commander of the supporting artillery battery, Major Reginald Wise, to finalise the details. In this meeting, Howden played a role akin to that of a chairman. Dwyer and Wise

'Ever the peace maker': Lieutenant Colonel Walter Howden, CO of the 2/8th Battalion (holding pointer), conferring with his officers on the slopes of Mount Shiburangu, Wewak area, New Guinea, 27 June 1945. Captain Mick Dwyer, OC of C Company, is to the immediate right of Howden, and Major Reginald Wise, commander of the supporting battery from 2/1st Field Regiment, is on the far right. (AWM Neg. 093473)

clashed about the extent of artillery support required and Howden, 'ever the peace maker', stepped in and led the two back through the details of the plan, and eventually the 3500 rounds of artillery support were supplied. During this conference, Dwyer also requested additional support for his company from within the battalion, including the attachment of an extra platoon, Vickers guns, mortars and two flame-throwers.[90]

An attitude of frustration at being left out of the action seems to have prevailed among several COs as a result of this more static role. Harry Dunkley of the 7th Battalion wrote home from Bougainville in May 1945: 'Life is, naturally, a little more interesting, but just. COs are horribly safe and even when things warm up I get most of my excitement through binoculars.' Not a single CO was killed or wounded as a result of enemy action in 1945. Dunkley's frustration continued throughout his time on Bougainville. On 5 August he wrote: 'I live an almost sheltered life . . . not the rough rootin' and shootin' life like the Salamaua days.'[91] Mick Dwyer's recollections reveal Walter Howden to be quite uncomfortable with his sedentary existence at battalion headquarters. During the fight for Shiburangu, he called Dwyer on the field telephone several times until Dwyer, who 'really had a battle on [his] hands', eventually told him 'if he came on the line again I would throw the bloody phone in the creek'. Dwyer conceded that Howden's 'concern for the company was genuine for he also hated casualties', but ultimately he 'showed his confidence in me and sweated it out in silence and no doubt frustration'.[92]

Although it is difficult to determine whether it resulted from frustration or relief, it is still illuminating to note a diary entry made by Geoff Matthews on 23 November 1944, following the issue of his orders for the 9th Battalion's deployment on the Numa Numa Trail in central Bougainville: 'I find I have nothing much to do today as all the planning is done.' Matthews did come to terms with his role. On 29 November his D Company attacked Little George Hill. He did not go forward but remained at his headquarters, believing that his 'presence would hamper the officers doing the job'.[93] In a static command post, however, COs needed to take care to ensure that they retained control of each of their subunits and mastery of the tactical situation as a whole. Discussing command in Bougainville, Brigadier Hammer reflected that while 'tropical warfare . . . is a "section, platoon or company commander's war" . . . it would be foolish for a brigade commander to place the tactical command of his formation in the hands of 108 section commanders'.[94] Some COs, Matthews among them, lost touch with the situation affecting their forward troops, with dire consequences. As Michael Howard has reflected: 'It is one thing to be able to inspire men to risk their lives and endure prolonged hardship if one is sharing those dangers and hardships. To be able to do so by remote control demands qualities of personality possessed by very few.'[95]

FIGHTING THE 'UNNECESSARY CAMPAIGNS'

In 1945 there was a widespread feeling in the AMF that the talents and lives of its soldiers were being squandered on operations of dubious strategic relevance.[96] Unaddressed, this feeling undermined morale and threatened to impede the army's operational proficiency. It also resulted in a high degree of interference in battlefield operations from higher commanders, which often contradicted basic tactical principles, thereby confusing and frustrating COs, their officers and their soldiers. They were encumbered with arbitrary geographic limits to their operations, 'no casualty' policies and continued directions to 'take your time. There's no hurry.'[97] This only complicated the task of commanding increasingly complex and protracted operations in terrain that was every bit as difficult as that encountered earlier in the war. The inherent difficulties of command in 1945 are illustrated by the fact that 15 battalion COs were removed from command between January and August 1945 and another three had to stand aside owing to medical conditions.

In many instances, the operations of 1945 presented COs with tactical conundrums with which they were not all equipped to deal. These were the result of a tactical doctrine that emphasised offensive action combined with muddled strategic direction that urged caution and sought to prevent Australian battalions becoming embroiled in operations that the AMF's logistics system could not sustain. In addition, COs had to face their own personal misgivings about the worth of their operations. In every campaign of 1945, there are signs of COs taking a more cautious approach. Peter Stanley notes the heavy emphasis placed on patrolling by the COs of the 26th Brigade during their first week on Tarakan.[98] While such an approach was characteristic of both 'Fag' Tucker and Bob Ainslie, restraint on the part of George Warfe is telling. Anecdotal evidence suggests that some COs made it known that the lives of 'originals' were not to be risked unnecessarily.[99] Before moving up for the first operation of the Aitape–Wewak campaign, Paul Cullen cautioned the 2/1st Battalion to 'take all precautions as you can, and ensure small casualties'.[100] He later recalled that he tried to 'protect everybody' as much as he could.[101]

Ironically, tentative action could often result in more rather than fewer casualties. Cullen still enjoined his battalion to 'do the task allotted to you to the best of your ability'.[102] As the 15th Brigade report on its Bougainville operations pointed out, it was essential to retain contact with the enemy and thereby retain the initiative on the battlefield. The brigade had to fight a costly 'patrol battle' after contact with the Japanese was lost during the relief of the exhausted 7th Brigade. Once contact was regained, the brigade was able to force the Japanese to 'conform to [its] moves'.[103] After the early fighting east of Aitape Major General Stevens similarly observed that 'vigorous action' was usually the 'cheapest way' to complete a task: 'We had two brief periods of very tentative action on the part of two units in the Division and in each case the troops concerned suffered greater relative casualties and a greater loss of morale than would have been the case had more vigorous leadership been in evidence.'[104] The two COs concerned, Hector Binks of the 2/11th and Nevis Farrell of the 2/4th, both commanding their battalions in action for the first time, were dispatched to training units in Australia. Binks was replaced by Charlie Green and Farrell by Geoff Cox, aged 30. Venturesome, brusque and driven, inclined to take calculated risks,[105] Cox would later be described as the 'outstanding CO of 6 Div[ision] in the Wewak Campaign'.[106]

An aggressive tactical posture, however, brought its own complications. Particularly in the early stages of the Aitape–Wewak and Bougainville campaigns, before Headquarters First Army had confirmed an offensive policy, almost arbitrary geographic constraints were imposed on COs. In early 1945 William Parry-Okeden was ordered to prevent Japanese movement along and east of the Ramu River in the vicinity of the Allied base at Hansa Bay in northern New Guinea. Parry-Okeden's diary reveals him to be forthright and aggressive with little tolerance for undue hindrance from above. He had acted in command of the 2/9th Battalion at Gona and Sanananda and, as CO of the 30th Battalion, was the only battalion commander in the 8th Brigade who retained his command following the advance along the Rai Coast in 1944. Like many Australian officers in 1945, he seems to have been committed to wiping out every remaining pocket of Japanese. The 30th Battalion operated on the Ramu between January and February 1945 and the constraints imposed on Parry-Okeden by First Army made him 'spit blood'. On 9 January he was ordered not to deploy any troops west of the Ramu and to limit his patrol action to Marangis and Bosman, two small villages on its west bank. Positions occupied on 7 January to hold the villages had to be withdrawn, thereby yielding initiative to the Japanese: 'This method is absurd and we are now allowing the Jap freedom of movement up to the Ramu and cannot prevent him from crossing. If only these bloody old fools would come and see for themselves.'[107] Parry-Okeden's anger was exacerbated when the troops defending Bosman were ambushed in the process of withdrawing and suffered three casualties.

By 20 January, 'determined to beat the little swine', Parry-Okeden had managed to persuade his brigade commander to authorise patrols west of the Ramu. He was well aware of the consequences: 'There will be a howl from 1st Army should my chaps get into a fight and have a casualty. I'm quite prepared to take a "bowler hat" for taking offensive action against the enemy.'[108] The patrols operated successfully until early February, locating and destroying many Japanese parties between the Sepik and Ramu Rivers. On 9 February, the inevitable orders to withdraw east of the Ramu once more arrived from First Army: 'This was a shock for we had him beaten.'[109] The 35th Battalion relieved the 30th on 15 February. Parry-Okeden was not sorry to be leaving: 'Am fed up and don't intend to spoil my men by remaining passive thereby killing the offensive spirit in the individual.'[110] Parry-Okeden's performance, however, does not appear to have harmed his career. In May he was told he was being considered for a I Corps (AIF) command, and in August he was nominated to command a BCOF battalion; characteristically, having recently been denied leave, he replied, 'They can stick the army of occupation up their rears.'[111]

Paul Cullen similarly argued against arbitrary geographic limits imposed on his early operations. He, however, seems to have been motivated more by the attitude that the application of sound tactical principles saved Australian lives rather than killed Japanese. In late January 1945 the 16th Brigade was ordered to take over the defensive line between Abau and Malin to prevent Japanese movement west towards Aitape. Its battalions were to patrol, but not establish outposts, east of this line. Cullen's 2/1st Battalion was allocated the coastal sector. Here, the terrain was dominated by Nambut Ridge (Hill), a long, high feature that lay just to the east of the Abau–Malin Line. Cullen assessed that its occupation would present a serious threat to the 2/1st's operations, particularly its patrols along the coast, and, finding it unoccupied, sought permission to do so. Owing to the limitations imposed by divisional headquarters, he was given permission only to establish an outpost on the northernmost knoll. The enemy occupied the rest on the night of 29 January. The battalion was subsequently required to clear the Japanese from the ridge. Despite several air strikes and artillery bombardments, and extensive patrolling to find approaches from the flanks, it took 16 days and 14 deaths to clear Nambut Ridge. Even during these operations, several knolls cleared of Japanese had then to be abandoned owing to orders precluding the advancement of the battalion's forward defensive positions; at least one was reoccupied. Cullen remained bitter about the experience and admitted to being under 'terrible strain' during the campaign, which he described as 'absurd'.[112] The strain was evident – Stevens commented to Gavin Long in 1945 that Cullen was 'very worried in action. He felt the casualties very deeply.'[113]

The moral challenges faced by Australian COs in the last year of the war represent a wider war-weariness afflicting Allied armies. Max Hastings has written of the bitterness felt by Robin Hastings, CO of the 6th Battalion, the Green Howards, after his battalion suffered heavily during a

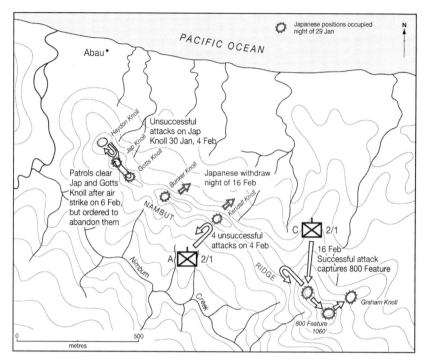

Map 8.2: 2/1st Battalion operations around Nambut Ridge, January–February 1945

poorly supported and ill-conceived attack at Crisot during the Normandy campaign. Hastings' observations on the attitudes of British COs have a familiar ring to them. The end of the war was in sight and, after 'bloody losses and failures', many British COs decided they would 'husband the lives of their men' and personally assess whether their objectives 'justified an all-out effort, regardless of casualties, or merely sufficient movement to conform with that of other units and satisfy higher headquarters'.[114] Echoing the cautious approach of Binks and Farrell, the 7th Armoured Division, which had been fighting since 1940, has been criticised for being so cautious in Normandy that it verged on being ineffective. Roger McElwain has similarly observed that by late 1944 the 2nd New Zealand Division was very tired and its COs reluctant to take risks.[115]

Australian COs in the South West Pacific Area, however, had an advantage over their British and New Zealand counterparts in Europe. As we have seen, there was not the same pressure on Australian COs from higher headquarters to pursue their operations at the high tempo required by the British in north-west Europe. The Australians had the time as well as the opportunity to patrol and seek alternative approaches to enemy positions, to employ extensive and prolonged fire support programs and, in most cases, to commit their troops to the assault in the knowledge that all possible precautions to reduce casualties had been taken. Rhoden's operations at Manggar unfolded over five days, which seems almost leisurely when compared to the frantic pace of operations in Normandy, where full-scale battalion assaults were mounted with only a few hours preparation.[116] The relative intensity of the operations also needs to be considered. In that one attack at Crisot, against dug-in panzer grenadiers, the 6th Green Howards suffered 250 casualties.[117] In the entire Aitape–Wewak campaign, the 2/1st Battalion suffered 105 casualties; in all of the campaigns of 1945 only four Australian battalions – 25th, 31/51st, 58/59th and 2/24th – suffered more than 200 casualties.[118]

But, as Gavin Long noted during a visit to the Aitape region, 'to the man in the section a pat[rol] in which three men are killed . . . is as serious as a full-dress attack by 3 divisions. The fact that 100 other sec[tion]s are or are not having similar experiences makes no difference to the "seriousness" of the fighting.'[119] Ultimately, Australian COs in New Guinea, Bougainville and Borneo faced the same challenges as their British counterparts in Europe: 'to persuade their battalions that the next ridge, tomorrow's map reference, deserved of their utmost.'[120] The combination of war weariness, the pervading sense of strategic futility and protracted operations in some of the filthiest terrain in the South West Pacific Area demanded the utmost of COs' leadership and man-management skills. There are numerous indications that the battalions were under significant strain in these last campaigns, which were largely glossed over in the official history. In March 1945 Long observed that the infantry of the 16th Brigade had 'lost none of their go', although a perception that the Aitape–Wewak campaign was 'unworthy of their quality as troops' had led to the 'drive for loot' being substituted for the 'drive for victory for its own sake'.[121]

The 2/3rd Battalion's medical report for the same month, however, reveals that some soldiers had in fact lost much of 'their go': 'Some long service soldiers continually and loudly complain that they have "had it" and this has undoubtedly upset some new soldiers, a few of whom developed a well marked neurosis owing to a feeling of insecurity among their mates.'[122] In the previous month Long had noted the breakdown of several officers on Bougainville, and Stanley recounts a number of similar incidents on Tarakan. Yet most battalions held together. Paul Cullen has reflected that in the Aitape–Wewak campaign he did not actively work to maintain morale but that the battalion just held itself together as the troops knew he was doing his best.[123] This simple, self-effacing remark is particularly insightful. A combination of unit pride, loyalty to comrades and trust in experienced commanders, including the CO, seems to have held many battalions together. A crude measure of the influence of experienced COs is provided by the fact that only one CO who had previously commanded his unit in action was replaced in 1945. Perhaps the best illustration of what these COs were doing right is provided by examining a vivid example of when COs got it wrong and battle discipline broke down: the Bougainville mutinies.

The first of these mutinies occurred in Geoff Matthews' 9th Battalion on 31 January 1945.[124] The 9th Battalion had already established a credible record on Bougainville, having captured Little George Hill and Artillery Hill in the central sector of the island. In late January it moved with the rest of the 7th Brigade to take over the tentative southern advance from the dysfunctional 29th Brigade, which was being relieved for further training. Originally ordered to defend the landing point at Mawaraka, Matthews was given permission on 25 January to advance along the axis of the Mosigetta Road. The coastal flats in this part of Bougainville were dominated by tidal swamps through which the infantry had to patrol by day. Night brought no respite; some overnight patrols were required to sleep in trees, while others on seemingly dry land had to occupy weapon pits that filled with water to just below ground level.[125] Japanese resistance was sporadic and fleeting at first, but a complex of well-sited bunkers halted C Company around Makotowa on 28 January. In addition the battalion was subject to artillery fire for the first time during the war. The next day, Matthews dispatched a second company, D, to outflank Makotowa and cut the road behind it. It subsequently blundered around in the swamps for a day and a half and was lost to battalion headquarters. It was located by a reconnaissance aircraft and eventually established itself astride the road.

Signs of strain in the battalion first became evident on 29 January. One C Company soldier refused to go on patrol, another shot himself in the foot. Matthews' attitudes to combat exhaustion seem ambivalent. Initially he snidely commented in his diary: 'Once cowardice was punished by death but now we give them medicine!!' but the following day he relieved C Company with B after his medical officer reported that it was suffering from numerous cases of diarrhoea and its morale was very low. On the morning of 31 January Matthews' command post was shelled. Via field

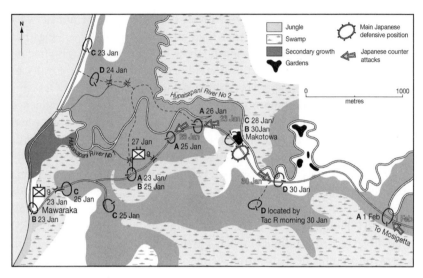

Map 8.3: 9th Battalion operations towards Mosigetta, 23 January – 1 February 1945

telephone, he ordered D Company to find the gun responsible and destroy it. The company commander refused, reporting that no man would leave his position as they were too tired, cut off from support and had no way of evacuating any casualties. Under further pressure from Matthews to get his company on the move, the company commander 'cracked up', and started to cry. Matthews blamed the incident on his 'jittery influence' but his own performance must be questioned. His control of his companies was lacking, and his diary reveals that he had little conception of the conditions his troops were encountering; there is no indication that he ventured beyond his headquarters from the time it opened at Mawaraka until D Company refused duty.

The mutiny was a wake-up call, and Matthews' subsequent actions are enlightening. He cancelled his orders to find the gun, and D Company was instead directed to push a patrol back to open the road through to B Company, which was pushing patrols forward to the same ends. The Japanese positions at Makotowa were found abandoned, and the battalion's pioneers were immediately set to repairing the road. Matthews was on the first jeep through and found D Company very tired and suffering from dysentery and trench foot. The YMCA jeep with its tea urn followed close behind, and D Company's packs, cooking gear and canteen supplies were rushed forward by nightfall.

Trouble continued in the battalion the next day with both troops and at least one company commander expressing a lack of confidence in Matthews' command. From this point, however, a change in Matthews' actions can be traced through his diary. He regularly visited his forward troops and often accompanied patrols, including that which located the gun that had sparked the trouble in the first place. Although still losing companies in the swamps, he appears to have been paying more attention to keeping his troops supplied and supported, and employed two companies forward, leapfrogging one through a secure foothold established by the other. One of the major difficulties he faced at this stage in the campaign was a lack of supporting arms but he argued vehemently for the liberal employment of artillery as it became available. The 9th Battalion was withdrawn to rest on 26 February and, although Matthews recorded further instances of jittery troops on its return to operations in April, there was no recurrence of the mass disobedience experienced on 31 January.[126]

The 9th Battalion was relieved by 'Bill' Dexter's 61st. It had commenced operations in the southern sector of Bougainville in early January 1945 and initially performed well.[127] Working further inland than the 9th, it did not encounter so much difficulty with swamps, and Dexter's earlier experience in New Guinea was apparent. He was leapfrogging his companies from the beginning, taking care to rotate platoons and companies regularly to provide them with opportunities to rest and maintain their fitness, and he took advantage of the good tracks to visit his troops regularly.[128] Gavin Long observed on 11 February that the battalion was 'as happy a crowd' as he had ever seen.

Around the same time, however, things began to go awry. Rains turned the tracks into 'a morass of mud', and supplies had to be airdropped and casualties evacuated on foot along a line of communication continuously subject to infiltration and ambush. Dexter emphasised the difficulties of casualty evacuation, which served only to heighten the men's awareness of the predicament. In late February Dexter began to worry about the health and morale of his troops. Men 'too frightened to get out of their trenches' were refusing duty. On 19 March he wrote to Brigadier Field stating that his battalion had reached 'such a stage of mental and physical strain' that it could no longer be regarded as an efficient striking force.[129] By this stage the battalion had been engaged in operations for more than nine weeks. A study conducted by American psychiatrists in northwest Europe found that after four weeks in action the performance of infantry soldiers began to degrade rapidly and after six signs of emotional exhaustion became widespread.[130] As Mark Johnston has pointed out, however, other Australian battalions had endured similarly extended periods of operational service.[131]

Dexter's influence must be considered. As a CO, he proved unequal to the challenges presented by his battalion's role in southern Bougainville. Confronted with scared, tired men, his own physical and emotional condition deteriorated. The standards of administration in the battalion fell away, Dexter stopped visiting the forward companies, which had a 'highly detrimental' effect on morale, and, struggling to rein in negative sentiments within the battalion, he made several erratic decisions regarding the promotion and dismissal of officers.[132] In addition, Dexter seems to have simply lacked empathy with his troops. Although he told Field they were 'not yellow', in his post-war writings and conversations he railed about the 'attitude of those conscripted for service' compared to men of the AIF and spoke disparagingly of 'rotten so-and-sos who just wouldn't fight'.[133] Dexter was evacuated to hospital on 4 April with neuritis and never returned to duty. The adverse report lodged by Brigadier Field stated that 'from war-weariness [Dexter] does not now possess the robust mentality and stamina needed to deal with a problem of this nature'.[134]

Yet Dexter had lacked neither robustness nor stamina in the past. Keith Picken reflected after the war that he had expected Bill Dexter would 'crack up in com[man]d of a b[attalio]n during a hard campaign': 'D. was one of those men who just couldn't relax and soon was tired and living on his nerves. His appearance of alertness and hardness was a cloak for the strain beneath. There were many like him who came out of every campaign, whether it lasted 2 days or 6 months, utterly done.'[135] Field had also noted that another of Dexter's faults was a failure to utilise his 2iC to ease the strain on himself. On 27 April, having spoken with Field, Matthews jotted in his diary: 'Dexter, late CO of 61 Bn will not be returning to his unit, he has shown his inability to comd. "Uneasy lies the head . . ."'[136] He should perhaps also have added: 'There but for the grace of God . . .'

Dexter was not the only CO to have worn out in 1945. As it was among the soldiers and officers, the very experience that contributed to the AMF's success was taking its toll among the COs. In 1945 at least nine were removed from command for what is best termed combat exhaustion. All but one had joined the AIF in 1939, and their decorations included three DSOs, an MC and three mentions in dispatches. Four were officially diagnosed with psychological conditions,[137] and the notes in their service records provide an inkling of the origin of their difficulties. One was

'Uneasy lies the head': Lieutenant Colonel Walter 'Bill' Dexter, CO of the 61st Battalion. Dexter struggled with the challenges presented by the Bougainville campaign despite prior impressive performance as a company commander. Dexter's brigade commander attributed his difficulties to war weariness. (AWM Neg. 075779)

described as 'unfit for duties involving responsibility for three months' while another was found 'unfit for undue mental strain or concentration'.[138] One of these officers was John Abbott of the 26th Battalion. He had been awarded the DSO for 'conspicuous gallantry and devotion to duty' leading a company of the 2/3rd Battalion at Tobruk on 21 January 1941.[139] Gavin Long was on Bougainville when Abbott was sacked: 'John Abbott, recognised by all as a fine soldier, is going out and is not expected back.' The word was that family problems back in Australia had pushed Abbott over the edge. 'I wonder how often this has happened,' Long mused.[140]

The most dramatic impact of exhaustion on a battalion CO was witnessed in Brunei in September 1945 when James Hendry shot himself in the head with his service revolver. Hendry was another 1939 enlistment who had been awarded an MC for his actions at Bardia and fought in Libya, Greece and British North Borneo. Aside from his personal file, official records make no mention of his death. The abbreviated findings of a court of inquiry convened at 9th Division Headquarters state that at the time of his death his 'mind and judgement were temporarily unbalanced by worry'.[141] Don Jackson, his replacement, described James Hendry as a 'casualty of war'.[142]

The overall rate of psychological casualties among COs for the entire war is difficult to determine. Exhaustion and psychological illness often occurred concurrently with other medical conditions, and diagnoses of the latter often disguised the former. Recorded cases constituted around 13 per cent of battle-related casualties. The highest recorded total for the AMF as a whole was 11 per cent on Tarakan, suggesting that the COs might have suffered psychological casualties at a greater rate than their men.

In 1945 the AMF was widely regarded as the most capable jungle fighting army in the world. It had soundly defeated the Japanese in every campaign in which it had confronted them since 1943, and its personnel were eagerly sought to pass on their experience to the British Army. In 1945 the AMF's infantry battalions reached the peak of their efficiency. This was the result of a conjunction between highly experienced and closely bonded command teams, a proliferation of specialist equipment and fire support, and the efficient communications and sound doctrine needed to employ it efficiently. The COs themselves were ultimately a validation of the philosophy underlying the reforms in command training and appointment practices in the late 1930s. The COs of 1945 were young, fit and the product of centralised training in doctrine developed specifically for the AMF, principally from its own experience.

In the battalions, command had become an increasingly technical pursuit, requiring COs to be tied to their command posts more than ever before and to rely on an increasingly large group of advisers from the other arms and services, as well as demanding even more from their own staffs. This meant that many of the functions of command were devolved to company commanders, a system that worked only when they and their CO shared experience, respect and trust. This, however, did not negate the CO's role as the 'father' of the battalion. He might have been only 30, but he was still the 'old man'. In the gruelling battlefield conditions of 1945 the CO's presence in times of trial, the sense that he genuinely cared about the interests of those under his command, was aware of their fears and privations, and was seen to be doing his best to address them, was still essential for the maintenance of morale and therefore fighting efficiency.

As had been the case for most of the war, 1945 clearly demonstrated that Australia was maintaining an expeditionary combat force much larger than the nation could comfortably sustain. The AMF was employing more divisions simultaneously than it had at any other point throughout the war, and the highly experienced cohort of officers and SNCOs on which its success was founded was stretched increasingly thin. For the AMF, the war ended just in time. Like the AIF in the First World War, it had reached its peak in the last year of operations. Had it been required to fight in 1946, significant reorganisation and retraining would have been necessary following the discharge of large numbers of experienced personnel, including COs and company commanders. Following such a loss, it is questionable whether the AMF would have exhibited the same proficiency as it had in 1945.

CONCLUSION

Decision-making is the core function of command. Successful decisions, which manage the risks of the battlefield and thereby optimise the chances of success, depend on judgement, and judgement is gained through a combination of training and experience. This book has charted the AMF's accumulation of command experience during the Second World War and the development of a centralised and universal system of training based on it. These were the foundations of the high level of command proficiency achieved in the latter years of the war.

The story of Australian battalion command in the Second World War is one of improvisation, adaptation and evolution. COs were just one part of a battalion command system. Where elements of this system were found to be ill-suited to the nature of operations they were seeking to control, COs had to take action to compensate. As a result, the manner in which command was exercised varied widely across the theatres in which Australian battalions fought. In the Mediterranean theatres of 1941 COs employed a variety of measures in an attempt to overcome the control difficulties presented by inexperienced soldiers and officers and inadequate communications. Their own inexperience often hampered their decisions, and standards were uneven. In 1945 command practice was much more uniform, as most COs had several years of active service experience behind them, and good communications and experienced troops facilitated strong control. In Malaya and Singapore COs struggled to employ supporting weapons effectively owing to a lack of experience and training; in 1945 these weapons were central to the Australian way of war.

The constant in the experience of Australian COs was the human aspect of war. Command, in all theatres, demanded immense reserves of physical and mental stamina, reflected in the steady decline of the age of COs between 1939 and 1945. Such endurance, in the face of prolonged hardship and extreme danger, had also to be elicited from the rest of the battalion. Successful leadership was an evolutionary process. Reflecting on the qualities of a good commander, Phil Rhoden observed:

> A commander has status, has his badges of rank on him, [is] supported by military law to get his wishes. But that's not enough; it never can be enough. He's got to have standing. He needs knowledge of his craft; be able to get on with people both upwards and downwards; have the power of communication; [be] seen to be interested in the job and those under his command and care . . . but above all must know his trade. If he has the opportunity to display that in battle and he comes out on top, it's a great plus for him.[1]

These remarks highlight the essence of leadership in the AMF during the Second World War. COs could not rely on the authority of their hierarchical status to make men carry out their orders. They needed to embrace the exercise of referent and expert power – win the trust and respect of their officers and men – else their ability to direct and control their actions would rapidly deteriorate.

Expert power was reliant on a CO's administrative and tactical skill, which further emphasises the role of training and experience in their proficiency on the battlefield. Referent power could be built up in a number ways, but it was usually the product of fair, consistent and often stern discipline, and of a CO whose own actions would exemplify the behaviour he expected of his men. Leadership styles were as much determined by experience and judgement as was tactical decision-making. A CO needed to judge when his presence was needed among the forward troops and when they would be better served by him remaining in his command post. COs needed to judge the demeanour of their men, weigh up their experience and determine what they as followers required – both physically and psychologically. Recall the wise counsel that urged 'Myrtle' Walker and Charles Anderson not to lead bayonet attacks; the shame-faced determination with which the 58/59th set out to prove 'Warfie' wrong, compared to the cynical incredulity his methods met in the 2/24th; the calm determination of Fred Chilton and Ralph Honner in the desperate stands at

Pinios and Isurava; and the vigorous personal aggression needed by 'Tack' Hammer and Charles Weir at El Alamein to coax one more attack from their tired and depleted battalions. Successful leadership in the AMF was truly situational.

Although there was no single, distinctive Australian style of command, a series of common characteristics can be identified: strong, coercive discipline; leadership based on trust and respect; tactical and administrative proficiency, with a commitment to aggression and offensive action wherever possible; a willingness to lead by personal example; and capable cohesive command teams with a high level of delegation, particularly in the South-West Pacific Area. Delegation, however, was a common feature only within battalions. As the conduct of operations in Malaya, Papua and New Guinea highlighted, COs often had to fight to be accorded any real scope for initiative by their superior commanders.

But the story of Australian COs is not unique. There are echoes of all facets of their experience in the conduct of infantry commanders in the armies of all of Australia's major allies. The NZEF was created from the same administrative chaos as was the AIF.[2] Australia, Canada and New Zealand all had tiny regular armies during the interwar period, and the majority of the original COs of each nation's expeditionary infantry battalions were drawn from the citizens forces and were First World War veterans in their forties.[3] In the Mediterranean theatres, the NZEF struggled to overcome inexperience and an inadequate command system. Its younger officers sat beside Australians in the classrooms at METS and subsequently succeeded the First World War generation. David Hingston, writing of the development of the NZEF in the Western Desert, tells a familiar tale. He writes of the action at Point 175 in November 1941: 'The New Zealanders failed to coordinate artillery fire and rushed their battle procedure at the cost of thorough tactical appreciations and clear orders.' By the time of Alamein, however, the New Zealand division was operating with 'well positioned tactical headquarters with good communications, readily available reserves and support weapons' guided by the 'smooth passage of orders'.[4]

The AMF in the Second World War was part of a larger process of command development and innovation that occurred in all of the Allied armies. The United States Army also had its own great purge of overage officers in 1942, and the success of the 82nd Airborne Division in Western Europe in 1944–45 has, like that of Australian battalions half a world away, been attributed to the cohesion and insight engendered in command teams as a result of shared experience.[5] Common features in the experience of successful British COs can also be identified. In *Infantry Colonel* George Taylor recounts how, in the fighting in Normandy, he was able to rely on brief verbal orders, reliable communications and well-trained troops[6] – just as Tom Daly was able to put his battalion into action with 'three paragraphs over the telephone'.

Similarly, Australian battlefield leadership was not unique. In his study of New Zealand battalion commanders Roger McElwain emphasises aggression, competence and courage as the foundation of battlefield leadership.[7] The ability to remain calm and collected in the most desperate of situations has been emphasised as a trait of successful British COs.[8] Geoff Hayes notes that there was no one style of acceptable leadership in the Canadian Army as long as an officer was competent in the field. In another echo of the Australian experience he states that officer–men relationships in the Canadian Army might have seemed 'informal relative to other armies, but Canadian enlisted men also expected that their officers not be "one of the men"'.[9]

Command and leadership also faltered in other Commonwealth armies in the same ways as they did in the AMF. A Canadian CO was killed while personally endeavouring to find the forward elements of his battalion after a breakdown in communications,[10] and Martin Lindsay of the Gordon Highlanders reflected that one of his greatest failings as a CO was a 'failure to recognise the need to rest men who were not wearing well, for which we should always have had a battalion short-leave centre. We should also have had a permanent jeep team visiting our wounded.'[11] It seems that the AMF did not exhibit a distinct Australian style of battlefield leadership but simply the varied characteristics common to leadership in broad-based volunteer armies.

Despite the professionalisation charted in this book, the AMF, at its heart, remained a citizens army. If in 1939 the AMF was an amateur force with a professional ethos, then perhaps a good way to describe the AMF of 1945 was a professional force with an amateur ethos. It never became a pure meritocracy. Even in 1945 command appointments and relationships were still subject to the vagaries of patronage and personality that had held sway in the interwar period. Further, the very small number of wartime COs who chose to continue serving on a permanent basis demonstrates the dominance of the citizen soldier ideal. They were 'warriors for the working day', as John Field entitled his memoirs of his war years. Less than 15 non-PMF COs volunteered for the post-war interim army. Two went to Japan with BCOF[12] and another two eventually ended up commanding battalions in the Korean War;[13] but as a group they had a negligible influence on the infant Australian Regular Army (which was formed in September 1947). It must be acknowledged that the plans for the post-war regular army were modest – a brigade group of three battalions – and offered only a limited range of opportunity for officers who had commanded in war. Tom Cotton sized up his chances: 'Yes, I'd remain in the army after the war, but I won't be given a chance. They'd say I was too old. Or else I'd be asked to drop to a captain and serve under some Staff Corps bloke who'd never seen any fighting and I wouldn't do that.'[14]

After the war, the majority of Australia's COs returned from whence they came – the urban middle class – with several rising to prominent positions in business and the public service. Ralph Honner returned to teaching and became Australian ambassador to Ireland in 1968. Fred Chilton assisted in the creation of the Joint Intelligence Organisation before becoming Assistant Secretary of the Department of Defence and ultimately head of the Repatriation Commission. Ivan Dougherty and George Warfe became heads of state civil defence organisations, although Warfe, perhaps characteristically, made a detour to Vietnam to help establish a counter-revolutionary warfare school. Charles Anderson and Robert Joshua became federal members of parliament, and three others became state parliamentarians. Others returned to pre-war callings: Phil Rhoden, John Wilmoth and Oscar Isaachsen to the law, Paul Cullen to business and Geoff Cooper to the family brewing concern.

Many retained their commitment to defending the society in which they lived and were prominent in the resurrection of the militia, as the Citizen Military Forces (CMF), in 1948. Several sought to rebuild their wartime battalions,[15] and the wartime CO cohort eventually provided the CMF with 11 major generals, ten of whom were divisional commanders; four would end their careers as the CMF member of the Military Board. The rise of professional soldiers, however, meant the CMF generals had little real influence on army policy. Ironically, the proficiency achieved by citizen soldiers during the war, and the training they had conducted to achieve it, had spelled the end of their role as Australia's first line of defence. The PMF COs, on the other hand, produced two Chiefs of the General Staff, including Tom Daly, who retired as a lieutenant general with a knighthood in 1971.[16]

Although he rose to the most senior post in the Australian Army, Tom Daly still described his period in command of the 2/10th Battalion as a highlight of his career – it had been his ambition for the entire war.[17] Fred Chilton reflected similarly on his time in command of the 2/2nd:

From Balikpapan to South Vietnam: Lieutenant General Sir Thomas Daly, Chief of the General Staff, talking to members of the 3rd Special Air Service Squadron, Nui Dat, South Vietnam, November 1969. Daly described command of the 2/10th Battalion during the Second World War as a highlight of his career, but it was only the first of a succession of senior command appointments. (AWM Neg. EKN/69/0176/VN)

'There's nothing like a fighting unit, an infantry battalion . . . the people are the salt of the earth.' Inherent in these reflections on what was only a very small part of long careers is the mutual respect, camaraderie and sense of purpose that were essential for an infantry battalion to succeed on the battlefield. With characteristic humility, Chilton observed that 'in a battalion in battle' the 'CO is just one of the team'.[18] The sense of responsibility that officers like Chilton and Daly felt to that team was deep and genuine. Daly 'loved' his soldiers and, despite the distance a CO

'The CO is just one of the team': Sir Frederick Chilton (right), aged 101, unveils a commemorative plaque for the 2/2nd Battalion, which he commanded between 1940 and 1941, at the Australian War Memorial, Canberra, 25 October 2006. (AWM Neg. PAIU2006/112.07)

needed to maintain, 'became very close to his men: 'because you know that some of them are going to die and it may be your fault that they do'.[19] The longevity of the bonds of mutual trust and respect developed between a successful CO and his men are a testament to their strength. In 2001, 56 years after the end of the Second World War, and following a distinguished legal career, Phil Rhoden reflected: 'One of the joys of being a commander is the friendship you have with these fellows now. They come into your home, and they're respectful; there's a genuineness about it. If you can achieve that you've done a lot in life.'[20]

> A leader is best
> When people barely know that he exists,
> Not so good when people obey and acclaim him,
> Worst when they despise him.
> 'Fail to honour people,
> They fail to honour you';
> But of a good leader, who talks little,
> When his work is done, his aim fulfilled,
> They will all say, 'We did this ourselves.'
>   Lao Tzu, *Tao Teh Ching*, Verse 17, 6 BC

APPENDIX I

THE DEMOGRAPHICS OF AUSTRALIAN BATTALION COMMANDERS

The immense quantities of records generated by twentieth-century military bureaucracies have proved a boon for historians. Personnel records in particular have allowed detailed analysis of the social composition of military units or specific cohorts of soldiers and officers. In Australia, the pioneering study in this field was Lloyd Robson's 'Origin and character of the First AIF'. Similar work has been used to good effect in Dale Blair's *Dinkum Diggers*, Joan Beaumont's *Gull Force* and Peter Henning's *Doomed Battalion*. As Pat Brennan has demonstrated in his short study of Canadian battalion commanders in the First World War,[1] the analysis of demographic data derived from personnel records can illuminate more than just the social composition of a military force. It can also highlight trends in promotion and the policies behind them, training practices, and even the nature and influence of the battlefield environment.

This book is underpinned by detailed statistical evidence derived from the personal service records of 276 officers who commanded Australian infantry battalions during the Second World War. This evidence clearly illustrates that Australian infantry battalion commanders were drawn overwhelmingly from a narrow social stratum – the educated middle class – and that, in keeping with the general attitudes of social responsibility of that class, most had served in the militia before the war. The service records reveal the citizen soldiers' continued dominance of command appointments throughout the war as well as highlighting promotion policies that kept regular soldiers at bay. The data relating to the officers' military education charts a steady professionalisation of the business of command, for both regular and citizen soldiers, resulting in a correlation between appointment to command and attendance at centralised schools of instruction. The strain of command on the battlefield is also readily apparent in the steadily declining age of battalion commanders and clusters of battalion commander casualties that coincide with the AMF's most trying operations.

The names of the 276 Australian infantry battalion commanders embraced by this book (listed in appendix 2) were derived from various appointment and promotions lists printed by the AMF during the war, from battalion records and from published battalion histories. The resulting list is comprehensive but not complete. Many discrepancies can be found within these sources. Moreover, between 1940 and late 1944 the AMF did not publish consolidated regimental and gradation lists. The archival record is particularly weak regarding command appointments to militia battalions between January 1940 and December 1941. The AMF had ceased publishing its regular officers lists, and these battalions do not appear to have maintained comprehensive unit records – at least any that have survived – until placed on a full war footing in December 1941.[2] It is estimated, however, that fewer than ten officers have been omitted, which equates, at most, to a little under 4 per cent of the sample.

Only officers who were formally appointed to command have been included. Periods of administering or acting command, particularly during operations, were not consistently documented in unit or personal records. Seeking to produce a comprehensive list of all officers who commanded in these capacities would be virtually impossible, and any partially complete list has the potential to cause significant statistical anomalies. In keeping with this approach, the dates of command used are also those promulgated in the appointments and promotions lists and recorded in an officer's service record. Again, this is principally for the sake of consistency as the dates of an officer's actual assumption of command were not consistently recorded. Promulgated dates of command have been amended only in limited instances where the examination of unit records has revealed significant discrepancies between the promulgated date and the actual assumption or relinquishment of command.

The main source of information on which this appendix is based are the personal service records, or dossiers, originally maintained by 2nd Echelon AMF Headquarters, which would later become known as the Central Army Records Office (CARO).[3] These records are now in the custody of the National Archives of Australia. The major components of each service record are Australian Army forms B103 and B199a, other ranks and officers records of service, respectively. These forms record postings, promotions, punishments, embarkations, disembarkations, attendance at courses of instruction, and instances of sickness, injury and wounding. While the records of service provide a summary of a soldier's career, it is the various versions of attestation forms that provide the richest source of information on his background. Most commonly, these forms include details of birthplace, religion, marital status, occupation, place of residence and next of kin. As a historical source for demographic analysis, however, the early versions of the attestation forms for militia mobilisation (AAF Mob. 1) and special service abroad (AAF A. 200) have several weaknesses. They do not list details of educational qualifications, prior military service or dependants. A later version of the special service form (AAF A. 203) did subsequently include this information, but it was found in only a minority of service records and therefore some of the observations presented below relate to a more limited sample.

In order to gauge just how representative the COs were of the men they commanded and the society they defended, the observations derived from the service records have been compared with wider studies. The most significant are the Commonwealth Census of 1933 and the AMF census of 1942–43. The 1933 Commonwealth Census[4] was the last conducted in Australia before the Second World War, and its complete statistical analysis was not published until 1940, although the preliminary statistician's report appeared in 1936. All further references made in this work to the demographics of the wider Australian population are drawn from that study. In most areas of analysis the census breaks data down into four-year – quinquennial – age ranges. Representing the closest fit to the age range for service in the AMF, most comparisons have been made with the 20–54 male age group.

The AMF Census was conducted among most units of the AMF, both at home and overseas, between early 1942 and mid-1943 to accumulate data for the placement of specialists within the army and to facilitate planning for its eventual demobilisation. It has been estimated that the census canvassed approximately 90 per cent of the AMF,[5] and a 10 per cent sample was presented in a tabulated summary at the end of 1943.[6] The observations that follow have been compared with this summary. Given its purpose, however, the range of data collected by the AMF Census was largely confined to educational and occupational details. As no sustained demographic study of the AMF as a whole has yet been completed, observations drawn from the studies of the 2/21st and 2/40th Battalions, completed by Beaumont and Henning respectively, have also been used for comparison.

Origins

An analysis of the origins of Australia's Second World War infantry commanding officers (COs) reveals that they were drawn from a relatively narrow section of society. When their residential origins, occupations and educational levels are examined, an image of a 'command class' within the AMF begins to form. At the most basic level of analysis, the COs were predominantly Australian-born and Australian-raised (see table A1.1). The most significant minority were the 16 born in the United Kingdom, who comprised a little over 5 per cent of the group and were therefore under-represented as 10 per cent of Australian male residents in 1933 were recorded as having been born in the United Kingdom.[7] Another eight COs were born overseas, and their countries of birth were all slightly over-represented in comparison to the community at large. There was little to differentiate the prior experience of overseas-born COs from their Australian-born counterparts, however. The majority had lived in Australia since at least young adulthood, most since childhood. Eleven had served with the AIF during the Great War, and others had completed the military obligations required of Australia's young men during the 1920s. Only three of the overseas-born COs appear to have arrived in Australia as mature adults, as would be expected given that immigration virtually ceased during the 1930s.[8]

The Australian-born COs were generally representative of the origins of the male population at large (see table A1.2). The only exception was those born in Western Australia, who were represented in twice the proportion in which they existed in the wider community. The coincidence between the years of their birth and the Western Australian gold rushes may offer a possible explanation. Just under a third of the Western Australians, however, lived in the eastern states by the time of the Second World War.

Demonstrating the universality of the Australian effort during the Second World War, the AMF closely replicated the distribution of the eligible Australian male population (see table A1.3). There was no more than 6 per cent difference between the relative male populations of each state, their enlistees for Second World War service, the number of infantry battalions they raised and maintained, and the number of COs they provided.

Table A1.1 Place of birth of COs

Place of birth	COs	Percentage	In 2/40 Bn (%)[a]	In Australian male population (%)
Australia	249	90.2	96	84.59
United Kingdom	16	5.8	3	10.42
New Zealand	3	1.0		0.70
China (Shanghai)	1	0.4		0.24
South Africa	1	0.4		0.09
United States	1	0.4	0.5	0.10
New Hebrides	1	0.4		0.05
At sea	1	0.4		0.03
Unknown	3	1.0	0.5	–
TOTALS	276	100	100	–

[a] Figures from Henning, *Doomed Battalion*, p. 43.

Table A1.2 State of birth of Australian-born COs

Place of birth	COs	Percentage	In Australian male population 20–54 (%)
Federal Capital Territory	0	0	<0.1
New South Wales	82	32.9	37.3
Northern Territory	0	0	<0.1
Queensland	25	10	13.1
South Australia	25	10	10.1
Tasmania	10	4	4.7
Victoria	85	34.1	29.9
Western Australia	22	8.8	4.1
TOTALS	249	100	100

The most populous states, New South Wales and Victoria, provided the largest number of COs: 179, or 65 per cent of the total.

This situation was a product of the AMF's practice of allocating units and recruiting quotas to each state on the basis of population. Early in the war there was a strong commitment to maintaining regional battalions, as was the practice in the Great War and the interwar period. For militia battalions this was a necessity as they remained based in their home states and trained for only short periods; for the battalions of the Second AIF it was a matter of state prestige and unit cohesion. The Australian

Table A1.3 State of residence of COs on enlistment for Second World War service

Place of residence	COs	Percentage	Infantry battalions raised or head-quartered in state	Total Australian infantry battalions during war (%)	In AMF (%)[a]	In Australian male population 20–54 (%)
Federal Capital Territory	4	1	1[b]	1	38[c]	<1
New South Wales	95	34	34	31		39
Northern Territory	1	<1	0	0	<1	<1
Queensland	29	10	17	16	14	15
South Australia	27	10	9	8	8	9
Tasmania	7	3	5	5	3	3
Victoria	84	30	33	30	28	27
Western Australia	29	11	7	6	8	7
All states	–	–	3	3	–	–
TOTALS	276	100	109	100	100	100

[a] Derived from figures in Long, *The Final Campaigns*, p. 635.
[b] The 3rd Battalion (Werriwa Regiment) was originally headquartered in the Federal Capital Territory but drew most of its strength from the towns and farms of the New South Wales Southern Highlands.
[c] Long does not provide separate figures for enlistees from the Federal Capital Territory, presumably because they were enlisted in the 2nd Military District and thus allocated service numbers with New South Wales prefixes: 'N' or 'NX'.

Table A1.4 State of origin of battalions and COs

Command and state of origin	COs in command Sep 1939–Aug 1945	(%)	COs appointed Sep 1939–Jan 1942	(%)	COs appointed Jan 1942–Jan 1944	(%)	COs appointed Jan 1944–Aug 1945	(%)
Commanded battalion from state of residence	243	68	115	85	77	55	13	31
Commanded battalion from other than state of origin	95	27	12	9	53	38	29	69
Commanded 'all states' battalion (2/31, 2/32, 2/33 Bn)	17	5	8	6	9	7	0	0
TOTALS	355	100	135	100	139	100	42	100

federation was only 40 years old, and parochialism was intense. The most vivid example was the determined public and political campaign waged in mid-1940 to have the 2/40th Battalion recruited as a solely Tasmanian unit.[9] Only three 'all states' battalions were formed for service in the Second World War – the 2/31st, 2/32nd and 2/33rd[10] – all of which were assembled in the United Kingdom from the 18th Brigade and auxiliary units of I Corps.

In keeping with this regional recruitment policy, the overwhelming majority of COs appointed before the outbreak of war with Japan hailed from the state in which their battalion was recruited (see table A1.4). This relationship, however, broke down as the war continued and is one marker of the increasing professionalisation of the AMF. By 1943 a new CO was just as likely to hail from a state other than that with which his battalion identified. By 1945 most newly appointed COs were not from their battalions' home states. There was no longer any official regard for parochial loyalties in the posting of officers, NCOs or reinforcements, although some battalions still clung to such identities, causing challenges for new COs.[11]

Table A1.5 Metropolitan/rural distribution – place of residence of COs on enlistment for Second World War service

Place of birth	COs	Percentage	In 2/10 Bn (%)[a]	In 2/21 Bn (%)[b]	In 2/40 Bn (%)[c]	In Australian male population 20–54 (%)
State capitals	186	67.4	} 45	52.4	} 26	44.7
Major provincial cities[d]	21	7.6		} 47.3		15.7
Rural	69	25	55		73	38.9
TOTAL URBAN	207	75	–	–	–	60.4

[a] Figures from Allchin, *Purple and Blue*, p. 364.
[b] Figures from Beaumont, *Gull Force*, p. 120.
[c] Figures from Henning, *Doomed Battalion*, p. 42.
[d] The 1933 Census provided a breakdown of those living in the capital cities, provincial urban areas and rural areas. A 'provincial urban area' was defined as 'those cities and towns which are not adjacent to the metropolitan areas, and which are incorporated for local government purposes'. A determination of which cities and towns this definition encompassed is beyond the scope of my research. For my purposes a provincial city is defined as one with a population of more than 20,000, which includes Ballarat, Bendigo, Broken Hill, Geelong, Launceston, Newcastle, Rockhampton, Toowoomba and Townsville (*Official Year Book of the Commonwealth of Australia, No. 32 – 1939*, pp. 523, 527).

Although the numbers of COs drawn from each state were broadly representative of the character of both AMF and the Australian male population, their specific places of residence were not. As table A1.5 shows, 75 per cent of COs lived in Australia's major urban areas before the war, the majority in the state capitals. The equivalent proportion of urban dwellers in the male working population was 60 per cent. Further, anecdotal evidence suggests that men from rural areas, often lacking mechanical skills that could be utilised in specialised units, were over-represented in the infantry battalions. The examples of three infantry battalions, for which such figures are available, tend to support this impression. In the 2/21st Battalion, an ostensibly urban battalion raised in Melbourne, rural dwellers were represented in a greater proportion than they were in the wider community. In the 2/10th and 2/40th Battalions, raised in South Australia and Tasmania, both states with much smaller urban populations, rural dwellers were in the majority.

Generally, the COs of Australia's infantry battalions were urban-dwelling, well-educated, middle-class, white-collar workers. In keeping with the nature of the AMF as a citizens' force, only a very small minority were professional military officers. On their attestation forms the COs listed 100 separate occupations, shown in table 3.6. Included among them was a banana inspector,[12] the proprietor of the *Argus* newspaper,[13] the deputy head of the Commonwealth Investigations Service,[14] a police constable[15] and a drapery manager.[16] When grouped into the broad occupational categories of the 1933 Census (see table A1.7), those engaged in public administration, clerical duties and the non-industrial professions predominate, followed by those employed in commercial or financial pursuits. These professions were represented among the COs in a proportion far in excess of that in the AMF or the wider male population. The public administration, professional and clerical group represented 45 per cent of the COs but only 15 per cent of AMF rank and file and 10 per cent of the Australian male population aged 20 to 54. It must be noted that this category does not include professionals working in industry; architects, for example, are included in the 'Building' category and industrial chemists in 'Industrial'. Men working in commerce and finance were also over-represented among the COs: 26 per cent as opposed to 16 per cent of the AMF and 12 per cent of its recruiting base. Conversely, the largest occupational group represented in the AMF, workers in industry, only produced 7 per cent of COs compared to 25 per cent of the rank and file.

The image of a distinct command class becomes even clearer when industrial boundaries are discarded and the COs' pre-war occupations are categorised to reflect their degree of inherent responsibility and relative socioeconomic standing in the community (see table A1.8). Grouped together, the professional men of all industries and callings were the most numerous, constituting 26 per cent of all COs. The clerks, a category embracing bank tellers and office workers, constituted a similarly large proportion: 22 per cent. When these two groups are combined with the high-level managers, company directors and large-scale employers, those in middle-level management and other supervisory roles, men previously employed in retail, sales and marketing,[17] and teachers and university lecturers, an overwhelmingly urban white-collar cohort emerges. These men constituted 77 per cent of COs while comprising only 24 per cent of the wider male population. Only 2 per cent of future COs stated their occupation as tradesmen on enlistment; no unskilled labourers would rise to command an infantry battalion. The professional backgrounds of

Table A1.6 Pre-war occupations of COs

1.	Accountant	11	34.	Drapery Manager	1
2.	Advertising	1	35.	Electrical and Mechanical Fitter	1
3.	AIC Officer	1			
4.	Architect	6	36.	Electrical Engineer	3
5.	Assistant Manager	1	37.	Farmer	9
6.	Assistant Valuer	1	38.	Forestry Inspector	1
7.	Assurance Inspector	1	39.	Fruit Merchant	1
8.	Auctioneer	1	40.	Fuel Merchant	1
9.	Auditor and Accountant	1	41.	Gold Refiner	1
			42.	Grazier/Pastoralist	9
10.	Banana Inspector	1	43.	Grocer	1
11.	Bank Inspector	1	44.	Headmaster	2
12.	Bank Manager	1	45.	Horticulturist	3
13.	Bank Officer	15	46.	Hospital Attendant	1
14.	Bank Teller	2	47.	Hotel Manager	1
15.	Barrister	3	48.	Inspector – Commonwealth Audit Office	1
16.	Branch Manager Pastoral Company	1			
17.	Business Manager	5	49.	Inspector of Food and Drugs	1
18.	Cabinet Maker	2			
19.	Chartered Accountant	2	50.	Insurance Agent	1
			51.	Insurance Inspector	2
20.	Chemical Salesman	1	52.	Insurance Official	3
21.	Chemist	3	53.	Investigations Officer	1
22.	Chief Clerk	1	54.	Jackeroo	2
23.	Civil Engineer	7	55.	Journalist	1
24.	Clerk	22	56.	Law Clerk	1
25.	Combustion Engineer	1	57.	Manager	2
26.	Commercial Traveller	4	58.	Managing Director	2
27.	Company Director (retired)	1	59.	Managing Law Clerk	1
28.	Company Manager	4	60.	Manufacturer	1
29.	Company Representative	2	61.	Manufacturer's Agent	2
30.	Conveyancer	1			
31.	Dairy Farmer	1	62.	Marine Engineer	1
32.	Dental Technician	1	63.	Master Printer	1
33.	Deputy Head – Commonwealth Investigations Service	1	64.	Mechanical Engineer	2
			65.	Medical Practitioner	1
			66.	Merchant	3

Table A1.6 (cont.)

67.	Newspaper Departmental Manager	1	85.	Share Broker	1
			86.	Solicitor	20
			87.	Staff Corps Officer	20
68.	Office Manager	1			
69.	Optometrist	1	88.	Station Master	1
70.	Orchardist	2	89.	Stock and Station Agent	3
71.	Plumbing Instructor	1			
72.	Police Constable	1	90.	Stock Buyer	1
73.	Proprietor, *The Argus*	1	91.	Sugar Cane Farmer	1
74.	Public Servant	8	92.	Superintendent	1
75.	Railway Official	1	93.	Telephone Traffic Officer	1
76.	Refinery Manager	1			
77.	Retail Departmental Manager	2	94.	Tobacco Grower	1
			95.	Train Controller	1
78.	Retired	1	96.	Trust Officer	1
79.	Sales Manager	4	97.	Trustee	1
80.	Salesman	5	98.	University Lecturer	1
81.	Saw Miller	1			
82.	School Teacher	10	99.	Wool Broker	1
83.	Secretary	4	100.	Wool Clerk	2
84.	Service Manager	2	101.	Unknown	1

Australian COs echoed those of their New Zealand counterparts: 78 per cent of the COs of the 2nd New Zealand Division were formerly urban white-collar workers.[18]

The dominance of urban white-collar workers is hardly surprising. They represented the best educated men in a community in which roughly half of all boys who attended primary school received no education beyond their fourteenth birthday, and only a tenth of the remainder would matriculate from high school.[19] This is borne out by an examination of the educational standards of the 45 COs for whom such information is available (see table A1.9). More than 60 per cent completed high school, 20 per cent had professional qualifications, and 42 per cent had completed a university degree. Fewer than 2 per cent of the AMF had university degrees, but this was still double the proportion of the wider male community.

Several former COs have reflected that many of the skills developed by a professional occupation were readily applicable to the military. Paul Cullen valued his professional training as an accountant for the order its

Table A1.7 COs grouped according to pre-war occupation

Professional category	Among COs	Percentage	In AMF (%)	In Australian male population 20–54 (%)
Fishing & Trapping	0	0	<1	<1
Rural	38	14	18	22
Forestry	1	<1	1	1
Mining & Quarrying	1	<1	2	3
Industrial	18	7	25	15
Building	16	6	6	10
Transport & Communications	4	1	10	9
Commerce & Finance	72	26	16	12
Public Administration, Professional & Clerical	123	45	15	11
Sports & Entertainment	0	0	1	1
Hospitality & Domestic	1	<1	3	2
Ill-defined & Other	0	0	3	10
Unemployed & Retired	1	<1	<1	4
Unknown	1	<1	–	–
TOTALS	276	100	100	100

processes imposed on his own innate exuberance. It allowed him to adapt to and readily function within the military's structured way of conducting its business.[20] Fred Chilton commented that his calling, law, was a 'very practical profession' one in which one dealt regularly with the problems of real people, which had to be solved by referring to basic principles. He likened this application of basic principles to the process of solving tactical problems on the battlefield.[21] Phil Rhoden, echoing Chilton's sentiments, reflected that the law developed an ordered way of thinking and an ability to grasp the nub of a problem quickly: 'If you're on your feet in a courtroom you have to learn to think pretty quickly . . . you've got to be quick and you've got to be right.'[22] Chilton and Rhoden's thoughts are difficult to discount. Barristers, solicitors and conveyancers constituted almost 9 per cent of infantry COs, yet they formed less than 0.2 per cent of the male working population aged between 20 and 54.[23]

The dominance of the urban white-collar cohort cannot be completely explained by level of education. The nature of the AMF as a citizen

Table A1.8 COs grouped according to pre-war occupation and level responsibility

Professional category	Among COs	Percentage	In Australian male population 20–54 (%)
Professional	71	26	2
Military Officers	21	8	<1
Pastoral	30	11	13
Clerical	61	22	6
Education/Academia	13	5	1
Retail/Sales/Marketing	26	9	9
Managerial – high level	18	7	2
Managerial – mid- to low level	23	8	4
Tradesman/Urban labourer[a]	6	2	47
Rural Labourer	3	1	10
Emergency Services	1	<1	<1
Medical Worker (not 'professional')	2	<1	<1
Unknown	1	<1	–
TOTALS	276	100	95

[a] As it is difficult to separate tradesmen from unskilled workers in the 1933 Census data, and as no COs were unskilled urban workers before the war, these two groups have been combined. The differing social status of these two groups, however, is recognised.

force meant that it relied heavily on officers of the pre-war militia to fill command appointments from company commanders up, and the costs associated with being an officer in the militia also often limited commissions to men of the middle class. Additionally, part-time soldiering was just one of many leisure-time activities competing for the attentions of young men. It was often those with strong notions of community service and a commitment to self-improvement who were attracted to the commissioned ranks; these were both attitudes that can be attributed to the middle class.

Although not a comprehensive sample, all but one of the nine former COs[24] interviewed for this book shared a remarkably similar background in keeping with the statistical portrait of COs previously discussed. Not all of the nine had privileged upbringings, but their families were sufficiently well off to allow them to complete a full high school education.

Table A1.9 Educational standards of COs

Highest qualification	Among COs	Percentage	In AMF (%)	In Australian male population (% – est.)
Primary only	1	2	78	
High school entrance	2	5		
Intermediate	3	7	15	N/A
Leaving/matriculation	9	20	6	
Diploma/professional course	9	20	N/A	
University degree	19	42	1	1[a]
Postgraduate degree	2	4	N/A	N/A
TOTALS	45	100	100	

N/A Figures not available.
[a] Calculated from the number of Australian males aged 16–29 attending university in 1933 (5854) compared to the wider male population aged 16–29 (822,965). The 1933 Census grouped all university attendees older than 29 into a single total – '30 and over' – of 398.

Of the eight who completed high school, six went on to complete university degrees, one an accountancy qualification, and Tom Daly, unable to gain a university scholarship, attended the Royal Military College, Duntroon. Fred Chilton was typical of the group. He completed secondary education at Sydney High School, gained a scholarship to Sydney University and subsequently graduated with degrees in arts and law. A commitment to society is evident from his service in the militia from 1923 onwards – he was a major by 1939 – and his membership of Toc-H, an early form of service club. As Chilton recalled, '. . . one took things seriously in those days.'[25]

Exceptional among those interviewed was Frank Sublet. He was the son of a Kalgoorlie gold miner and Great War veteran. He attended public schools and left secondary school with a Junior Certificate at the age of 14. His first job was with a manufacturer in Perth. Despite his different origins, Sublet, however, was not far removed from the other men discussed. He subsequently secured a position, then promotion, in the Western Australian public service, and his penchant for the books of the United Services Institute (referred to in chapter 1) demonstrates a definite commitment to self-improvement.

But what of the men from rural areas, who are so dominant in popular conceptions of Australian military prowess? A quarter of the infantry

COs hailed from the bush. The composition of this cohort in some ways reflected the trends seen among the urban COs. Landowners and members of the rural business community, grouped under the 'pastoral' category in table A1.8, were the most numerous, constituting 11 per cent of COs, roughly in proportion to their numbers in the male working population aged 20–54. Generally, these men enjoyed status and influence in their communities, and many of them would have shared what could be loosely described as a conservative world-view with their urban white-collar comrades. Maintaining socioeconomic status and levels of responsibility as categorisation criteria, rural labourers were excluded from the pastoral group. Only three rural labourers, two jackeroos and one farmhand, employed by his father, rose to command infantry battalions during the war.

The socioeconomic status of the pastoral group, however, needs to be treated with caution. Land was not necessarily an indicator of wealth nor of social status. From the brief annotations made in personal service records it is difficult to differentiate between the owner of a large and productive holding, employing multiple stockmen and other labourers, and the owner of a struggling small holding. There is one small linguistic clue in the records as to the socioeconomic status of the rural landowners. Some are listed as 'graziers' while others are listed simply as 'farmers'. Assuming that the use of the term 'grazier' implies ownership of a large landholding and the social status that accompanied it, and not just the whim of an AMF clerk, the proportion of COs from the middle strata of Australian society increases. Combined, the graziers, the land, stock and produce agents, and the urban white-collar occupational groups represented 81 per cent of the commanding officers.

There are further indications that the COs represented a command class with a narrow social base. As shown in table A1.10, those who stated their religion on enlistment as Church of England, the establishment religion in interwar Australia, comprised 55 per cent of COs as opposed to 43 per cent of the 2/21st Battalion or 39 per cent of the wider male population. In both the 2/21st and 2/40th Battalions Catholics were represented in proportion to their numbers in the wider population (17–18 per cent), yet among the COs they were under-represented by almost half. The Catholics are even further marginalised if the 32 per cent of COs adhering to other Protestant faiths are grouped with the Anglicans. This is not to suggest that there was active discrimination against Catholics in the selection of COs, but in Australian society at the time religion was a social marker. Catholicism was more predominant among the urban working

Table A1.10 Religious adherence of COs

Religion	Adherents among COs	Percentage	In 2/21 Bn (%)[a]	In 2/40 Bnb (%)[b]	In male Australian population 20–54 (%)
Anglican	151	54.7	43.0	51	38.71
Other Protestant denominations[c]	87	31.5	36.7	25	23.45
Baptist	3	1.1		–	1.42
Roman Catholic	27	9.8	18.6	18	16.72
Jewish	1	0.4	–		0.40
No religion stated	4	1.5	–	56	–
Unknown	3	1.1	–		–
TOTALS	276	100	98.3	–	–

[a] Figures from Beaumont, *Gull Force*, p. 28.
[b] Figures from Henning, *Doomed Battalion*, p. 45.
[c] The other Protestant denominations noted among the COs were: Congregational, Lutheran, Methodist, Presbyterian and Salvation Army.

class, while the various Protestant faiths, particularly Anglicanism, were dominant among the middle and upper classes.

In proportion with the size of the Australian Jewish community, there was only one Jewish battalion commander, Paul Cohen of the 2/1st Battalion. Cohen, interestingly, changed his name to Cullen while serving in the Middle East in 1941. At first glance, this action raises a question of whether there was institutionalised prejudice in the AMF. Cullen, however, denied there was any such prejudice. After narrow escapes from the Germans in Greece and Crete he decided that he would prefer a more anglicised name on his identification discs before he risked capture again. In any case, Cullen had renounced his faith by the end of the war. After all he had seen, he reflected that religion, of any sort, 'all sounds like bullshit'.[26]

Prior military experience

The most common shared experience in the backgrounds of Australian COs during the Second World War was service with the militia during the 1920s and 1930s. Those who had been citizen soldiers numbered 245 (89 per cent), and another 21 had been PMF officers posted to militia units. For just over half these men it was their only military experience before 1939; the rest had fought in the Great War. By any measure, the Great War must be seen as the defining experience of any military career, although 17 per cent of the COs had served in the citizen forces before 1914. Only five COs had seen no military service before the Second World War.

Proportionately, the militia's two university regiments – the Melbourne University Rifles (MUR) and the Sydney University Regiment (SUR) – produced the largest number of wartime COs. Twenty-three COs had served in either of these two units – 8 per cent of the total from units that constituted only 4 per cent of the militia's battalions, were not part of its field force, and were maintained at much smaller strengths than the line battalions. The success of these officers further supports the notion of a command class in the wartime AMF. The university regiments were originally established in the early twentieth century to provide an opportunity for male students to undertake military training. They soon developed an elite ethos, and in the 1920s and 1930s the universities and major private schools still provided the majority of the units' soldiers, NCOs and junior officers. Although officially designated as standard infantry battalions, MUR and SUR had an unofficial role of grooming candidates for commissions.[27]

It can be argued that the reason so many ex-university regiment officers rose to command is that, like the middle-class professionals discussed earlier, these men were well educated and imbued with a notion of social responsibility. In addition they had been introduced to the military in a unit that reinforced notions of their social and intellectual superiority, and sought to maintain standards in keeping with its elite reputation.[28] Patronage networks might also have been at play. The university regiments' old boys' networks have long had a reputation in the wider army for furthering the interests of their own.[29] Fred Chilton, for instance, was encouraged to apply for appointment to the 2/2nd Battalion by Major Ian Campbell, former adjutant of SUR and newly appointed brigade major of the 16th Brigade.[30] Direct links between a brigadier, the COs he had a hand in appointing, and previous service with SUR can be drawn in only three cases, however. In two of them the brigadier was William Windeyer, CO of SUR between July 1937 and May 1940. The COs he appointed were Colin Grace of the 2/15th Battalion, in May 1943, and John Broadbent of the 2/17th Battalion, in February 1944. Both had served under Windeyer's command at SUR. These connections, however, are hardly conclusive.

Table A1.11 Maximum rank attained by COs before 1939

Rank	Militia	PMF	No interwar service
Lieutenant colonel	70	0	0
Major	52	9	1[a]
Captain	67	8	2[b]
Lieutenant	47	4	2[c]
Warrant officer	0	0	0
Sergeant	3	0	0
Corporal[d]	1	0	0
Private	1	0	0
Unknown	4	0	0
None	0	0	5
TOTALS	245	21	10

[a] Listed on the Reserve of Officers following his return from the First World War.
[b] Listed on the Reserve of Officers following their return from the First World War.
[c] Listed on the Reserve of Officers following their return from the First World War.
[d] Lance corporal was not a rank at the time but an appointment made by the CO.

The highest ranks achieved by the COs during their pre-war militia service are shown in table A1.11. Three-quarters had attained the rank of captain or above, which generally meant they had had some prior experience of unit or sub-unit command. Although, as we have seen, promotion in the pre-war militia was affected by socioeconomic factors, the AMF's commitment to the Great War policy of promotion from the ranks is readily apparent in the COs' military backgrounds. From their service records, it is possible to determine the precommissioning experience of 195 of them.[31] Only 38 were commissioned with no experience in the ranks of a militia unit; 20 of these were graduates of RMC, and at least another 10 had been officers in cadet units. The rise from private to lieutenant colonel was a long journey for many, in some cases beginning with the introduction of compulsory training in 1911, and including service in two world wars. There were only 10 men[32] who rose all the way from the ranks to battalion command during the Second World War itself. All enlisted for service with the Second AIF in late 1939 and had been commissioned by the middle of 1940, and therefore saw no active service in the ranks.

Caution must be exercised in considering the significance of pre-war militia experience. The structure, doctrine and training of the militia all suffered from serious deficiencies that affected the proficiency of its officers. Nevertheless, the experience gave them a firm grounding in military practice and procedure, taught them much about man management, and honed skills of improvisation and adaptation. As a military organisation, the pre-war militia failed in many ways, but it generally fulfilled its primary aim of providing a base of trained officers on which to found an expanded wartime army.

In command

Once appointed, an Australian CO in the Second World War would command his battalion for an average of 13 months before he was posted elsewhere, wounded or killed (table A1.12). The average, however, does not tell the full story. Some COs remained in command for several years, while others did not see out a month. The longest-serving CO was Thomas Bartley, who took command of the 40th Battalion in December 1937 and still held the appointment in July 1945. In that month the 40th was amalgamated with the 12th Battalion; Bartley assumed command of the new unit, took it to Timor as part of the reoccupation force, and relinquished command only in May 1946. Bartley's longevity, however, was no doubt assisted by the fact that neither the 40th nor the 12/40th saw active service.

Militia COs like Bartley enjoyed the longest periods of command, but this was due to the benign military environment that existed in Australia until Japan's entry to the war. Of the 75 militia COs who held their commands for two years or longer, 56 were appointed before December 1941. Only 26 AIF COs served for longer than two years, further demonstrating the effect of active service on command tenure. Once the two elements of the AMF were both fighting in the South-West Pacific theatre, the maximum duration of command tenure became remarkably similar (see table A1.13 and table A1.14) The statistics relating to command tenure also demonstrate the effect of the great purge of militia COs and their replacement by AIF veterans between mid-1942 and mid-1943. While 1942 and 1943 appear to have been a period of command consolidation in AIF battalions, in which the average command tenure rose from 17 months to

Table A1.12 Tenure of AMF COs

Duration of command	COs in command Sep 1939–Aug 1945	COs appointed Sep 1939–Jan 1942	COs appointed Jan 1942–Jan 1944	COs appointed Jan 1944–Aug 1945
Average duration of command (months)	13	15	15	9
Longest duration of command	92	50	43	24
Shortest duration of command	<1	1	<1	2

Table A1.13 Tenure of militia COs

Duration of command	COs in command Sep 1939–Aug 1945	COs appointed Sep 1939–Jan 1942	COs appointed Jan 1942–Jan 1944	COs appointed Jan 1944–Aug 1945
Average duration of command (months)	13	14	13	8
Longest duration of command	92	50	43	21
Shortest duration of command	<1	2	<1	1

23 months, in the militia battalions, the average command tenure hardly rose between September 1939 and December 1943.

The posting paths of COs further illustrate promotion policy in the AMF. The postings of COs immediately before assuming command are shown in tables table A1.15 and table A1.16. The standard route to command for officers in both militia and 2nd AIF battalions was to command a platoon and a company, then serve for a time as a battalion 2iC, although it was not uncommon to be appointed CO straight from a company command. Roughly three-quarters of AMF COs followed this

Table A1.14 Tenure of COs of original 2nd AIF battalions

Duration of command	COs in command Sep 1939–Aug 1945	COs appointed Sep 1939–Jan 1942	COs appointed Jan 1942–Jan 1944	COs appointed Jan 1944–Aug 1945
Average duration of command (months)	17	16	23	12
Longest duration of command	43	42	43	24
Shortest duration of command	<1	1	<1	2

route to command. In the AIF, 30 COs who had commanded a battalion previously were appointed. Demonstrating a hierarchy between the militia and the AIF, most AIF COs who commanded multiple battalions commanded a militia unit first, then one from the AIF; only five officers commanded two AIF battalions during the war; most went up or out.[33] Only four officers commanded militia battalions after having been AIF COs;[34] all four had been relieved of their AIF commands – three owing to perceived command failures and one being over-age. A proportionate number of militia COs commanded more than one battalion, but this figure is much lower than it appears because about a third of the multiple command appointments can be attributed to the amalgamation or separation of two battalions.

Small numbers of officers were appointed to battalion commands from training, staff and other non-infantry postings, but the most notable group are the 9 per cent appointed to AIF battalions from staff positions on operational headquarters. What is informative in this case is that two-thirds were PMF officers. The other significant group of officers appointed to command from outside the battalions was the 24 who were appointed to command militia battalions from the retired or unattached lists or from the reserve of officers. Their presence reinforces the notion that the militia was regarded as a second-string army; only a single AIF CO was not actively serving in the AMF when appointed. The appointment of these officers is evidence of the strain placed on the militia by the formation and maintenance of the 2nd AIF; more than half of the men

Table A1.15 Posting before battalion command appointment – COs of original 2nd AIF battalions

Previous posting	Before first wartime command[a] No.	%	Before second wartime command No.	%	Before third wartime command No.	%	Before fourth wartime command	Totals No.	%
CO of other battalion	0	0	29	88	1	100	–	30	23
2iC in battalion (at least)[b]	23	24	0	0	0	0	–	23	18
Regimental service in battalion	7	7	0	0	0	0	–	7	5
2iC in other battalion	17	18	1	3	0	0	–	18	14
Regimental service in other battalion	21	22	0	0	0	0	–	21	16
OC independent company	0	0	0	0	0	0	–	0	0
CO of other combat unit	5	5	0	0	0	0	–	5	4
CO of support unit	0	0	0	0	0	0	–	0	0
Regimental service in other unit	0	0	0	0	0	0	–	0	0
Operational staff	12	13	0	0	0	0	–	12	9
Training	5	5	2	6	0	0	–	7	5
Base or lines of communication	3	3	1	3	0	0	–	4	3
Brigade commander	1	1	0	0	0	0	–	1	1
UL/RL/RoO	1	1	0	0	0	0	–	1	1
Ceremonial duties	1	1	0	0	0	0	–	1	1
TOTALS	96	100	33	100	1	100	–	130	100

[a] Figures include incumbent militia COs in September 1939.
[b] Figures for battalion 2iCs are not comprehensive as appointments below unit command were rarely noted in an individual's service record, and are not consistently recorded in unit records.

Table A1.16 Posting before battalion command appointment – COs of militia battalions

Previous posting	Before first wartime command[a] No.	%	Before second wartime command No.	%	Before third wartime command No.	%	Before fourth wartime command No.	%	Totals No.	%
CO of other battalion	11	6	31	84	7	100	1	100	50	22
2iC in battalion (at least)[b]	22	12	0	0	0	0	0	0	22	10
Regimental service in battalion	33	18	0	0	0	0	0	0	33	15
2iC in other battalion	8	4	0	0	0	0	0	0	8	4
Regimental service in other battalion	42	23	2	5	0	0	0	0	44	19
OC of independent company	1	<1	0	0	0	0	0	0	1	<1
CO of other combat unit	4	2	0	0	0	0	0	0	4	2
CO of support unit	2	1	0	0	0	0	0	0	2	1
Regimental service in other unit	4	2	0	0	0	0	0	0	4	2
Operational staff	12	7	0	0	0	0	0	0	12	5
Training	10	6	3	8	0	0	0	0	13	6
Base or lines of communication	7	4	0	0	0	0	0	0	7	3
Brigade commander	0	0	1	3	0	0	0	0	1	<1
UL, RL or RoO	24	14	0	0	0	0	0	0	24	11
Ceremonial duties	0	0	0	0	0	0	0	0	0	0
TOTAL	180	100	37	100	7	100	1	100	225	100

[a] Figures include incumbent militia COs in September 1939.
[b] Figures for battalion 2iCs are not comprehensive as appointments below unit command were rarely noted in an individual's service record, and are not consistently recorded in unit records.

Table A1.17 Regimental origins – COs of original 2nd AIF battalions

Battalion commanded	COs appointed Sep 1939–Jan 1942		COs appointed Jan 1942–Jan 1944		COs appointed Jan 1944–Aug 1945	
	No.	%	No.	%	No.	%
Commanded their previous battalion	13	20	19	35	3	27
Commanded battalion they had never served in	51	80[a]	36	65	8	73
TOTALS	64	100	55	100	11	100

[a] The large proportion of COs appointed to AIF battalions in which they had never served early in the war is accounted for by the formation of the 36 battalions of the 2nd AIF.

recalled to duty to take command were appointed in the period October 1939–February 1942 when the AIF was formed and sent abroad.

Posting paths also demonstrate that, despite pre-war practice, there was no long-term commitment to a regimental system in the wartime AMF: more than 70 per cent of battalion COs were appointed to command battalions in which they had not served. Only small heed was paid to regimental loyalties in the AIF throughout the war, although in its early years some attempts do seem to have been made to preserve a regimental system, with a small majority of COs being appointed to battalions in which they had served (see tables table A1.17 and table A1.18). In the AIF, 73 per cent of COs were appointed to command battalions in which they had never served. The last vestiges of a regimental system, however, were blown away once AIF combat veterans began to be appointed to militia commands from early 1942 onwards. Between 1944 and 1945 only a single militia officer would be promoted to command from within a battalion.

Analysis of the fate of COs at the conclusion of their period of command (see tables table A1.19 and table A1.20) further illustrates promotion trends in the AMF as well as providing a crude measure of the COs' success. As would be expected, the fate of AIF COs reflected their more extensive operational service. Although the number of command appointments made to AIF battalions represented a little more than half

Table A1.18 Regimental origins – COs of militia battalions

Battalion commanded	COs appointed Sep 1939–Jan 1942		COs appointed Jan 1942–Jan 1944		COs appointed Jan 1944–Aug 1945	
	No.	%	No.	%	No.	%
Commanded their previous battalion	36	59	19	24	1	3
Commanded battalion they had not served in	25	41	60	76	33	97
TOTALS	61	100	79	100	34	100

of those made to militia battalions, five times as many former AIF COs were immediately promoted to brigadier, and four times as many were appointed to the staffs of operational headquarters. Presumably both were roles for which considerable operational experience was a prerequisite. This experience came at a cost, however. Proportionally, four times as many AIF COs were killed, and ten times as many were taken prisoner.

Similarly, the fate of militia COs reflects the experience and characteristics of militia battalions. Proportionally, five times more militia COs would proceed directly to a second or third battalion than their counterparts in the AIF. This figure (65), however, is reduced significantly if the COs appointed from militia battalions to raise AIF battalions (34) and those who succeeded to the command of amalgamated or split battalions (17) are excluded. Readily demonstrating the higher ages of militia COs, particularly early in the war, their retirement rate was three times that of their AIF counterparts; a larger proportion were also transferred to garrison battalions and the Volunteer Defence Corps – both the domain of over-age soldiers and officers.

What of the cohort as a whole? Can the statistical analysis of their fates provide a crude measure of their competence? Ultimately, 38 would be promoted to brigadier and seven to major general and above, three during the war. Sixty-eight were awarded the Distinguished Service Order, five on two separate occasions. Extraordinary achievements stand out, but plain competence is a little more difficult to discern. The outcomes of 162 command appointments suggest competent performance in that they ended with promotion, appointment to a further command, a period of administering command at battalion or brigade level, or demobilisation

Table A1.19 Fate following command – COs of original 2nd AIF battalions

Fate	Following first wartime command		Following second wartime command		Following third wartime command		Following fourth wartime command		Totals
	No.	%	No.	%	No.	%	No.	%	
Brigade command	14	15	10	30	0	0	–	24	18
Training	13	14	6	18	0	0	–	19	15
Demobilised at end of war	13	14	4	12	1	100	–	18	14
Base or lines of communication	8	8	5	15	0	0	–	13	10
Prisoner of war	10	10	3	9	0	0	–	13	10
Operational staff	10	10	1	3	0	0	–	11	8
Dead	7	7	2	6	0	0	–	9	7
Another battalion command	8	8	0	0	0	0	–	8	6
RL or RoO	4	4	2	6	0	0	–	6	5
CO combat unit	4	4	0	0	0	0	–	4	3
Admin. brigade command	3	3	0	0	0	0	–	3	2
Admin. battalion command	1	1	0	0	0	0	–	1	1
Garrison battalion or VDC	1	1	0	0	0	0	–	1	1
Voluntary demoted (AIF)	0	0	0	0	0	0	–	0	0
Demoted	0	0	0	0	0	0	–	0	0
CO support unit	0	0	0	0	0	0	–	0	0
RL or RO – recalled to duty	0	0	0	0	0	0	–	0	0
TOTAL	96	100	33	100	1	100	–	130	100

Table A1.20 Fate following command – COs of militia battalions

Fate	Following first wartime command[a] No.	%	Following second wartime command No.	%	Following third wartime command No.	%	Following fourth wartime command No.	%	Totals No.	%
Another battalion command	55	31	9	24	1	14	0	0	65	29
RL or RoO	26	15	9	24	2	29	0	0	37	16
Base or lines of communication	25	14	6	16	1	14	0	0	32	14
Training	22	12	5	13	0	0	0	0	27	12
Demobilised at end of war	12	7	4	11	2	29	1	100	19	8
Brigade command	8	4	1	3	0	0	0	0	9	4
Garrison battalion or VDC	7	4	0	0	0	0	0	0	7	3
Voluntary demoted (AIF)	5	3	0	0	0	0	0	0	5	2
CO combat unit	5	3	0	0	0	0	0	0	5	2
Admin. brigade command	3	2	1	3	0	0	0	0	4	2
Operational staff	4	2	0	0	0	0	0	0	4	2
Killed	2	1	1	3	1	14	0	0	4	2
Demoted	2	1	0	0	0	0	0	0	2	1
Admin. battalion command	1	<1	1	3	0	0	0	0	2	1
RL or RO – recalled to duty	2	1	0	0	0	0	0	0	2	1
CO support unit	1	<1	0	0	0	0	0	0	1	0
Prisoner of war	0	0	0	0	0	0	0	0	0	0
TOTAL	180	100	37	100	7	100	1	100	225	100

[a] Figures include incumbent militia COs in September 1939.

at the end of the war. This figure, however, translates to a competency rate of only 46 per cent – surely too low.

The COs' service records indicate that only 33 (12 per cent) were removed from command for incompetence. This figure is probably also too low. Command failures in the AMF were often dealt with according to a code of gentlemanly discretion; command in battle was a difficult pursuit, and failure was not necessarily the result of an inherent weakness that would prevent an officer from performing capably in another role. Hence, the removal of an officer from command was not routinely recorded on his service record. It cannot be assumed, however, that every training or rear area appointment indicates a command failure. Some officers had skills more applicable to these appointments, and they were also used to provide wounded, over-age or medically unfit, but otherwise competent, COs with meaningful employment. A conservative estimate of COs removed directly as a result of command failure is approximately 45 (16 per cent), leaving a competency rate of approximately 84 per cent. It is easy, however, in the snug confines of a scholarly office to speak glibly of command failure. The cause of several command failures was sheer physical and mental exhaustion.

CO CASUALTIES

Just as the sensitivities of the AMF make it difficult to determine the precise number of COs removed for command failure, they also prevent an accurate assessment of the mental strain of command. It has been possible to identify positively only seven COs who broke down; six were removed from command, another was evacuated but eventually able to return to duty. Six of these cases could be linked to operational service, but one occurred before the officer concerned, William Jeater of the 2/20th Battalion, had seen action. Exhaustion and other psychological conditions were routinely disguised by other medical diagnoses.[35] Therefore it is likely that the rate of psychological casualties among the COs was higher than recorded.

Hard data about the rate of psychological casualties in the AMF is scarce. At the very least, COs succumbed to the mental strain of operational service in a similar proportion to the men they commanded, and it is likely that they suffered psychological casualties at a higher rate than their troops, but this is difficult to confirm. The overall rate of recorded CO psychological casualties was a little under 7 per cent. In *Middle East and Far East*, Allan Walker states that during one three-month period in Tobruk, 1.13 per cent of infantry soldiers admitted to the 2/4th AGH for all causes were psychological casualties. From one battalion the rate was as high as 3 per cent, and it was the highest among anti-aircraft units (9 per cent). The infantry figures for Tobruk seem remarkably low when rates of between 6 and 7 per cent were experienced in Papua and New Guinea, and up to 11 per cent during the Tarakan campaign of 1945.[36] The CO psychological casualties that could be linked to operational service constitute 13 per cent of battle-related casualties. At first glance this figure would seem to point to an added strain imposed by command, but it must also be considered that psychological casualty rates in British and Dominion forces during the war ranged as high as 25 per cent.[37]

Altogether, just under half (45 per cent) of Australian infantry COs became casualties during the Second World War. The battle casualties constituted 37 per cent (see table A1.21), and the rest (see table A1.22) included the victims of accident, illness, a self-inflicted wound and an execution. Fifteen COs died, ten as a result of enemy action. The circumstances of the battle casualties reflect both the AMF's experience in the different theatres of war and the actions of its COs. Of the COs who saw active service, 6 per cent were killed, 11 per cent wounded and 7 per cent taken prisoner. These casualties are clustered around three of the AMF's most desperate and hard-fought campaigns (see table A1.23): 15 per cent occurred in the vicious face-to-face fighting along the Kokoda Track and at the Japanese beachheads; 17 per cent during the ill-fated defence of Malaya and Singapore; and 24 per cent in the prolonged slug-out at El Alamein. All but one of the battle deaths occurred during this fighting. These casualties provide a pointer to the role COs were playing on the battlefield. Desperate, confused fighting required COs to be at the forefront of battle to try to stay in control of the situation and rally their troops, thus exposing them to great danger. The lack of battle fatalities and relatively low casualty rates among Australian COs from the Lae landings onwards are similarly indicative of a less hostile battlefield and a type of warfare that had moved COs further away from harm. The COs of the 2nd New Zealand Division, which spent the entire war fighting in the Western Desert and Italy, suffered much higher casualty rates: 13 per cent killed and 48 per cent wounded.[38] These figures provide a further indication that the experience of COs in the South-West Pacific was much different from those who served in the high-intensity warfare of the European and Mediterranean theatres.

Table A1.21 CO battle casualties

Casualty	Number of COs	Percentage of COs	Percentage of CO battle casualties	Percentage of CO casualties	Percentage in AMF Infantry
Wounded in action (remained on, or returned to duty)	10	4	22	8	43.5

Summary: Bourne, 2/12 Bn (Shaggy Ridge); Colvin, 2/13 Bn (El Alamein); Cotton, 2/33 Bn (Balikpapan); Cummings, 2/9 Bn (Buna); Hammer, 2/48 Bn (El Alamein); Marson, 2/25th (Balikpapan); Porter, 2/31 Bn (Syria); T. Scott, 2/32 Bn (Lae); Stevenson, 2/3 Bn (Kokoda Track); Wood, 2/6 Bn (Aitape–Wewak)

Casualty	Number of COs	Percentage of COs	Percentage of CO battle casualties	Percentage of CO casualties	Percentage in AMF Infantry
Wounded in action (did not return to duty)	8	3	17	6	

Summary: Balfe, 2/32 Bn (El Alamein); Burrows, 2/13 Bn (Tobruk); Cooper, 2/27 Bn (Gona); Honner, 2/14 Bn (Ramu Valley); Lamb, 2/3 Bn (Greece); Ogle, (El Alamein); Weir, 2/24 Bn (El Alamein); Wrigley, 2/5 Bn (Bardia)

Casualty	Number of COs	Percentage of COs	Percentage of CO battle casualties	Percentage of CO casualties	Percentage in AMF Infantry
Killed in action or died of wounds	9	3	20	7	22

Summary: R. Anderson, 2/32 Bn (Tobruk); Assheton, 2/20 Bn (Singapore); Boyes, 2/26 Bn (Singapore)[a]; Magno, 2/15 Bn (El Alamein); Owen, 39 Bn (Kokoda Track); Robertson, 2/29 Bn (Malaya); Turner, 2/13 Bn (El Alamein); Wall, 2/23 Bn (Lae); Ward, 55/53 Bn (Kokoda Track)

Casualty	Number of COs	Percentage of COs	Percentage of CO battle casualties	Percentage of CO casualties	Percentage in AMF Infantry
Prisoner of war	13	4	28	10	35

Summary: Campbell, 2/1 Bn (Crete); Key, 2/14 Bn, (Kokoda Track)[b]; Leggatt, 2/40 Bn (Timor); Marlan, 2/15 Bn (Mechilli); McCarter, 2/28 Bn (El Alamein); Oakes, 2/26 Bn (Singapore); Pond, 2/29 Bn (Singapore); Ramsay, 2/30 Bn (Singapore); J. C. Robertson, 2/19 Bn (Singapore); W. E. Robertson, 2/20 Bn (Singapore); W. Scott, 2/21 Bn (Ambon); Spowers, 2/24 Bn (El Alamein); Walker, 2/7 Bn (Crete)

Table A1.21 (cont.)

Casualty	Number of COs	Percentage of COs	Percentage of CO battle casualties	Percentage of CO casualties	Percentage in AMF Infantry
Psychological casualty (returned to duty)	1	<1	2	<1	2–11 per cent on top of WIA figures
Summary: Wain, 2/43 Bn (El Alamein)					
Psychological casualty (did not return to duty)	5	2	11	4	
Summary: Abbott, 26 Bn (Bougainville); Dexter, 61 Bn (Bougainville); Embrey, 37/52 Bn (New Britain); Lovell, 55/53 Bn (Bougainville); Miell, 19 Bn (New Britain)					
TOTALS	46	17	100	c. 37	–

[a] Boyes was actually killed leading 'X' Bn, a scratch unit of reinforcements, several days after being removed from command of 2/26 Bn.
[b] Key is believed to have been taken PoW by the Japanese and subsequently executed.

The strain of prolonged military service is also evident among the non-battle casualties. In the region of 70 per cent can be attributed to the rigours of operational service. The effect of the tropical environment is particularly evident, half of all sickness cases being due to its oppressive nature, including 28 cases of malaria and a death from scrub typhus. The medical statistics also confirm that the AMF was lacking command talent early in the war. A fifth of the medical casualties leading to the loss of command were due to pre-existing conditions, including deafness, emphysema and osteoarthritis, which should have been detected on enlistment and barred the individuals concerned from active service; all but one were appointed to their terminal wartime command before the end of 1940.

A steady decline of the average age of COs throughout the war (see figure A1.1 and tables table A1.24 and table A1.25) provides further evidence of the physical and mental demands of command. The average age of AMF COs appointed in 1939 was 43; by 1945 it had dropped to 35.

Table A1.22 CO non-battle casualties

Casualty	Number of COs	Percentage of COs	Percentage of CO non-battle casualties	Percentage of CO casualties
Executed while prisoner of war	1	<1	1	<1

Summary: Key, 2/14 Bn (Papua)

Suicide	1	<1	1	<1

Summary: Hendry, 2/28 Bn (British North Borneo)

Psychological illness	1	<1	1	<1

Summary: Jeater, 2/20 Bn (Malaya)

Died of sickness	3	1	4	2

Summary: Brock, 31/51 Bn (Atherton); Miller, 2/31 Bn (Papua) T[a]; W. E. Robertson, 2/20 Bn (Malaya)[b]

Sickness or medical downgrade (did not return to command)	30	11	38	24

Summary: Anderson, 2/19 Bn (Singapore) T[c]; Arnold, 2/12 Bn (Australia) T; Bartley, 12/40 Bn (Australia) T; Carr, 2/22 Bn (Australia) T; Carstairs, 22 Bn (New Britain); Conran, 39 Bn (Papua) T; Conroy, 2/5 Bn (Australia); Coombes, 47 Bn (Australia); Davidson, 42 Bn (Australia) T; Dobbs, 2/10 Bn (Papua) T; Foster, 19 Bn (New Guinea); Galleghan, 2/30 Bn (Singapore)[d]; Goldrick, 35 Bn (Australia); Horley, 16 Bn (New Britain) T; Lee, 2/9 Bn (Balikpapan); Lillie, 5 Bn (Australia); Loughrey, 2/28 Bn (Australia); Matthewson, 22 Bn (New Guinea); McGregor, 4 Bn (Australia); McEwin, 10/48 Bn (Australia); Melville, 11 Bn (New Britain) T; Miles, 25 Bn (Papua); Montgomery, 47 Bn (New Guinea) T; Proud, 28 Bn (Australia); Rowan, 37/52 Bn (Australia); Spotswood, 12/50 Bn (Australia); Starr, 58/59 Bn (New Guinea) T; Teele, 38 Bn (Australia); Verrier, 2/10 Bn (Tobruk); Wade, 20/34 Bn (Australia)

Sickness (returned to duty)	35	11	45	28

Table A1.22 (cont.)

Casualty	Number of COs	Percentage of COs	Percentage of CO non-battle casualties	Percentage of CO casualties

Summary: Amies, 15 Bn (Papua & New Guinea) T; Colvin, 2/13 Bn (New Guinea) T; Cooper, 2/27 Bn (Papua); Corby, 2/33 Bn (Syria); Cox, 2/4 Bn (New Guinea); Cullen, 2/1 Bn (Papua & Australia) T; Dexter, 61 Bn (Australia) T; Eather, 2/1 Bn (Middle East); Edgar, 2/2 Bn (Papua & Australia) T; Farrell, 2/4 Bn (New Guinea) T; Geard, 2/10 Bn (Australia) T; Grace, 2/15 Bn (Australia) T; Guinn, 2/7 Bn (New Guinea) T; Hammer, 2/28 Bn (Middle East); Honner, 39 Bn (Papua) T; Howden, 2/8 Bn (New Guinea) T; Hutchison, 2/3 Bn (Australia) T; Isaachsen, 36 Bn (New Guinea) T; Kessels, 49 Bn (Australia) T; Loughrey, 2/28 Bn (Middle East); Lovell, 55/53 Bn (Bougainville); Marson, 2/25 Bn (Australia) T; Marston, 57/60 Bn (Australia); Matthews, 9 Bn (New Guinea) T; Meldrum, 61 Bn (Papua) T; Montgomery, 8 Bn (New Guinea) T; Norman, 2/28 Bn (New Guinea) T; O'Connor, 22 Bn (New Guinea) T; Ogle, 2/15 Bn (Tobruk); Parbury, 31/51 Bn (New Guinea); Robson, 2/31 Bn (Australia) T; Scott, 2/32 Bn (Middle East & New Guinea) T; Simpson, 2/43 Bn (Australia) T; Warfe, 58/59 Bn (New Guinea) T; Withy, 2/25 Bn (Syria)

Sickness (returned to another command)	2	1	3	2

Summary: Buttrose, 2/33 Bn (Cairns) T; Farquhar, 27 Bn (Australia)

Killed accidentally	1	<1	1	<1

Summary: Norris, 7 Bn, (Vella Lavella)

Accidentally injured, did not return to duty	2	1	3	2

Summary: Robson, 2/31 Bn (Balikpapan); Smith, 24 Bn (Australia)

Accidentally injured, returned to duty	2	1	3	2

Summary: Amies, 15 Bn (Papua); Caldwell, 14/32 (New Guinea)

TOTALS	78	32	100	63

[a] 'T' represents those conditions resulting from tropical service.
[b] W. E. Robertson died while a prisoner of war.
[c] C. G. W. Anderson was subsequently taken prisoner on the surrender of Singapore, but was not in command of the 2/19th Battalion at the time.
[d] F. G. Galleghan was subsequently taken prisoner on the surrender of Singapore, but was not in command of the 2/30th Battalion at the time.

Table A1.23 Distribution of CO battle casualties (including psychological casualties in theatre of operations)

Campaign	Casualties	Percentage
El Alamein	10	24
Malaya and Singapore	8[a]	17
Papua	7	15
New Guinea 1943–44	4	9
Bougainville and New Britain 1944–45	4	9
Greece and Crete	3	7
'Benghazi Handicap' and siege of Tobruk	3	4
'Bird Force' operations	2	4
Borneo	2	4
Cyrenaica offensive	1	2
Lebanon and Syria	1	2
Aitape–Wewak	1	2
TOTALS	46	c. 100

[a] Neither C. G. W. Anderson nor F. G. Galleghan are included in this figure for, although taken prisoner at the fall of Singapore, sickness had already caused both to relinquish their commands.

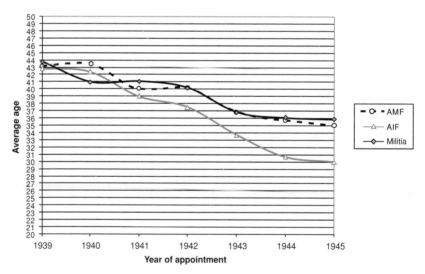

Figure A1.1: Decline in average age of COs

Table A1.24 Average age of 2nd AIF COs on appointment

Year of appointment	First wartime command (years)	Second wartime command (years)	Third wartime command (years)	Fourth wartime command (years)	Overall average age (years)
1939	43	42.6	–	–	42.8
1940	41.2	43.5	–	–	42.4
1941	39	39	–	–	39
1942	36.4	38.6	–	–	37.5
1943	33.6	33.6	34	–	33.7
1944	30.7	30.6	–	–	30.7
1945	30	30	–	–	30

Table A1.25 Average age of militia COs on appointment

Year of appointment	First wartime command (years)	Second wartime command (years)	Third wartime command (years)	Fourth wartime command (years)	Overall average age (years)
1939	43.8	–	–	–	43.8
1940	45	37	–	–	41
1941	44.2	41.1	38	–	41.1
1942	38	40.3	42.5	–	40.2
1943	32	38.3	40	–	36.8
1944	33.6	35.7	39	–	36.1
1945	32.2	41.5	35	35	35.9

Among the AIF battalions, reflecting their more extensive operational experience, the drop was more pronounced. During 1941 the average age of AIF CO appointees dropped below 40; in the wider army this did not happen until 1943 and by the end of the war the average had decreased another ten years to 30.

The rate of decline of the average age of militia CO appointees was more in keeping with that of the AMF as a whole, although the figures relating specifically to those appointed to command for the first time are instructive. The average age of first-time militia COs actually rose during 1940, from 43.8 to 45, and was the only significant movement against the downwards trend during the entire war. This phenomenon supports previous comments made about the nature of officers appointed to militia commands during this period. The sequel to this story came with the influx of young AIF veterans into the militia: between 1942 and 1943 the average age COs appointed to their first commands in militia battalions dropped from 38 to 32.

Three rough cohorts of COs can be discerned. The first is the initial group of officers who were in command of militia battalions in 1939. The original COs for the Second AIF battalions were selected from these men. Most had served in the First World War, and all had extensive experience in the militia; very few were younger than 40. With the maximum age for a lieutenant colonel in the 2nd AIF being 45, not many of these men would carry their commands beyond 1941, and having raised and trained their battalions some even missed the opportunity to lead them in action. In home-based militia units, men of this cohort remained in command longer, but Japan's entry into the war, and the impending deployment of their battalions to New Guinea, also saw them replaced.

The members of the second cohort also had considerable experience in the militia, and were generally enlisted into the 2nd AIF as majors. Being in their mid-thirties, however, they were younger than their predecessors and had not served in the First World War. They began replacing the original COs from late 1940 onwards and had the benefit of attending the British Army schools in the Middle East. It was mostly these men who led Australian battalions into action in North Africa, Greece, Crete and Lebanon. Some did not rise to command in the Middle East but, returning to Australia, were posted back to militia battalions as COs.

The third cohort began to take command from 1942 onwards. These men were younger still, aged in their late twenties to early thirties. The promotion of such young men reflected the physically demanding nature of the war being fought by Australian troops in the South-West Pacific Area, the breadth of experience now found in the AMF and the development of a promotion system based on merit. This third cohort had generally served in the 2nd AIF's early campaigns, as lieutenants and captains. Pre-war militia experience was not as dominant in this cohort as the first two, as most had been old enough to see only a few years' service before the war. These men were arguably the most capable of all COs as their ideas of warfare were founded in the current war, honed by their experience and refined by the latest developments in British and Australian military education.

APPENDIX 2

Periods of command

The list that follows contains 276 individual officers officially appointed to command Australian infantry battalions during the Second World War. For the reasons discussed in the introduction and appendix 1 it does not include those who administered command or those who acted in command on the battlefield. Hence, in some cases, there appear to be quite large gaps between two COs. During these periods it should be assumed that an officer, or a series of officers, administered command of the battalion pending the formal appointment of the next incumbent.

This list is comprehensive but not complete. Weaknesses in the archival record relating to militia battalions in the period September 1939–December 1941 have meant that the COs of some battalions during this period cannot be determined; it is estimated that less than ten officers have been thus omitted.

The dates of command listed are those contained in each officer's service record (Australian Army Form B. 199, Officer's Record of Service), which were promulgated in the AMF's various appointments and promotions lists. Where examination of unit records – primarily war diaries – has revealed significant discrepancies, the dates have been amended accordingly.

PMF officers are indicated with an asterisk.

1st Battalion
Douglas, William Huggett   23 October 1939–22 July 1940
Unknown
Casier, Leon William   9 December 1941–11 August 1942

1/45th Battalion

Casier, Leon William	11 August 1942–26 May 1943
Hosking, Frank	16 August 1943–28 July 1944

2nd Battalion

Jeater, William David	1 March 1937–1 July 1940
Owens, William Thomas	6 March 1942–14 July 1942
Morris, Alfred George	11 August 1942–18 September 1942
Phillips, George William	18 September 1942–31 July 1944

3rd Battalion

Paul, Albert Thomas	1 August 1938–17 October 1942
Cameron, Allan Gordon	17 October 1942–5 July 1943

4th Battalion

Parsons, Percival Augustus	1 October 1936–13 October 1939
Crawford, John Wilson	15 December 1939–26 April 1940
Smith, William John	26 April 1940–7 October 1942
Crosky, Percy William	4 February 1942–19 August 1944
McGregor, Wilfred Robert	19 August 1944–6 October 1944
Neville, Cilve Henry	13 November 1944–14 December 1945
Jackson, David Arion Collingwood	14 December–18 April 1946

5th Battalion

Cook, Thomas Page	18 May 1937–13 October 1939
Knox, George Hodges	15 December 1939–1 May 1940
Johnston, Robert Alexander Bold	2 May 1940–11 August 1942
Caldwell, William Blythe	11 August 1942–5 November 1942
Lillie, Cyril McEachern	5 November 1942–30 October 1943
Goble, Norman Leopold	12 November 1943–14 September 1944

6th Battalion

McArthur, Neil Alexander	31 March 1938–29 April 1942
Evans, John Granville	29 April 1942–6 August 1942
Egan, Eugene William	6 August 1942–19 August 1944

7th Battalion

Goucher, Frederick Thomas Henry	15 December 1936–1 February 1940
Conran, Hugh Marcell	1 February 1940–22 June 1940
Goucher, Frederick Thomas Henry	Unknown–1 September 1941

Sadler, Rupert Markham 17 September 1941–25 March 1942
Wilmoth, John Alfred 14 April 1942–18 October 1943
Norris, Geoffrey Moore 18 October 1943–8 October 1944
Dunkley, Howard Leslie Ewen 20 October 1944–7 November 1945
Webster, Peter Glynn Clifton 4 December 1945–29 March 1946

8th Battalion
Reed, Alfred Effingham 15 November 1943–4 July 1940
Unknown
Wallis, Robert Jonathan 1 February 1942–25 April 1942
Montgomery, Keith Hamilton 14 April 1942–22 July 1943
Berry, Bernard Nash 18 October 1943–1 June 1945
Loughran, Lewis John* 1 June 1945–5 November 1945

9th Battalion
Kessels, Owen Andrew 14 October 1940–27 January 1942
Morgan, Harold David* 13 April 1942–4 September 1942
Matthews, Geoffrey Ronald 5 September 1942–4 October 1945

9/49th Battalion
Kessels, Owen Andrew 1 August 1939–14 October 1940

10th Battalion
Veale, William Charles Douglas 1 October 1936–1 June 1940
Hill, John Holroyd 1 June 1940–1 June 1942
Farquhar, Lindsay Keith 1 June 1942–22 July 1942

10/48th Battalion
McEwin, Kenneth John 11 August 1942–9 January 1943
Davies, Thomas Francis 9 January 1943–29 October 1945

11th Battalion
Hubbard, Stanley Harold 18 December 1936–28 September 1941
James, Herbert 20 October 1941–1 August 1942
Buntine, Martyn Arnold 11 August 1942–18 October 1943
Melville, William Sydney 12 November 1943–28 August 1945

12/50th Battalion
Youl, Geoffrey Arthur Douglas 1 July 1937–1 July 1940
Unknown

Spotswood, Henry Edward Tregaskis 5 February 1941–28 September 1943
Chilton, Henry Humfrey Marsden 18 August 1943–17 May 1945

12/40th Battalion
Bartley, Thomas William 14 July 1945–10 May 1946

13th Battalion
Patterson, Gordon Andrew 1 July 1939–11 August 1942

13/33rd Battalion
Chilton, Henry Humfrey Marsden 11 August 1942–18 August 1943
Joshua, Robert 28 February 1944–19 November 1945

14th Battalion
Steele, Clive 1 July 1933–13 October 1939
Roach, Leonard Nairn 9 November 1939–1 July 1940
Unknown
Lillie, Cyril McEachern 1 March 1942–11 August 1942

14/32nd Battalion
Caldwell, William Blythe 5 November 1942–23 July 1945

15th Battalion
Fleming, Patrick Vincent Osborne 1 August 1939–1 May 1941
Edmonds, Frederick Gabriel 2 July 1941–11 August 1942
Amies, Jack Lowell 11 August 1942–14 September 1944
McDonald, Herbert Hector 14 September 1944–4 December 1945

16th Battalion
Louch, Thomas Steane 1 December 1936–13 October 1939
Lloyd, John Edward 4 November 1939–17 June 1940
Sanderson, William Lauchlan 15 August 1940–1 December 1942
Horley, Ronald John 1 December 1942–4 December 1945

17th Battalion
Galleghan, Frederick Gallagher 1 March 1937–17 October 1940
Morgan, Charles Eric 5 November 1940–14 June 1944

18th Battalion
Cameron, Claude Ewen	1 October 1936–30 April 1940
Adcock, Garnet Ingamells	30 April 1940–18 September 1942
Neville, Clive Henry	18 September 1942–13 November 1944

19th Battalion
Unknown	June 1941–April 1942
Berry, Bernard Nash	14 April 1942–10 January 1943
Foster, Edwin Peter	10 January 1943–18 April 1944
Miell, Lindsay de Lisle	18 April 1944–29 March 1945
Barnes, Eric	29 March 1945–28 August 1945

20th Battalion
Ward, Kenneth Harry	30 June 1941–29 September 1941
Wade, Frederick Bede	23 October 1941–2 December 1943

20/19th Battalion
Ward, Kenneth Harry	1 September 1939–30 June 1941

20/34th Battalion
Wade, Frederick Bede	2 December 1943–14 June 1944

22nd Battalion
Birch, Herbert Ralph	1 August 1939–23 July 1940
Matthewson, Eric Clive Tate	27 August 1940–3 July 1943
O'Connor, John Christian Watson*	3 July 1943–14 March 1944
Carstairs, James de Mole	18 April 1944–28 August 1945
Barnes, Eric	28 August 1945–23 May 1946

23/21st Battalion
Godfrey, Arthur Harry Langham	1 August 1935–13 October 1939
Roberston, John Charles	16 November 1939–17 October 1940
Hall, Joseph Rex	8 February 1941–5 June 1942
Ainslie, Robert Inglis	5 June 1942–22 July 1943

24/39th Battalion
Walker, Theodore Gordon	24 August 1939–13 October 1939
Spowers, Allan	1 January 1940–5 May 1940
Falconner, Alexander Hugh	1 November 1940–7 October 1941

24th Battalion
Falconner, Alexander Hugh	7 October 1941–9 July 1943

Smith, George Frederick   9 July 1943–20 October 1944
Anderson, Arthur Jeffrey   20 October 1944–14 January 1946

25th Battalion
Hatton, Neville Gordon   1 October 1934–22 May 1940
Unknown
Miles, Edward Samuel   19 June 1941–11 January 1944
McKinna, John Gilbert   11 January 1944–16 October 1945

26th Battalion
Murray, Henry William   21 July 1939–11 August 1942
Abbott, John Noel   11 August 1942–8 February 1945
Callinan, Bernard James   8 February 1945–14 December 1945

27th Battalion
Best, Francis Mayfield   1 May 1934–19 February 1940
Farquhar, Lindsay Keith   27 March 1940–14 April 1942
Pope, Alexander   14 April 1942–22 October 1945

28th Battalion
Anketell, Michael Joseph   1 February 1937–5 November 1940
Proud, Alfred Joseph   16 December 1940–5 June 1942
Brennan, James Gerald   13 August 1942–9 August 1945
Chilton, Henry Humphrey Marsden   14 July 1945–28 February 1946

29th Battalion
Barber, Charles Stanley   1 August 1939–24 June 1940
Hosking, Frank   24 June 1940–12 April 1942

29/46th Battalion
Cusworth, Kenneth Stanley   11 August 1942–19 November 1945

30th Battalion
Russell, James William   1 July 1939–4 April 1942
Anderson, Eugene Leroy   1 June 1942–8 September 1943
Parry-Okeden, William Nugent 17 November 1943–4 December 1945

31st Battalion
North, Francis Roper   1 October 1933–1 August 1940
Beattie, Oscar Alexander   Unknown
Cardale, Philip Heathcote George*   May 1942–February 1943

31/51st Battalion

Brock, Geoffrey Hutton	13 April 1943–19 December 1943
Parbury, Philip Kingsmill	21 January 1944–20 November 1944
Kelly, Joseph Lawrence Andrew*	20 November 1944–17 January 1946

32nd Battalion

Gordon, Robert James Moody	11 June 1939–3 October 1941
Watson, Frederick Robert	15 March 1942–11 August 1942

33rd Battalion

Dougherty, Ivan Noel	1 December 1938–13 October 1939
Assheton, Charles Frederick	16 December 1939–1 May 1940
Hague, Albert George	19 November 1940–3 October 1941

34th Battalion

Milne, Edmund Osborn	1 June 1937–13 December 1939
Anschau, John	13 December 1939–28 September 1943
Corby, John Armstrong	28 September 1943–12 December 1943

35th Battalion

Varley, Arthur Leslie	13 September 1939–1 July 1940
Goldrick, Robert Austin	5 August 1940–31 May 1942
Rae, Douglas Frank	1 June 1942–19 August 1944
Egan, Eugene William	19 August 1944–17 May 1945
Armstrong, Frederick Henry Montgomery	17 May 1945–28 February 1946

36th Battalion

Burrows, Frederick Alexander	15 August 1938–3 March 1940
Purser, Muir	4 March 1940–13 March 1942
Brown, Arnold	18 May 1942–2 September 1942
Isaachsen, Oscar Cedric	2 September 1942–28 August 1945

37th Battalion

Stewart, Albert William John	24 August 1939–29 January 1942
Minogue, John Patrick	14 May 1942–2 June 1942
Rowan, John George	23 June 1942–5 September 1942

37/52nd Battalion

Rowan, John George	6 September 1942–22 December 1944

Embrey, Frederick James 22 December 1944–25 July 1945
Pitt, Douglas Davison* 25 July 1945–19 February 1946

38th Battalion
Swatton, John William 9 November 1936–29 September 1940
Langley, George Furner 30 September 1940–16 February 1942
Teele, Athol Dyring 20 March 1942–12 June 1942
McGregor, Wilford Robert 20 November 1942–19 August 1944

39th Battalion
Conran, Hugh Marcell 1 October 1941–24 May 1942
Owen, William Taylor 6 June 1942–28 July 1942
Honner, Ralph 1 August 1942–5 July 1943

40th Battalion
Bartley, Thomas William 19 December 1937–14 July 1945

41st Battalion
Barham, Lindley de Lisle* 16 January 1939–7 November 1939
Moyes, John 15 December 1939–1 January 1942
Foster, Edwin Peter 30 March 1942–10 January 1943
Berry, Bernard Nash 10 January 1943–18 October 1943

41/2nd Battalion
Corby, John Armstrong 2 December 1943–3 May 1944
Hosking, Frank 14 June 1944–26 November 1945

42nd Battalion
Martin, James Eric Gifford 1 August 1937–13 October 1939
Heron, Alexander Robert 1 January 1940–16 March 1942
Davidson, Charles William 6 September 1942–14 September 1944
Byrne, Joseph Henry 14 September 1944–4 October 1945

43rd Battalion
Verrier, Arthur Drummond 1 July 1939–13 October 1939
Pope, Alexander 1 March 1940–1 May 1940
Unknown
Baldock, Albert 9 July 1941–20 July 1942
Frith, Clifford 10 September 1942–22 June 1944

44th Battalion
McKenzie, Eric George Henderson 1 January 1936–31 December 1939

Unknown
Garner, Walter Brendon 1 June 1940–17 July 1944

45th Battalion
Rena, John Alfred 1 June 1937–2 July 1940
Morris, Alfred George 3 July 1940–11 August 1942

46th Battalion
Carr, Howard Hammond 1 November 1938–26 July 1940

47th Battalion
Paterson, Eric Ewen 1 November 1936–23 August 1942
Rowan, John George 23 August 1942–6 September 1942
Tasker, Henry McKean 18 September 1942–22 July 1943
Montgomery, Keith Hamilton 22 July 1943–31 August 1944
Goble, Norman Leopold 14 September 1944–19 January 1945
Coombes, Charles John Allen 19 January 1945–6 April 1945
Fry, William Gordon 16 April 1945–7 December 1945

48th Battalion
Moten, Murray John 1 July 1939–26 April 1940
McEwin, Kenneth John 29 August 1940–11 August 1942

49th Battalion
Oliver, William 5 March 1941–29 January 1942
Kessels, Owen Andrew 29 January 1942–3 July 1943

51st Battalion
Harris, Hubert Cochrane 1 November 1936–16 March 1942
Brock, Geoffrey Hutton 18 September 1942–12 April 1943

52nd Battalion
McCormack, Patrick John 1 April 1937–20 February 1940
Symons, Arno Edgar 21 February 1940–11 August 1942

53rd Battalion
Ward, Kenneth Harry 29 September 1941–27 August 1942
Cameron, Allan Gordon 2 September 1942–17 October 1942

55th Battalion
Brackpool, Albert Arnold 1 October 1941–1 March 1942

Lovell, David John Hamilton   11 August 1942–31 October 1942

55/53rd Battalion

England, Vivian Theophilus	1 August 1937–13 October 1939
Brackpool, Albert Arnold	15 December 1939–1 October 1941
Lovell, David John Hamilton	31 October 1942–27 March 1945
Shanahan, Patrick Mannix	21 May 1945–30 January 1946

56th Battalion

Bundey, Eric William*	14 December 1938–2 October 1939
Maxwell, Duncan Stuart	25 November 1939–1 July 1940
Downing, Reginald George	15 July 1940–2 January 1943
Bryen, Robert Thomas	10 February 1943–9 June 1944

57/60th Battalion

Stevens, Jack Edwin Stawell	1 July 1935–13 October 1939
Evans, Bernard	15 December 1939–1 July 1940
Moss, Richard George	20 August 1940–23 May 1942
Hale, Frederick William Harvey	11 August 1942–6 September 1942
Creed, Sydney Stewart	6 September 1942–19 February 1943
Marston, Robert Raymond	28 April 1943–27 February 1945
Webster, Peter Glynn Clifton	13 March 1945–4 December 1945

58th Battalion

Cannon, William George	1 October 1938–1 May 1940
Hale, Frederick William Harvey	18 May 1940–11 August 1942

58/59th Battalion

Whalley, Rupert Percy	11 August 1942–28 January 1943
Starr, Patrick Daniel Sarsfield	28 April 1943–10 September 1943
Warfe, George Radford	10 September 1943–4 January 1945
Mayberry, William Maurice	4 January 1945–4 October 1945

59th Battalion

Hill, Ernest Purnell	1 October 1938–25 June 1940
Whalley, Rupert Percy	9 August 1940–11 August 1942

61st Battalion

Grant, John MacDonald	15 October 1938–26 May 1941
Meldrum, Alexander	26 May 1941–18 August 1943

Wiles, Harold Joseph	19 August 1943–1 April 1944
Dexter, Walter Roadknight	1 April 1944–13 May 1945
Farrell, Terrance Joseph	13 May 1945–24 October 1945

62nd Battalion

Haupt, Alexander Graham Keith	1 January 1943–27 July 1944

2/1st Battalion

Eather, Kenneth William	13 October 1939–17 March 1941
Campbell, Ian Ross*	8 April 1941–30 May 1941
Eather, Kenneth William	31 August 1941–27 December 1941
White, Tom Warren*	1 January 1942–10 June 1942
Cullen, Paul Alfred	11 June 1942–28 August 1945

2/2nd Battalion

Wootten, George Frederick	13 October 1939–13 December 1940
Chilton, Frederick Oliver	1 December 1940–1 November 1941
Edgar, Cedric Rupert Vaughan	16 November 1941–16 June 1943
Cameron, Allan Gordon	5 July 1943–14 December 1945

2/3rd Battalion

England, Vivian Theophilus	13 October 1939–13 March 1941
Lamb, Donald James	18 March 1941–25 June 1941
Stevenson, John Rowlstone	25 June 1941–31 March 1943
Hutchison, Ian	26 April 1943–8 February 1946

2/4th Battalion

Parsons, Percival Augustus	13 October 1939–3 July 1940
Dougherty, Ivan Noel	19 August 1940–24 March 1942
Farrell, Nevis William Parkington	21 April 1942–29 March 1945
Cox, Geoffrey Souter	29 March 1945–11 December 1945

2/5th Battalion

Cook, Thomas Page	13 October 1939–17 December 1940
Wrigley, Hugh	17 December 1940–17 January 1941
King, Roy*	17 January 1941–1 November 1941
Wood, Frederick George	16 November 1941–14 January 1942
Starr, Patrick Daniel Sarsfield	14 January 1942–22 March 1943
Conroy, Thomas Mayo	23 March 1943–27 July 1944

Buttrose, Alfred William — 28 July 1944–12 November 1945

2/6th Battalion
Godfrey, Arthur Harry Langham — 13 October 1939–20 January 1941
Wrigley, Hugh — 22 March 1941–14 January 1942
Wood, Frederick George — 14 January 1942–28 July 1945

2/7th Battalion
Walker, Theodore Gordon — 13 October 1939–1 June 1941
Guinn, Henry George — 5 June 1941–20 November 1944
Parbury, Philip Kingsmill — 20 November 1944–24 September 1945

2/8th Battalion
Mitchell, John Wesley — 13 October 1939–16 June 1941
Winning, Robert Emmet — 7 July 1941–18 February 1943
Howden, Walter Stace — 19 February 1943–24 October 1945

2/9th Battalion
Martin, James Eric Gifford — 13 October 1939–27 December 1941
Cummings, Clement James — 27 December 1941–31 July 1944
Lee, Arthur James — 12 August 1944–12 August 1945

2/10th Battalion
Verrier, Arthur Drummond — 13 October 1939–1 August 1941
Bruer, Lionel Gregory — 25 July 1941–7 January 1942
Dobbs, James Gordon — 7 January 1942–5 January 1943
Geard, Charles John — 5 January 1943–8 August 1944
Daly, Thomas Joseph* — 14 October 1944–11 December 1945

2/11th Battalion
Louch, Thomas Steane — 13 October 1939–20 April 1941
Sandover, Raymond Ladais — 20 April 1941–1 May 1943
Binks, Hector McLean — 8 May 1943–28 February 1945
Green, Charles Hercules — 9 March 1945–12 November 1945

2/12th Battalion
Field, John — 13 October 1939–8 May 1942
Arnold, Arthur Sydney Wilfred — 7 June 1942–1 August 1943
Bourne, Charles Cecil Francis — 1 September 1943–1 November 1945

2/13th Battalion

Burrows, Frederick Alexander	26 April 1940–26 December 1941
Turner, Robert William Newton	27 December 1941–23 October 1942
Colvin, Geoge Edward	27 October 1942–30 October 1945

2/14th Battalion

Cannon, William George	26 April 1940–14 January 1942
Key, Arthur Samuel	14 January 1942–30 August 1942
Challen, Hugh Bradbury	30 September 1942–5 July 1943
Honner, Ralph	5 July 1943–23 December 1943
Rhoden, Phillip Edington	21 October 1943–8 October 1945

2/15th Battalion

Marlan, Robert Francis*	26 April 1940–6 April 1941
Ogle, Robert William George	13 June 1941–2 September 1942
Magno, Charles Keith Massy	9 September 1942–29 October 1942
Barham, Raymond James*	7 November 1942–11 March 1943
Grace, Colin Henry	11 May 1943–14 December 1945

2/16th Battalion

Baxter-Cox, Alfred Richard	26 April 1940–28 January 1941
MacDonald, Alexander Bath*	20 February 1941–12 August 1941
Potts, Arnold William	12 August 1941–6 April 1942
Caro, Albert Edward	16 April 1942–13 December 1942
Sublet, Frank Henry	30 April 1943–8 October 1945

2/17th Battalion

Crawford, John Wilson	26 April 1940–15 January 1942
Fergusson, Maurice Alfred	19 January 1942–7 March 1942
Simpson, Noel William	7 March 1942–27 February 1944
Broadbent, John Raymond	28 February 1944–8 February 1946

2/18th Battalion

Varley, Arthur Leslie	1 July 1940–12 February 1942
O'Brien, Charles Brian	12 February–15 February 1942

2/19th Battalion

Maxwell, Duncan Stuart	1 July 1940–1 August 1941
Anderson, Charles Groves Wright	1 August 1941–7 February 1942

Robertson, Andrew Esmond   7 February 1942–15 February 1942

2/20th Battalion
Jeater, William David   1 July 1940–4 August 1941
Assheton, Charles Frederick   4 August 1941–9 February 1942

2/21st Battalion
Roach, Leonard Nairn   1 July 1940–13 January 1942
Scott, William John Rendel   13 January 1942–2 February 1942

2/22nd Battalion
Carr, Howard Hammond   1 July 1940–25 March 1942

2/23rd Battalion
Evans, Bernard   1 July 1940–1 November 1942
Wall, Reginald Ernest   7 November 1942–4 September 1943
Tucker, Frederick Alfred George   13 October 1943–8 February 1946

2/24th Battalion
Spowers, Allan   1 July 1940–12 July 1942
Weir, Charles Gladstone   22 July 1942–7 November 1942
Finlay, Charles Hector*   7 November 1942–8 April 1943
Gillespie, Andrew Basil   11 May 1943–4 January 1945
Warfe, George Radford   4 January 1945–4 February 1946

2/25th Battalion
Marshall, Norman   1 July 1940–17 October 1940
Withy, Charles Burton   17 October 1940–26 October 1942
Marson, Richard Harold   27 October 1942–22 October 1945

2/26th Battalion
Boyes, Arthur Harold*   17 October 1940–6 February 1942
Oakes, Roland Frank   7 February 1942–15 February 1942

2/27th Battalion
Moten, Murray John   26 April 1940–27 December 1941
Cooper, Geoffrey Day Thomas   14 January 1942–1 December 1942
Bishop, John Ackland   11 March 1943–17 November 1943
Picken, Keith Sinclair   17 November 1943–8 October 1945

2/28th Battalion
Lloyd, John Edward   1 July 1940–7 March 1942

PERIODS OF COMMAND

Loughrey, Jack	7 March 1942–22 July 1943
McCarter, Lewis	23 July 1942–27 July 1942
Norman, Colin Hugh Boyd	27 July 1943–30 July 1945
Hendry, James Gordon	30 July 1945–29 September 1945

2/29th Battalion

Robertson, John Charles	17 October 1940–18 January 1942
Pond, Samuel Austin Frank	25 January 1942–15 February 1942

2/30th Battalion

Galleghan, Frederick Gallagher	17 October 1940–9 February 1942
Ramsay, George Ernest	9 February 1942–15 February 1942

2/31st Battalion

Strutt, Horace William	26 June 1940–16 September 1940
Garrett, Alwyn Ragnar*	16 September 1940–26 February 1941
Porter, Selwyn Havelock Watson Craig	27 February 1941–25 March 1942
Dunbar, Colin Campbell	16 April 1942–5 November 1942
Miller, James	5 November 1942–14 December 1942
Byrne, Joseph Henry	31 March 1943–25 May 1943
Robson, Ewan Murray	26 May 1943–8 October 1945

2/32nd Battalion

Sparkes, Alonzo Sydney Clive	26 June 1940–14 April 1941
Anderson, Raymond Keith	15 April 1941–13 June 1941
Conroy, Thomas Mayo	13 June 1941–27 February 1942
Whitehead, David Adie	27 February 1942–9 September 1942
Balfe, John	10 September 1942–7 November 1942
Scott, Thomas Henty	7 November 1942–4 December 1945

2/33rd Battalion

Bierwirth, Rudolph*	26 June 1940–3 June 1941
Monaghan, Raymond Frederic	1 June 1941–18 June 1941
Corby, John Armstrong	18 June 1941–23 April 1942
Buttrose, Alfred William	23 April 1942–12 April 1943
Cotton, Thomas Richard Worgan	2 May 1943–3 December 1945

2/40th Battalion

Youl, Geoffrey Arthur Douglas	1 July 1940–25 October 1941

Leggatt, William Watt   25 October 1941–22 February 1942

2/43rd Battalion
Crellin, William Wauchope*   1 July 1940–3 November 1941
Wain, William John   16 November 1941–22 July 1943
Joshua, Robert   22 July 1943–27 February 1944
Simpson, Noel William   28 February 1944–21 March 1945
Jeanes, Mervyn Roderick   31 March 1945–16 December 1945

2/48th Battalion
Windeyer, William John Victor   1 July 1940–6 January 1942
Hammer, Howard Heathcote   8 January 1942–18 June 1943
Ainslie, Robert Inglis   22 July 1943–12 August 1945

Notes

Introduction
1. Long, notebook 86, p. 49, AWM 67, item 2/86.
2. This account is drawn from: '2/10 Aust Inf Bn Report on Operations – Oboe Two', para. 1–6, WD 2/10 Bn, Aug 1945, AWM 52, item 8/3/10.
3. Daly, interview, 26 Oct 2000.
4. Long, notebook 86, p. 49.
5. Ibid.
6. Ibid.; *Infantry Training*, Part I: *The Infantry Battalion. 1943*, p. 1; 'LHQ Tactical School précis Inf 1: Org of an Aust Inf Bn', p. 1, AWM 54, item 881/1/4, part 2.
7. French, *Raising Churchill's Army*, p. 70.
8. Griffiths-Marsh, *The Six Penny Soldier*, p. 1.
9. Ibid.
10. Compiled from figures in AWM 54, item 171/1/13.
11. Chilton, interview, 25 Oct 2000.
12. Similar observations are made about the significance of New Zealand battalion commanders in McElwain, 'The commanding officers of the infantry battalions of 2nd New Zealand Division', in Harper & Hayward (eds), *Born to Lead?*, pp. 177–8.
13. Until October 1943 various promotions lists were issued by separate approving authorities within the AMF. These included the CiC AMF, the Offices of the Adjutant-General and the Military Secretary and the GOCs of AIF Middle East, AIF Malaya, First and Second Australian Armies, New Guinea Force and III Australian Corps. These were consolidated into the 'Australian Military Forces – Appointments, Promotions, etc., List', the first of which was issued on 28 Oct 1943.
14. Butler, 'The youngest CO: Charles Green', Butler, Argent & Shelton, *The Fight Leaders*, pp. 47–8.
15. Peter Brune's reverence for Honner is born of many years of close association and is well known among the Australian military history community. Bill Edgar 'met Brigadier Potts and was inspired by his dignified vitality' (Edgar, *Warrior of Kokoda*, back cover). Green's biography was written by his widow, Eather's by a distant relation but principally informed by Eather's stepson, (Jones (Eather's great niece), interview, 5 Mar 2004) and Galleghan's by one of his sergeants.
16. Clift, *War Dance*.

17. Rhoden, interview.
18. Budge, 'Tribute to a CO', p. 11.
19. Savige, 'Notes on Volume V. Chapter 3.' (comments on draft of *The New Guinea Offensives*), p. 24; AWM 67, item 3/348, part 2.
20. CARO Dossier VX104143, Falconer, A. H.
21. Corfield, *Hold Hard, Cobbers*, pp. 193–4, 208, 221–2, 224–5, 236, 246, 249, 261–3.
22. Bean, *The Australian Imperial Force in France, 1916*, pp. 707–8. An example of Bean's treatment of another CO – Owen Howell-Price of the 3rd Battalion – can be found on pp. 578–9 of the same volume.
23. This discussion has been partly informed by consultation with Professor Jeffrey Grey, who has conducted extensive research into the production of Second World War official histories in Australia, Canada, Great Britain, New Zealand, South Africa and the United States.
24. Oakes, 'Chapter 13' (comments on draft of Wigmore's *Japanese Thrust*), pp. 1–2, AWM 67, item 3/291.
25. Louch to Long, 22 Jan 1950, pp. 1–2, AWM 67, item 3/223.
26. Jeanes, 'Chapter 19. Pages 18–20', p. 1, AWM 93, item 50/2/23/633.
27. Long, *The Final Campaigns*, p. 461.
28. Wootten to Long, 5 April 1948, p. 4, AWM 67, item 3/435.
29. Windeyer, 'Comments on draft of chapter 8 – "The Pacific Front"', pp. 2–3, AWM 93, item 50/2/23/250.
30. Dexter, *The New Guinea Offensives*, p. 229.
31. An award recognising such distinguished conduct that it is made immediately following an action, rather than the periodic lists in which such decorations are usually promulgated.
32. Galleghan, 'History of the War in Malaya. Comments on draft chapters [sic] 11 by Brigadier F.G. Galleghan', pp. 8–9, AWM 67, item 3/140.
33. Evans, 'The tyranny of dissonance: Australia's strategic culture and way of war 1901–2005', p. 52.
34. Barrett, *We were There*, pp. 338, 366.
35. Johnston, *At the Frontline*, pp. 151–4.
36. Barter, *Far Above Battle*, pp. 54, 55.
37. DSO citation, NX231, Chilton, F. O., CARO dossier NX231, Chilton, F. O.
38. Beaumont, *Gull Force*, p. 36.
39. Beaumont, pers. comm.
40. Stanley, *Tarakan*, pp. 62, 90, 196.
41. For example, see the discussion of the situation faced by Heathcote 'Tack' Hammer on the night of 30 October 1942 in Johnston and Stanley, *Alamein*, p. 220.
42. Brune, *Those Ragged Bloody Heroes* (1992); *Gona's Gone!* (1994); *The Spell Broken* (1997); *A Bastard of a Place* (2003).
43. For example, see the discussion of the decisions made by James Dobbs, CO 2/10th Battalion, before the fight for KB Mission in Brune, *A Bastard of a Place*, pp. 320–3; or that of the actions of Geoff Cooper, CO 2/27th Battalion, at Gona in Brune, *Those Ragged Bloody Heroes*, pp. 232–5.
44. Brune, *Gona's Gone!*, pp. 93–4.

45. See, for example, the discussion of Lieutenant Colonel Robert Ainslie at Sattelberg: Coates, *Bravery Above Blunder*, p. 219.
46. Collins, 'Tactical command', in Brown and Matthews (eds), *The Challenge of Military Leadership*.
47. AWM 52, 2nd AIF and CMF Unit War Diaries, 1939–45 War.
48. In the case of units taken prisoner the diaries were compiled months later in prisoner of war camps, although, particularly in Singapore, considerable efforts were made to interview battalion personnel to ensure an accurate and cohesive narrative.
49. Thomas, interview with Long; AWM 67, item 1/4, p. 59; McAllester, interview, 1 Jun 2003. After meeting John Treloar, CO of the War Records Section during the First World War, Jim McAllester maintained a very detailed war diary while IO of the 2/14th Battalion. He recalled, however, that his CO, William Cannon, was not happy with some of what he wrote and made many changes.
50. Thyer to Wigmore, 2 May 1952, p. 1, AWM 93, item 50/2/23/480; Travers, interview Long, 30 Oct 1944, AWM 67, item 2/105, pp. 4–5.
51. Sweeting cited in Maclean, *A Guide to the Records of Gavin Long*, p. 15.
52. AWM 67, *Official History, 1939–45 War: Records of Gavin Long, General Editor*, subseries 2.
53. Thomson, *Anzac Memories*, p. 228.
54. Ibid., pp. 229–34.
55. Van Creveld, *Command in War*, p. 5.
56. Sheffield, 'Command, leadership and the Anglo-American experience', in Sheffield (ed.), *Leadership and Command*, p. 1.
57. Australian Army, *The Fundamentals of Land Warfare*, p. 31.
58. 'Spartacus', 'Command and leadership: A rational view', p. 24.
59. *Army Doctrine Publication*, Vol. 2, *Command*, pp. 1.2–1.3.
60. Van Creveld, *Command in War*, p. 6.
61. *Command*, pp. 1.4–1.5
62. Klein, *Sources of Power*, pp. 31–3, cited in Graves, 'Choosing the best: Battalion command and the role of experience'.
63. Sheffield, 'Command, leadership and the Anglo-American experience', p. 2.
64. Van Creveld, *Command in War*, p. 10.
65. Checklists of characteristics or qualities needed for successful command are utilised by Glyn Harper in *Kippenberger*, David Horner in *The Commanders* and, to some extent, by Peter Pedersen in *Monash as Military Commander*. An indicative list of the qualities of command appears in *The Fundamentals of Land Warfare* (1993), pp. 31–2, and includes: leadership, intellect, robustness, courage and resolution, boldness, professional knowledge, judgement, decisiveness and flexibility, integrity and imagination.
66. Van Creveld, *Command in War*, p. 11.
67. Gabriel, *Military Incompetence*, p. 6.
68. Watson, *War on the Mind*. A similar point is made in Keegan, *The Mask of Command*, p. 325.
69. Burns, *Leadership*, p. 2, cited in Rocke, 'Trust: The cornerstone of leadership', p. 32.

70. 'Spartacus', 'Command and leadership: A rational view', p. 27.
71. Howard, 'Leadership in the British Army in the Second World War: Some personal observations', in Sheffield (ed.), *Leadership and Command*, p. 117.
72. 'Leadership', Human Service Department, Flathead Valley Community College, http://www.fvcc.edu/academics/dept_pages/human.services/leadership.htm, accessed 12 Jan 2005.
73. Larison, 'Course Syllabus, BA 321, Principles of Management, Eastern Oregon University School of Education and Business Programs', cited in Ambur, 'Reconsidering the higher-order legitimacy of French and Raven's bases of social power in the Information Age', http://users.erols.com/ambur/French&Raven.htm, accessed 20 Dec 2001, p. 3; Nicholson, 'Thoughts on followership', Part 2, p. 41.
74. Segal, 'Leadership and management: Organization theory', in Buck & Korb (eds), *Military Leadership*, pp. 42–3; Sherman & Schmuck, 'Kurt Lewin's contribution to the theory and practice of education in the United States', http://www.users.muohio.edu/shermalw/Lewin_Conference_ PaperV3.doc, accessed, 20 Apr 2005, p. 14.
75. Broedling, 'The psychology of leadership', in *Military Leadership*, p. 86.
76. 'Key leadership theories in the "Stimulating Leaders" Report', Manufacturing Foundation of the United Kingdom, http://www.manufacturingfoundation.org.uk/upload/leadership_theories.pdf, accessed 24 Apr 2005.
77. Most notable among the leadership models that provide templates or matrixes to follow in given situations are those of Fred Fiedler: 'Contingency theory'; Charles Handy: 'Motivational calculus'; Victor Vroom and Phillip Yetton: 'Normative decision making theory'; and Paul Hersey and Kenneth Blanchard: 'Situational leadership'. *Leadership Theory and Practice* (Australian Army, 1973), pp. 6–14–6–15, also provides advice on different styles of leadership to adopt in different military situations.
78. Broedling, 'The psychology of leadership', p. 73.
79. Ibid., p. 78. For instance, the landmark 'Hawthorne Studies' conducted at a US electricity plant between 1927 and 1932 found that individual workers would forgo personal benefits offered by management in order to remain accepted by their peers.
80. Holmes, *The Firing Line*, p. 47.
81. Nicholson, 'Thoughts on followership', Parts 1 & 2.
82. French & Raven, 'Bases of social power', in Cartwright (ed.), *Studies in Social Power*, pp. 150–67.
83. Bartlett, *Psychology and the Soldier*, cited in Sheffield, 'Command, leadership and the Anglo-American experience', p. 10.
84. *The Fundamentals of Land Warfare* (1993), p. 32.
85. Larson & Lafasto, *Teamwork: What Must Go Right, What Can Go Wrong*, pp. 85–7, cited in Rocke, 'Trust: The cornerstone of leadership', p. 30.
86. Rocke, 'Trust: The cornerstone of leadership', p. 33.
87. Baldwin, 'The leadership formula', p. 85.
88. Wilson, 'Theory X, Theory Y won't take the hill', p. 36; Stokesbury, 'Leadership as an art' in *Military Leadership*, p. 38; Nicholson, 'Thoughts

on followership', Part 2, p. 39; *Leadership Theory and Practice*, pp. 9–2; Daly, interview; Rhoden, interview.
89. Wilson to Kiggell, 15 Sep 1914, cited in Nicholson, 'Thoughts on followership', Part 2, p. 41.
90. Keegan, *The Mask of Command*, p. 329.
91. Wakin, 'The ethics of leadership', in Buck and Korb, *Military Leadership*, p. 95.
92. Keegan, *The Mask of Command*, p. 329.
93. Ibid., p. 11.
94. Baldwin, 'The leadership formula', p. 85.
95. *Leadership Theory and Practice*, pp. 6–13–6–15, for instance, defines three styles of leadership – authoritative, participative and free rein – derived from the work of Lewin, Lippitt and White. An authoritative leader retains complete control of all planning and decision-making and tasks; a participative leader consults his subordinates, involves them in decision-making and delegates authority to allow them to act as they see fit, in accordance with set guidelines and standard; and a leader giving his subordinates free rein allowed them to act at will with no direction or standards.
96. Nicholson, 'Thoughts on followership', Part 2, p. 40.
97. Keegan, *The Mask of Command*, p. 316; Johnston, *At the Front Line*, p. 71.
98. Kirkland, 'Combat leadership styles: Empowerment versus authoritarianism', p. 61.
99. Burns, *Leadership*, p. 443, cited in Funnell, 'Leadership: Theory and practice', p. 13.
100. Burns, *Leadership*, p. 19, in Funnell, 'Leadership: Theory and practice', p. 12.
101. Laffin, *Digger*, pp. 170–1.
102. Bridge, Harper, Spence, 'The Australian way of war: Some thoughts', p. 9.
103. Johnston, *At the Front Line*, p. 71.
104. Bridge, Harper & Spence, 'The Australian way of war', p. 13.
105. Hayes, 'A search for balance: Officer selection in the wartime Canadian Army, 1939–1945', p. 26.
106. Harper, 'A New Zealand way of war and a New Zealand style of command?', in Harper & Hayward, *Born to Lead?*, pp. 33–5.
107. For example, Faris Kirkland's discussion of German command practice in the Second World War emphasises trust across all ranks, the decentralisation of authority and the education of junior leaders (Kirkland, 'Combat leadership styles: Empowerment versus authoritarianism', pp. 61–72). Dominick Graham and Shelford Bidwell's comments on the German Army also have a familiar ring to them: 'it was also the duty of every man, private first class or general, to *lead* if the situation demanded it' (Graham & Bidwell, *Fire-Power*, p. 215).
108. Harper, 'A New Zealand way of war and a New Zealand style of command?', p. 36.
109. Hay, *Nothing Over Us*, p. x.

1 'Completely untrained for war': Battalion command in the pre-war army

1. Grey, *A Military History of Australia*, p. 138.
2. 'Basis of expansion for war', *Australian Army Journal*, No. 12, May 1950, p. 7; Pratten, 'Under rather discouraging circumstances'; Neumann, 'Australia's citizen soldiers, 1919–1939', p. 212; Grey, *A Military History of Australia*, pp. 136–40; Palazzo, *Defenders of Australia*, pp. 68–87; Wilcox, *For Hearths and Homes*, pp. 92–5, 101–2, 106–7.
3. Transcript of recruiting speech delivered at Kew (Victoria) Town Hall on 24 Aug 1936, p. 2, AWM 49, item 144.
4. 'First Report, 1938, by Lieutenant General E. K. Squires, Inspector General of the Australian Military Forces' (Squires Report), p. 6, AWM 1, item 20/11.
5. Grey, *A Military History of Australia*, p. 140.
6. *Australian Military Regulations and Orders 1927* (AMR&O), para. 114; MBA WN9/1939 'Extensions of Periods of Commands', NAA A2653, item 1939, vol. 1.
7. T. W. Bartley, CO 40 Bn, May 1938–Jul 1945; H. R. Birch, CO 22 Bn, Aug 1939–Jul 1940; R. H. Browning, CO 54 Bn, Oct 1922–Dec 1925, Sep 1939–Sep 1940; C. E. Cameron, CO 18 Bn, Jul 1933–Jun 1935, Oct 1936–May 1940, CO 18/51 Bn, Jul 1935–Oct 1936; T. P. Cook, CO 5 Bn, May 1937–Oct 39; P. V. O. Fleming, CO 15 Bn, Aug 1939–May 1941; F. G. Galleghan, CO 2/41 Bn, Aug 1932–Jan 1934, CO 2/35 Bn, Jan 1934–Mar 1937; A. H. L. Godfrey, CO 23 Bn, Mar 1927–Mar 1932, CO 23/21, Aug 1935–Oct 1939; R. J. M. Gordon, CO 32 Bn, Aug 1939–Oct 1941; N. G. Hatton, CO 25 Bn, Mar 27–Jul 1930, Oct 1934–May 1940; W. D. Jeater, CO 2 Bn, Mar 1937–Jul 1940; J. E. G. Martin, CO 42 Bn, Aug 1937–Oct 1939; N. A. McArthur, CO 6 Bn, Mar 1938–Apr 1942; E. G. H. McKenzie, CO 44 Bn, Jan 1936–Dec 1939; M. J. Moten, 43/48 Bn, Dec 1936–Jul 1939, CO 48 Bn, Jul 1939–Apr 1940; P. A. Parson, CO 4 Bn, Jul 1934–Oct 1939; E. E. Patterson, CO 47 Bn, Nov 1936–Aug 1942; J. W. Russell, CO 30 Bn, Jul 1939–Apr 1942; J. W. Swatton, CO 38/7 Bn, Jul 1931–Nov 1936, CO 38 Bn, Nov 1936–Sep 1940; T. G. Walker, CO 24 Bn, Jun 1935–Aug 1939, CO 24/39 Bn, Aug 1939–Oct 1939; and K. H. Ward, CO 20/54 Bn, Jul 1937–Aug 1939, CO 20/19 Bn, Sep 1939–Sep 1941.
8. T. G. Walker (CO 24 Bn, Jun 1935–Aug 1939; CO 24/39 Bn, Aug 1939–Oct 1939) enlisted in 2/24 Bn in 1918.
9. *Defence Act 1903–41*, sect. 20A.
10. Both Howard Carr (CO 46 Bn, Nov 1938–Jul 1940) and Hubert Harris (CO 51 Bn, Nov 1936–Mar 1942) were aged 40 in 1939 and would have been 19 in 1918. This age, however, was still too young to enlist without parental consent.
11. Derived from a survey of command selection files from the interwar period contained in NAA MP 508/1.
12. This total excludes the Melbourne University Rifles and the Sydney University Regiment.
13. Pratten, 'Under rather discouraging circumstances', p. 11.
14. AMR&O, para. 155.

15. *Defence Act, 1903–41*, sect. 27.
16. Fuller, *Generalship, Its Diseases and Their Cure*, p. 32.
17. Compiled from *Australian Imperial Force, Gradation Lists of Officers and Lists of Honours and Awards for Service in the Field: July, 1918* and *Australian Imperial Force, Staff and Regimental Lists of Officers: August, 1918*.
18. Squires Report, Appendix 2, p. 2.
19. Ibid.
20. H. W. Murray, CO 4 MG Bn, Mar 1918–20, CO 26 Bn, Jul 1939–Aug 1942.
21. Francis North was the AMF's longest serving CO in 1939, having been first appointed to command the 31st Battalion, as a captain, in December 1924. North would remain in command until promoted to brigadier in August 1940.
22. Compiled from *The Army List of the Australian Military Forces*, Part 1: *Active List* (1935).
23. 3rd Div Confidential Minute 10/39, 'Command: 24th Battalion', 17 Feb 1939, NAA B1535, item 709/11/310.
24. *Defence Act 1903–41*, sect. 21A (2).
25. The 21A Course was named after the section of the Defence Act that prescribed it.
26. 'Tactical exercise without troops: 21A course for promotion to Lt Col, Camden, May 1930', AWM 1, item 17/11.
27. *Defence Act 1903–41*, sect. 21A (2).
28. Fitness to command certificate, CARO dossier VX32, Walker, T. G.
29. Lavarack, 'Training of commanders and staff: Establishment of Command and Staff School', pp. 2–3, NAA A9787/1, item 46; Squires Report, p. 8.
30. 'Tactical exercise without troops: 21A course for promotion to Lt Col, Camden, May 1930', AWM 1, item 17/11.
31. Jackson, 'Autobiography', p. 57.
32. Neumann, 'Australia's citizen soldiers, 1919–1939', p. 212.
33. 'Army Headquarters Exercise, Duntroon, October, 1937', AWM 1, item 17/11.
34. 'Armour', in Dennis, Grey, Morris, Prior (eds), *The Oxford Companion to Australian Military History*, p. 50; Hopkins, *Australian Armour*, pp. 15–23.
35. Lavarack, 'Training of commanders and staff: Establishment of Command and Staff School', p. 3.
36. French, *Raising Churchill's Army*, pp. 60–1.
37. Lavarack, 'Training of commanders and staff: Establishment of Command and Staff School', p. 3.
38. Lavarack, 'Training of commanders and staffs', MBA 88/1935, pp. 1–2, NAA A2653/1, item 1935, Vol. 1.
39. Sublet, interview, 12 Sep 2000.
40. Field, 'Warriors for the working day', p. 3.
41. 'The basis of expansion for war', p. 7.
42. Personnel Return – 24 Bn, Mornington, 16 Feb 1938; Personnel Return – 37/39 Bn, Mornington, 9 Feb 1938, AWM 62, item 27/4/749.

43. 'Senior Officers' Exercise – Glenbrook Area. 17–22 Feb 36', p. 1, AWM 62, item 27/6/1130.
44. Between 1931 and 1935 the turnover rate of personnel in the militia reached 149 per cent. In December 1935 only 20 per cent of militia infantry had more than three years experience: Pratten, 'Under rather discouraging circumstances', p. 26; 57/60 Bn 38/128, 'Annual Training Report: 57/60 Bn: 1937/38', 21 May 1938, p. 1, AWM 62 27/4/737.
45. 29/22 Bn 37/142/36, 'Annual Training Report Pt I', 25 May 1938; 'Part III. Statistical Training Report. For year ending 30th June 1938', AWM 62, item 27/4/737.
46. Lavarack, 'Training of commanders and staffs', pp. 2–3.
47. Ibid., p. 2.
48. See, for example, Wilcox, *For Hearths and Homes*, pp. 102–3.
49. Lavarack, 'Training of commanders and staffs', p. 3.
50. Lavarack to Military Board, 12 Oct 1936, NAA A2653, item 1936, Vol. 1.
51. Lavarack, 'Training of commanders and staff', p. 3.
52. MBA 84/1938, NAA A2653, 1938, Vol. 1; Military Board Instruction (MBI) 44/1938.
53. Lavarack, 'Training of commanders and staff', pp. 2, 4.
54. Chilton, 'A life in the twentieth century', p. 15.
55. MBI 44/1938, 59/1938.
56. NAA MP 70/6, item a42 cited in Neumann, 'Australia's citizen soldiers, 1919–1939', p. 205.
57. Neumann, 'Australia's citizen soldiers, 1919–1939', p. 207.
58. AWM 62, 27/1/267, 'One sided outdoor skeleton force tactical exercises'.
59. Lloyd, interview, Jul 1994.
60. Palazzo, 'The way forward: 1918 and the implications for the future', p. 71.
61. French, *Raising Churchill's Army*, pp. 62–3.
62. Deverell to Liddell Hart, 29 Dec 1936, LHCMA Liddell Hart MSS 1/232/16 in French, *Raising Churchill's Army*, p. 63.
63. Dennis, *The Territorial Army 1906–1940*, pp. 149, 218–19.
64. Ibid., p. 151.
65. French, *Raising Churchill's Army*, p. 63.
66. *Regulations for the Territorial Army (including the Territorial Army Reserve) and For County Associations, 1936*, pp. 416–27.
67. Ibid., p. 127.
68. Various 'Annual Confidential Report on Officers' and AAF A8, 'Recommendation for appointment' held in AWM 49, CARO personal dossiers, and NAA B1535, MP 508/1 and SP 1048/6.
69. Blamey cited in 'Growth of army: Public interest needed', Argus, 25 Jun 1935, p. 15.
70. 3 Div Confidential 42/39, NAA B1535, item 709/11/343.
71. 1 Div 407/4/889, NAA B1535, item 709/10/336.
72. Major Eric Bundey (grad. RMC 1919) took command of the 56th Battalion, headquartered at Cootamundra, on 14 Dec 38, and Major Lindley Barham (grad. RMC 1921) took command of the 41st Battalion, headquartered at Lismore, on 16 Jan 1939: 2 Div Confidential 2/1/63, 'Command 41 Bn', NAA SP1048/6, item C2/1/76.

73. MC Citation, Lt D. S. Maxwell, CARO dossier NX12610, Maxwell, D. S.
74. Maxwell, interview with Long, 15 Feb 1953, p. 1, AWM 67, item 2/261.
75. 'Confidential Report on Officers, Major P. D. McCormack, 22 Mar 1937', NAA MP 508/1, item 251/704/37.
76. 1 Div 407/4/889, NAA B1535/0, item 709/10/336.
77. 'Proceedings of Promotion and Selection Committee', 11 Feb 1938, NAA MP508/1, item 251/704/25.
78. HQ 4 MD, Confidential 16/1934, 'Command – 27th Battalion', NAA MP508/1, item 251/704/25.
79. Peterson, interview, 10 Feb 2001.
80. Morgan, 'Old Scotch Collegians Platoon – 39th Battalion', AWM 49, item 144; Clarebrough to Butler, 22 Nov 1938, AWM 49, item 146.
81. Morgan, 'Old Scotch Collegians Platoon – 39th Battalion'.
82. Clarebrough to Butler, 22 Nov 1938.
83. Neumann, 'Australia's citizen soldiers, 1919–1939', pp. 182–4.
84. Pratten, 'Under rather discouraging circumstances', pp. 13–14.
85. R. E. Pratten, interview, 1992 in 'Under rather discouraging circumstances', p. 14; Corfield, *Hold Hard, Cobbers*, p. 23. Britain's Territorial Army experienced difficulties in recruiting enough officers owing to the cost of being an officer, and there is little doubt cost also imposed a de facto class restriction on its officer corps. See Dennis, *The Territorial Army*, pp. 154–5.
86. Pratten, 'Under rather discouraging circumstances', p. 14.
87. The membership of the League of National Security (White Army) in Victoria and the Old Guard in New South Wales included Major Generals Thomas Blamey and Gordon Bennett, one-time commanders of the 3rd and 2nd Divisions respectively; Colonel Frank Derham, commander of the 3rd Division's artillery; Edmund Herring, CO of several field artillery brigades; Colin Simpson, CO of the 39th Battalion; Jack Clarebrough, CO of the Melbourne University Rifles (MUR) and the 39th Battalion; Neil McArthur, CO of the 6th Battalion, Major Leonard Roach, 2iC of the 39th Battalion, Major Donovan Joynt, VC, 2iC of the 6th Battalion; Major William Scott, a former CO of the 46th Battalion; Major George Wootten, who, although on the Unattached List for most of the 1930s, took command of the 21st Light Horse Regiment in 1937; and Lieutenant John Wilmoth, a platoon commander in MUR (Cathcart, *Defending the National Tuckshop*, pp. 55–6; Moore, *The Secret Army and the Premier*, pp. 3, 89, 92; Horner, *Blamey*, pp. 96–7; 'White Army' (notes), AWM 3DRL 6224, item 1; Coulthard-Clark, *Soldiers in Politics*, p. 174).
88. *Australian Concise Oxford Dictionary of Current English*, p. 38.
89. MBA 9/1936, 'Training of Commanders and Staffs', p. 3, NAA A2653, item 1936, Vol. 1.

2 The foundations of battalion command: Forming the 2nd AIF, 1939–40

1. Austin, *Bold, Steady, Faithful*, p. 147.
2. Allen, interview with Long, 27 Dec 1944, p. 1, AWM 67, item 2/66. This officer was David Whitehead, CO of the 1st Machine Gun Regiment.
3. Blamey to Allen, 9 Oct 1939; AWM 3DRL 4142.

4. Stevens, 'A personal story of the service, as a citizen soldier, of Major-General Sir Jack Stevens, KBE, CB, DSO, ED', unpublished MS, AWM 3DRL 3561 (hereafter 'A personal story'), p. 21.
5. Including SUR.
6. Pratten, 'The "old man": Australian infantry battalion commanders in the Second World War', p. 95. The Australian Overseas Base was the administrative organisation that would support the AIF once it deployed to the Middle East.
7. Allen, interview, pp. 1–2.
8. Blamey to Allen, 9 Oct 1939; Long, notebook 66, p. 2a, AWM 67, item 2/66.
9. 'Confidential Report on Officers, Lieutenant Colonel K. H. Ward, 9 Sep 1941', NAA SP1048/6, item C2/1/339.
10. Arneil, *Black Jack*, p. 58. There were at least five lieutenant colonels on the Active List of the AMF in 1939 who were more senior to Galleghan, who was promoted to this rank in August 1932. They included (year in brackets indicates date of promotion to lieutenant colonel): W. H. Douglas (Mar 1932); A. H. L. Godfrey (1927); G. H. Knox (1918); J. W. Mitchell (1918); P. J. McCormack (1927); F. R. North (1927).
11. Arneil, *Black Jack*, p. 58; Campbell, 'A model for battalion command', p. 40.
12. Pratten, 'The "old man"', p. 98.
13. Savige, interview, p. 71. The connection between Blamey is unclear, but the fact that Mitchell, at 48, was oldest battalion CO in the 6th Division and would remain in command despite being charged with embezzlement in 1940 suggests it was strong.
14. Field, diary, 17 Sep 1939, AWM 3DRL 6937.
15. Pratten, 'The "old man"', p. 100.
16. Stevens, 'A personal story', p. 29.
17. Ibid.
18. Calculated from the number of Australian males aged 16–29 attending university in 1933 (5854) compared to the wider male population in this age group (822,965). The 1933 Commonwealth Census, the last completed before the Second World War, grouped all university attendees older than 29 into a single total – '30 and over' – of 398; Census of the Commonwealth of Australia, 30th June, 1933 (hereafter Commonwealth Census 1933), p. 1152.
19. All figures calculated from Commonwealth Census 1933. For full breakdown of figures see appendix 1.
20. Based on 'Australia. Schooling – Persons receiving instruction according to age', Commonwealth Census 1933, p. 1152.
21. J. L. A. Kelly, CO 31/51 Bn, Nov 1944–Jan 1946.
22. Lodge, *Lavarack: Rival General*; Horner, 'Staff Corps versus militia', pp. 14–15.
23. Horner, 'Staff Corps versus militia', p. 14.
24. Long, *To Benghazi*, p. 44.
25. Lodge, *Lavarack*, p. 78.

26. Coulthard-Clark, *Duntroon*, p. 142.
27. 'Raising of a second division for service overseas, and the dispatch of army co-operation squadrons RAAF for service overseas', appendix B2, NAA A5954, item 261/8.
28. In their original forms, the 7th and 8th Divisions included only one regular officer each, although it must be noted that Major General Gordon Bennett, GOC of the 8th Division, was an outspoken critic of the command abilities of the Staff Corps: R. F. 'Spike' Marlan, CO 2/15 Bn, Apr 1940–Apr 1941, and A. H. 'Sapper' Boyes, CO 2/26 Bn, Oct 1940–Feb 1942. At its formation the 9th Division included two PMF COs: R. Bierwirth, CO 2/33 Bn, Jun 1940–Jun 1941, and W. W. Crellin, CO 2/43 Bn, Jul 1940–Nov 1941.
29. 'History of the Activities of the Office of the Military Secretary during the War 1939–1945', p. 85; Horner, 'Staff Corps versus militia', p. 21.
30. Horner, 'Staff Corps versus militia', p. 21.
31. MacArthur, letter of commendation, CARO dossier NX65, Loughran, L. J.
32. Potts to Military Secretary, 1 Mar 1946, CARO dossier NX65, Loughran, L. J.
33. Marshall, diary, AWM 54, item 255/4/12, pp. 7–8.
34. Johnston, *At the Front Line*, p. 104.
35. AMR&O, para. 1423.
36. The Defence Act, Sect. 58; Regulations and Orders for the Australian Military Forces, 1927 (AMR&O), paras 1316, 1423–1433.
37. AMR&O, para. 300.
38. AMR&O, Regs 237–256.
39. NAA MP 742/1, item 85/1/612 cited in Johnston, *At the Front Line*, p. 152.
40. Godfrey, 'Anzac Day', WD 2/6 Bn, Jul 1940, AWM 52, item 8/3/6.
41. Parry-Okeden, diary, 20 Sep–15 Oct 1944, AWM PR, item 00321.
42. Smith, '2/6th Aust. Inf. Bn AIF', p. 44, AWM PR, item 85/223. Arthur Key took similar action against his troops after a spate of absences without leave when the 2/14th Battalion docked in Fremantle in August 1942 (Russell, *The Second Fourteenth Battalion*, pp. 104–5).
43. For example, John Field had bandsmen in the 2/12th Battalion confined to barracks for seven days for accepting glasses of beer proffered from the crowd during a march through Sydney in March 1940. Similarly, Vivian England of the 2/3rd earned a reputation for particularly harsh sanctions in Palestine in 1940 by routinely dispatching leave defaulters to the 1st Australian Detention Barracks in Jerusalem, an institution with a reputation for military discipline of the most strict kind that often verged on sadism (Field, diary, 28 Mar 1940, AWM 3DRL 6937, item 2; Clift, *War Dance*, p. 30; Wahlert, *The Other Enemy?*, pp. 135–7).
44. Chilton, interview; Daly, interview; Isaachsen, interview, 6 Feb 2001; Rhoden, interview, October 2000; Wilmoth, interview, 27 Sep 2000; Parry-Okeden, diary, 20 Sep–15 Oct 1944, AWM PR, item 00321.
45. Chilton, interview.
46. Daly, interview.
47. Johnston, *At the Frontline*, pp. 151–2.
48. AMR&O, paras 300, 1433.

49. Chilton, interview.
50. Rhoden, interview.
51. Givney, *The First at War*, p. 52; Burfitt, p. 29; Daly, interview; Johnson, *The 2/11th (City of Perth) Australian Infantry Battalion, 1939–45*, p. 2; Clift, *War Dance*, p. 30; Louch, *From the CO's Notebook*, p. 2; Arneil, *Black Jack*, p. 67.
52. Daly, interview.
53. Givney, *The First at War*, p. 14; Eather, *Desert Sands, Jungle Lands*, pp. 13–14.
54. Austin, *Let Enemies Beware*, p. 12.
55. Brigg, *The 36th Australian Infantry Battalion 1939–1945*, p. 12.
56. Fred Chilton reflected on the expectations of the men in the 2/2nd Battalion: 'A lot of the troops were pretty wild characters but they expect a lot better from their officers ... The things they do they don't expect their officers to do' (Chilton, interview).
57. Daly, interview; Chilton, interview; Rhoden, interview; Wilmoth, interview.
58. Lambert, *The Twenty Thousand Thieves*, pp. 30–1, 45–6, 73–4, 177, 264–9.
59. Rhoden, interview.
60. Lambert, *The Twenty Thousand Thieves*, p. 232; R. Cowie, interview, 15 Feb 2001.
61. Rhoden, interview; Wilmoth, interview; Hamilton, interview, Jul 2001.
62. Daly, interview.
63. Cowie, interview.
64. Circular No. 1, 'Notes on allotment of duties in an inf bn: regimental duties', p. 1, WD 27 Bde, Nov 1940, AWM 52, item 8/2/27.
65. *Field Service Regulations*, Vol. 1: Organization and Administration, 1940, p. 8.
66. *Infantry Training: Training and War*, 1937, p. 60.
67. Dwyer, 'Sergeants are snakes', *Sydney Morning Herald*, 2 Dec 1944.
68. Dunlop, 'Notes for Commanding Officers', p. 30, AWM 3DRL 6850, item 98.
69. Jackson, 'Autobiography', p. 100.
70. Ibid., p. 100.
71. Jackson, 'Autobiography', p. 219.
72. Hay, *Nothing Over Us*, pp. vii, 6–7, 43, 220–1.
73. Uren, *A Thousand Men at War*, p. 5.
74. Burns, *The Brown and Blue Diamond at War*, p. 3.
75. Daly, interview.
76. One of the 2/10th's former sergeants has commented that Verrier 'wasn't held in very great esteem as a soldier' (Peterson, interview, 10 Feb 2001).
77. Cundell, pers. comm., 3 Jan 2004.
78. Matthews, diary, 10 Aug 1940, AWM PR, item 89/79, folder 1.
79. Peterson, interview. Verrier's popularity with many of his troops is confirmed by several sources: Cundell, pers. comm., 3 Jan 2004, p. 2; Allchin, *Purple and Blue*, p. 189; Daly, interview.
80. Matthews, diary, May–Dec 1941.
81. Chilton, interview, 25 Oct 2000.
82. *Instructions for Medical Officers (Australia)*, 1942, p. 5.

83. AMR&O, para. 1425.
84. Braithwaite, 'The Regimental Medical Officer', p. 2, AWM 3DRL 6937, item 6; *Instructions for Medical Officers (Australia), 1942*, p. 4.
85. Dunlop, 'Notes for Commanding Officers', p. 34.
86. Rhoden, interview; Brune, *We Band of Brothers*, p. 150.
87. Matthews, diary, 1 Jun, 30 Oct, 3 Nov 1943; Wilmoth, interview; Rhoden, interview.
88. Chilton, interview.
89. Rhoden, interview.
90. 'NX—','All about officers', in *Jungle Warfare*, p. 141.
91. Routine Orders – Part 1, p. 2, WD 2/8 Bn, Feb 1940, AWM 52, item 8/3/8.
92. 'Statement of Numbers of Officers Volunteered for 2 AIF', WD HQ 17 Bde, Oct 1939, AWM 52, item 8/2/17. Conversely, the numbers of officers who volunteered for the AIF from militia units whose COs who had not volunteered or who had been rejected was quite low. There were suggestions that several of these COs actively discouraged their officers from volunteering ('Appointment of Officers, 2nd AIF: Military Board letter 39625 of 8 Nov 1939', WD 6 Div GS, Nov 1939, AWM 52, item 1/5/12; Field, diary, 17 Oct 1939; Allen, interview, pp. 1a, 2).
93. Hamilton, interview.
94. Fearnside (ed.), *Bayonets Abroad*, p. 2.
95. AMR&O, para. 311.
96. Johnson, *The 2/11th (City of Perth) Australian Infantry Battalion, 1939–1945*, pp. 4–5.
97. Dwyer, 'Sergeants are snakes'.
98. AMR&O, paras 1431–1432.
99. 'Training to 30 Nov 39', pp. 2–3, WD HQ 17 Bde, Nov 1939, AWM 52, item 8/2/17.
100. Field, diary, 24 Jan 1940. Morshead appears to have had some strange ideas regarding the type of fighting in which his brigade would be involved. The 18th Brigade war diary records that in an exercise in April 1940 he has his battalions exercising against several 'tribes' (WD HQ 18 Bde, 19 Apr 1940, AWM 52, item 8/2/18).
101. Field, diary, 31 Jan 1940.
102. Ibid., 5 Feb 1940.
103. Allen, interview, p. 31.
104. 'Questions asked of Bn Comds by Bde Comd in bivouac area at 2030 hrs. 14 Aug 40', WD HQ 21 Bde, 16 Aug 1940, AWM 52, item 8/2/21.
105. WD HQ 21 Bde, 16 Aug 1940, AWM 52, item 8/2/21.
106. 'No. 14 Course – Company Commanders (2nd AIF) Held at the Command and Staff School, from 4th to 22nd December, 1939', NAA MP 508/1, item 323/752/23.
107. 'No. 15 Course – Company Commanders (2nd AIF)', NAA MP 508/1, item 323/752/24.
108. 'Report on No. 14 Company, etc. Commanders (2nd AIF) Course – Command and Staff School (25 Jan 1940)', NAA MP 508/1, item 323/752/23.
109. WD HQ 18 Bde, 19 Apr 1940, AWM 52, item 8/2/18.

110. Unit History Editorial Committee, *White over Green*, p. 12.
111. WD HQ 16 BDE, 6 Jun 40, AWM 52, item 8/2/16.
112. Ibid., 17 Jul 40.
113. Stevens, 'A personal story', p. 24.
114. CARO Dossier, NX5, Parsons, P. A.
115. WD HQ 19 BDE, Dec 40, AWM 52, item 8/2/19. Louch, *From the CO's Notebook*, Nov 1940, p. iii. Louch discussed the capture of Cook by his companies: 'One Bn HQ which was sited in a rather conspicuous position on the railway line was raided twice by patrols from B and C companies; and when it was raided a third time by another patrol from C company the Bn commander complained that he had been assured by Capt. Honner that he would not be captured again. Anticipating that the 17th Bde might be sending out similar patrols I planted myself and my Bn HQ in the middle of the reserve company with instructions not to let anyone through on any account.'
116. Pratten, 'The "old man"', pp. 122–3.
117. Field, diary, 20 Oct–13 Dec 1941. 'Middle East Tactical School. Programme of work – Senior Wing – Seventh Course', AWM 54, item 941/1/13.
118. Field, 'Warriors for the working day', unpublished MS, chapter 8, p. 22, AWM PR, item MSS00785; WD 6 Royal Tank Regt, 16 Feb 1940; www.warlinks.com/armour/, accessed 19 Oct 2008.
119. Eather to sister (no further reference) in Eather, *Desert Sands, Jungle Lands*, pp. 22–3; Cullen, interview; Field, diary, 20 Oct–13 Dec 1941.
120. AWM 54, item 941/1/13; AWM 3DRL 6599, items 51–52. See, for example, the notes discussing lightly armed 'Jock' Columns as the answer to a 'properly organized and active defence': Middle East Tactical School – Senior Wing précis, 'Notes on a brief survey of tactics', p. 7, AWM 3DRL 6599, item 52; or the nebulous discussion of infantry-armour cooperation that concluded: 'A new method of tactical employment will therefore probably have to be worked out. The method outlined above should therefore not be accepted in any way as concrete': Middle East Tactical School – Senior Wing précis, 'Armoured Troops', p. 10; AWM 54, item 877/4/2.
121. Field, 'Warriors for the working day', chapter 8, p. 22.
122. Field wrote of Hamilton, 'His analysis of solutions propounded by students was incisive but gave due weight to the fact that since war is not a science but an art, a given situation might admit of more than one answer offering success. However, there could be no tampering with tested principles, however much the permutations and combinations of them offered variants of action' (Field, 'Warriors for the working day', chapter 8, p. 10).
123. French, *Raising Churchill's Army*, p. 233.
124. Long, *To Benghazi*, p. 126; Martin, interview with Long, n.d., AWM 67, item 1/2, p. 77; Allen, interview with Long, 27 Dec 44, AWM 67, item 2/66, pp. 19–20; Long, diary, 26 July 1943, AWM 67, item 1/2, p. 49.
125. '"Div" Exercise. 20/23 Nov 40', p. 2, WD HQ 17 Bde, Nov 40, AWM 52, item 8/2/17.
126. Eather interview with Long, 7 Aug 44, AWM 67, item 2/56, p. 20.

127. Jackson, 'Autobiography', p. 110. Jackson's comments are supported by the recollections of David Hay, an officer of the 2/6th Battalion. He stated that in 'general terms' the battalion was ready for action in late 1940 and had hit a 'peak of training', but continued that the battalion was not well prepared for the actual tasks it would confront in its first battles: 'It was a slight reflection on the training that went on beforehand' (Hay, interview with Welch in Pratten and Harper (eds), *Still the Same*, p. 78). Henry Gullett, a sergeant in the 2/6th, similarly commented: 'It was all fairly simple stuff, we didn't do any exercises with armour attached or under command, or working in co-operation with them. Neither did we do any exercises involving air support' (Gullett, interview with McClelland in Pratten and Harper (eds), *Still the Same*, p. 40).

3 'We were learning then': The AIF's Mediterranean campaigns, 1941

1. AIF Order No. 89, 1 Nov 1940.
2. Lambert, *The Twenty Thousand Thieves*, pp. 72–3.
3. Blamey, 'Organization of the 6th Division', 6 Div 68/3, 10 Nov 1939, p. 1, AWM 54, 721/2/6.
4. 2 AIF WE II/1940/12B/1, 'An Infantry Battalion', AWM 3DRL 6850, item 60; AMF WE II/1933/11/6, 'An Infantry Battalion', AWM 54, 327/6/6; J. C. W. Baillon, 'The New Infantry Organisation', p. 2, AWM 54, 721/8/11.
5. The rifle companies were originally identified by the first four letters of the British phonetic alphabet: Ace, Beer, Charlie and Don. In the interests of inter-allied communication, a new phonetic alphabet was adopted by the British Commonwealth forces in 1942; 'A', 'B' and 'D' becoming Able, Baker and Dog. This change caused a little disquiet in the Australian battalions, particularly among members of Don companies, owing to the negative connotations of the word 'dog' and generally companies continued to be known by their old designations.
6. Stanley, *Tarakan*, p. 34.
7. Baillon, 'The New Infantry Organisation', pp. 4–5; *Infantry Section Leading (Modified for Australia)*, 1939, p. 16.
8. The original British organisation for the 1938 pattern battalion specified eight-man sections, but when the 6th Division was reorganised it included nine-man sections. In 1940 Australian organisations were further amended to allow sections of up to 11 men (2 AIF WE II/1940/12B/1).
9. *Infantry Training: Training and War*, 1937, p. 3; *The Infantry (Rifle) Battalion, Part 1: Organization and Characteristics – The Carrier Platoon*, 1940, p. 5; Baillon, 'The New Infantry Organisation', p. 2; *Infantry Section Leading (Modified for Australia)*, pp. 31, 68.
10. English, *On Infantry*, pp. 52, 104–5; French, *Raising Churchill's Army*, p. 39.
11. *Infantry Section Leading (Modified for Australia)*, p. 66.
12. Baillon, 'The New Infantry Organisation', p. 6; *The Infantry (Rifle) Battalion, Part I: Organization and Characteristic – The Carrier Platoon*, p. 3.

13. Baillon, 'The New Infantry Organisation', pp. 6–7; *The Infantry (Rifle) Battalion*, Part I: *Organization and Characteristic – The Carrier Platoon*, pp. 8–12; 'The Carrier Pl (AIF Establishment)', p. 1, WD HQ 4 Div (GS), May 1942, AWM 52, item 1/5/8.
14. Many of these support and administrative personnel, such as the truck drivers, cooks and stretcher-bearers, were attached to the rifle companies.
15. Myatt, *The British Infantry 1660–1945*, pp. 200–1.
16. An example of a typical LOB group is provided by the 2/48th Battalion on 3 July 1942. The battalion's LOB Group consisted of 7 officers (1 major and 6 lieutenants), 5 sergeants (1 per company), 9 NCOs and 47 ORs (11 for HQ Company, 9 per rifle company) (H. H. Hammer, field notebook, AWM 3DRL 2642, item 1).
17. *Field Service Regulations*, Vol. 2: *Operations – General*, 1935, pp. 25, 61–2.
18. *Infantry Training: Training and War, 1937*, pp. 63–4.
19. Dunlop, 'Notes for Commanding Officers', p. 34, AWM 3DRL 6850, item 98; LHQ Tactical School précis, 'Part II – Control in Battle', p. 2, AWM 54, item 881/1/4, part 2; Matthews, 'Lecture to students at NGF School of Intelligence 22 June 43: What a CO expects from his 'I' Sec', AWM 3DRL 6937, item 19.
20. LHQ Tac School précis, 'Intelligence', p. 2, AWM 54, 881/1/4, part 1.
21. Liaison officers were intended to compliment the signals system and convey orders, instructions and information when time or the tactical situation prevented the use of normal channels of communication. They were also employed to provide an insight into the situation as it was perceived by the two liaising headquarters, a function described in a LHQ Tactical School précis as 'conveying local colour': LHQ Tactical School précis, 'Org of an Aust Inf Bn', p. 4, AWM 54, item 881/1/4, part 2; 'Circular 69. Instructions for Training. Training of Liaison Officers', WD HQ 24 Bde, Oct 1940, AWM 52, item 8/2/24.
22. *Infantry Training: Training and War, 1937*, p. 61; Dunlop, 'Notes for Commanding Officers', p. 24.
23. Dunlop, 'Notes for Commanding Officers', p. 24.
24. LHQ Tactical School précis, 'Part II – Control in battle', p. 1.
25. *Field Service Regulations*, Vol. 2, p. 26.
26. *Infantry Training: Training and War, 1937*, p. 57.
27. LHQ Tac School précis, 'Org of an Aust Inf Bn', p. 2.
28. LHQ Tactical School précis, 'Battle Procedure: Control in Battle: Infm and Recce. Part 1 – Battle Procedure', p. 1, AWM 54, item 881/1/4, part 2.
29. *Infantry Training: Training and War, 1937*, p. 57; *Infantry Training*, Part 1: *The Infantry Battalion, 1943*, p. 4; METS précis, 'Recce within the unit' (with notations by H. H. Hammer), AWM 54, item 941/1/13.
30. METS précis, 'Battle procedure and sequence of events', AWM 54, item 941/1/13.
31. *Infantry Training: Training and War, 1937*, p. 60; *Infantry Training*, Part I: *The Infantry Battalion, 1943*, p. 6; *Field Service Regulations*, Vol. 2, p. 29.
32. LHQ Tactical School précis, 'Part II – Control in battle', p. 1.
33. *Field Service Regulations*, Vol. 2, p. 25.

34. 'Copy of report on Lessons in Organization, Staff Duties and Minor Tactics from British Expeditionary Force', AWM 54, item 721/1/22; 'War Office London Secret letter – containing various recommendations under consideration for the reorganization of Army – resulting from experience gained in the recent campaign in French, 13 October 1940', AWM 54, item 721/12/28.
35. 'Notes on recently expressed concepts of tactics', p. 2, AWM 54, item 923/1/6; Chilton, 'Report on Operations of 2/2 Aust. Inf. Bn at Penios Gorge', p. 8, WD HQ 16 Bde, Mar–Apr 1941, AWM 52, item 8/2/16; *White Over Green*, p. 139.
36. Several conferences of former BEF COs recommended that there be no less than six 3-inch mortars in the battalion, and preferably eight. An independent Australian committee, convened by COs of the 18th Australian Brigade in the UK in 1940, also concluded that the numbers of 3-inch mortars in a battalion needed to be increased ('Consideration of Recommendations of Bartholomew Committee', p. 3, AWM 54, item 213/4/8; 'Lessons in Organisation, Staff Duties, and Minor Tactics from the BEF', p. 8, AWM 54, item 721/1/22).
37. One of the principal uses envisaged for the 2-inch mortar was to generate a smoke screen during an attack, so its first-line ammunition was 27 smoke bombs and only nine high-explosive bombs. A report compiled from several conferences convened among British Army COs after the campaign in France in 1940 stated: '2-in. mortars equipped only with smoke were useless though some units amused themselves by throwing smoke at the enemy. Recommended 80% HE bombs in future' ('Lessons in Organisation, Staff Duties, and Minor Tactics from the BEF', p. 8; Baillon, 'The New Infantry Organisation', p. 6; *Infantry Section Leading (Modified for Australia)*, p. 16; 2 AIF WE II/1940/12B/1, p. 6).
38. Glenn, *Tobruk to Tarakan*, p. 75.
39. 'Extracts from Draft Report by 30 Corps dated 21November 42', p. 1, Appendix A to '9 Aust Div Report on Operations El Alamein, 23 October 42–5 November 42', AWM 54, item 527/6/1A.
40. Baillon, 'The New Infantry Organisation', p. 3; *Infantry Training: Training and War, 1937*, p. 2.
41. *White over Green*, p. 139.
42. On one occasion, Chilton had to dispatch his carrier platoon in an effort to bring Bren gun fire to bear on an enemy party in the process of outflanking his position (Long, *Greece, Crete and Syria*, pp. 113–14). Reports from the campaign in Lebanon similarly discussed the need for the Vickers gun within infantry battalions ('Reports from the Syrian Campaign', WD 2/14 Bn, Aug 1941, AWM 52, item 8/3/14).
43. French, *Raising Churchill's Army*, p. 88.
44. Hay, *Nothing Over Us*, p. 102.
45. Contrary to much scholarship on the war in the Middle East, an HE shell was available for the 2-pounder anti-tank gun, although it was never issued to tanks (Hogg, *British and American Artillery of World War 2*, p. 75).

46. The term 'bunker buster' may seem anachronistic but in fact has its origins in the Second World War. See 'Tank Attack Guns – Infantry Battalions (Tropical Scale)', AWM 54, item 905/23/11.
47. Daly, interview, in Pratten and Harper (eds), *Still the Same*, p. 14.
48. 'Discussion on Signal Communications in the Field, held at "A" Mess on 1 Jul 41', p. 2, AWM 3DRL 6850, item 64; Comments of F. O. Chilton on chapter 17 of draft of *Greece, Crete and Syria*, p. 2, AWM 67, item 3/68; 'Report on Signal Operations in Greece', p. 1, AWM 54, item 425/6/92; Serle, *The Second Twenty-Fourth*, p. 59; Barker, *Signals*, p. 169; McAllester, *Men of the 2/14 Battalion*, p. 49; Chilton, 'Report on Operations of 2/2 Aust. Inf. Bn at Penios Gorge', p. 8, WD HQ 16 Bde, Mar–Apr 1941, AWM 52, item 8/2/16.
49. Chilton in Wick, *Purple over Green*, p. 63. A conference of 6th Division officers, chaired by the GOC, Major General Iven MacKay, in July 1941 concluded that 'with the speed of modern warfare and the distances involved, cable would not be practicable in many instances' ('Discussion on Signal Communications in the Field, held at "A" Mess on 1 Jul 41', p. 7).
50. Chilton in Wick, *Purple Over Green*, p. 60; Chilton, 'A Life in the Twentieth Century', unpublished MS, p. 27.
51. Long, *To Benghazi*, pp. 164–5; Eather, interview with Long, 7 Aug 44; AWM 67, item 2/56, p. 3.
52. Magill, pers. comm. undated, 2003; WD 2/4 Bn, 21–22 Jan 1941, AWM 52, item 8/3/4; WD 2/8 Bn, 21–22 Jan 1941, AWM 52, 8/3/8, WD HQ 19 Bde, 21–22 Jan 1941, AWM 52, item 8/2/19; Long, *To Benghazi*, pp. 225–31.
53. Louch, *From the CO's Notebook*, 'Bardia', p. i.
54. Wick, *Purple over Green*, pp. 62–3; Givney, *The First at War*, p. 99; Marshall, diary, AWM 54, item 255/4/12, pp. 86–90.
55. Marshall, diary, pp. 86–90.
56. Savige, 'Chapter VIII: The Battle of Bardia' (comments on draft of *To Benghazi*), AWM 67, item 3/348, part 2, p. 16.
57. 'Lessons from Tobruch' (24 Bde), p. 3, AWM 54, item 523/7/30.
58. For instance, the largest field telephone network operated by the 2/43rd Battalion at Tobruk consisted of 29 telephones, connected with 127 kilometres of line; battalion frontages at Tobruk were between five and eight kilometres long ('Bn Operations at Tobruch' (2/43 Bn), p. 2, AWM 54, item 523/7/9; 'Report on operations – 9 Aust Div in Cyrenaica, March–October 1941', chapter IV, p. 3, AWM 3DRL 2632, item 24).
59. Serle, The Second Twenty-Fourth, p. 59; Glenn, *Tobruk to Tarakan*, p. 34.
60. Chilton, comments on draft of chapter 17 of *Greece, Crete and Syria*, p. 2.
61. Long, *Greece, Crete and Syria*, pp. 107–8.
62. See 'Report on ops of 2/24 Bn 28 Apr–2 May 41', WD 2/24 Bn, May 1941, AWM 52, item 8/3/24.
63. Maughan, *Tobruk and El Alamein*, pp. 250–2, 256, 257.
64. See 'Night action of 2/12th Bn AIF night 3/4 May 41', WD 2/12 Bn, May 1941, AWM 52, item 8/3/12; '2/23 Bn – Report on action morning 17 May 41', WD 2/23 Bn, May 1941, AWM 52, item 8/3/23.

65. Daly, interview, in Pratten and Harper (eds), *Still the Same*, pp. 14, 19. Barton Maughan sums up the almost passive role of COs during the attacks mounted at Tobruk while discussing the role of Bernard Evans of the 2/23rd Battalion on the morning of 17 May: 'Evans moved back to his headquarters to await whatever good or sad news the day would bring' (Maughan, *Tobruk and El Alamein*, p. 233).
66. 'Night action of 2/12 Bn AIF, night 3/4 May 41', p. 4, WD 2/12 Bn, May 1941, AWM 52, item 8/3/12; J. Field, diary, 4 May 1941, AWM 3DRL 6937, item 3.
67. WD 2/23 Bn, 3 May 1941; Field, diary, 4 May 1941.
68. Daly, interview, in Pratten and Harper (eds), *Still the Same*, p. 14; '2/23 Bn – Report on action morning 17 May 41', p. 2.
69. Field, diary, 4 May 1941.
70. 'Report on operations – 9 Aust Div in Cyrenaica March–October 1941', p. 66.
71. Daly, interview, 26 Oct 2000; Daly, interview, in Pratten and Harper (eds), *Still the Same*, p. 14; Maughan, *Tobruk and El Alamein*, pp. 233, 259–62, 322, 326.
72. Stevens, 'Extracts from private diary', AWM 54, item 531/1/22, p. 2.
73. Long, *Greece, Crete and Syria*, p. 382.
74. Barker, *Signals*, p. 233.
75. 'Report on air co-operation during the battle of Damour', p. 1, AWM 54, item 531/1/2.
76. 'Report on W/T Set Patt No 108' (2/14 Bn), 'Ref. Originator's No. JW1' (2/16 Bn), 'Report on 108 R/T Sets 2/27 Bn 18/8/41', AWM 54, item 425/8/8; METS précis, 'Characteristics of wireless sets', AWM 3DRL 6599, item 51.
77. 'Report on 108 R/T Sets 2/27 Bn 18/8/41', AWM 54, 425/8/8.
78. 'Wireless Sets No 108' (1 Aust Corps), p. 1, AWM 54, 425/8/8/; Barker, *Signals*, p. 233.
79. Porter, 'The Capture of Jezzine', p. 5, AWM 67, item 3/317.
80. Porter, 'The Counter-attack', p. 11, AWM 67, item 3/317.
81. '2/23 Bn – Report on action morning of 17 May 41', p. 1; Field, 'Warriors for the working day', unpublished MS, pp. 14–17, AWM MSS 785, item 1.
82. WD 2/27 Bn, 1–2 Jul 1941, AWM 52, 8/3/27.
83. Chilton in Wick, *Purple over Green*, p. 58.
84. 'Report of action of 2/48 Bn on 1 May 41', p. 1, WD 2/48 Bn, May 1941, AWM 52, item 8/3/36; Field, diary, 3–4 May 1941. Although Barton Maughan states company commanders were able to conduct a reconnaissance on the afternoon of 3 May (*Tobruk and El Alamein*, p. 230), Field clearly states in his diary that no such reconnaissance was conducted by officers of the 2/12th, and there is no indication of it in the 2/9th's accounts of the attack. Tom Daly, BM of the 18th Brigade at the time, has also recalled that only the 2/9th's CO reconnoitered the ground the battalion was to attack over (Daly, interview, in Pratten and Harper (eds), *Still the Same*, p. 16).
85. Field, diary, 5 May 1941.

86. 'Report of action of 2/48 Bn on 1 May 41'; 'Report on 9 Bn operation at Tobruch – Night 3/4 May 1940', WD 2/9 Bn, April–June 1941, AWM 52, item 8/3/9; 'Night action of 2/12 Bn AIF night 3/4 May 41', Daly, interview, in Pratten and Harper (eds), *Still the Same*, pp. 3, 17–18.
87. Long, *To Benghazi*, p. 204.
88. 'Report on 9 Bn Operation at Tobruch – Night 3/4 May 1940', p. 3.
89. 'Trace "A" Objectives, Assembly Areas (Issued with 6 Aust Div OO No. 6 dated 1 Jan 41)', WD GS 6 Div, AWM 52, item 1/5/12.
90. Long, *To Benghazi*, pp. 171–2.
91. Allen, interview with Long, 27 December 1944, AWM 67, item 2/66, p. 19.
92. Hill, 'Lieutenant-General Sir Leslie Morshead: Commander, 9th Australian Division', in Horner, *The Commanders*, p. 183.
93. Field, 'Warriors for the working day', p. 16.
94. Maughan, *Tobruk and El Alamein*, p. 257.
95. '2/23 Bn – Report on action morning 17 May 41', p. 2; 'Action Diary 16–17 May 1941', p. 1, WD 2/23 Bn, Appendix 9, AWM 52, 8/3/23, Appendix. 9.
96. Maughan, *Tobruk and El Alamein*, p. 252; Simpson, interview with Dexter, 10 Dec 1951, AWM 172, item 5.
97. 'Report on ops of 2/24 Bn 28 Apr–2 May 41', p. 2, WD 2/24 Bn, May 1941, AWM 52, 8/3/24.
98. Savige, 'Chapter VIII: The Battle of Bardia' (comments on draft of *To Benghazi*), p. 15. Godfrey's commitment to an assault right across the wadi was partly founded on a concern for the welfare of his troops, but this concern was based on a fundamental misunderstanding of his role. On the morning of 2 January he noted in his diary: '. . . tomorrow's the big day. Our role is unsatisfactory. We will be chopped about if we don't get the cover of the N bank of the Muatered' (Godfrey, diary, 2 Jan 1940, AWM 54, item 523/7/30). A demonstration by fire, however, required him only to occupy covered positions on the south side of the wadi and fire on the Italian positions, thereby conveying the impression that an assault was to take place. By not attacking across the wadi, he had no need for the cover of the north bank.
99. Gullet in Charlton, *The Thirty-Niners*, p. xvii.
100. Only two D Company platoons were employed in the assault on Post 11; the third, with a detachment of medium machine guns, was used to provide fire support (Long, *To Benghazi*, p. 186).
101. There are indications that Godfrey may even have been considering a further attack at dawn on 5 January. Savige's recollections of the battle make reference to Godfrey abandoning two attack plans. One was for a night attack on 4 January that was cancelled after a reconnaissance just before dark. This corresponds to references made in the battalion war diary to orders being issued at 1700 on 4 January and for the OC of A Company being sent for at 1715. Savige makes a subsequent reference to Godfrey planning an 'all out' attack on Post 11 with all of his company commanders, which was cancelled when he received an order from Savige

not to attack any further. The recollections of George Smith, OC HQ Company, make no mention of a night attack, but they do speak of a conference planning a dawn attack involving all of the company commanders (Savige, 'Chapter VIII: The Battle of Bardia', p. 15; Signals log, WD 2/6th Bn, 4 Jan 1941; Smith, '2/6 Battalion in action – 1940–41 Bardia', p. 9; Smith, '2/6 Aust Inf Bn AIF', pp. 27–8, AWM PR 85/223).
102. Smith '2/6 Aust. Inf. Bn AIF', pp. 27–8; Smith note to 'Keith', AWM PR 85/223, item 1.
103. Twenty-two killed, 51 wounded.
104. Windeyer in Maughan, *Tobruk and El Alamein*, p. 217; 'Report on action of 2/48 Bn on 1 May', p. 1; WD 2/48 Bn, May 1941, AWM 52, item 8/3/52.
105. Evans, interview with Dexter, 13 Dec 1951, p. 2, AWM 172, item 13.
106. Daly, interview.
107. Stevens, 'A personal story', p. 61.
108. Ibid.
109. Long cited in Dennis et al., *Oxford Companion to Military History*, p. 566.
110. WD 2/31 Bn, AWM 52, item 8/3/31, 8 Jun 1941.
111. Porter, 'The Attack on Merdjayoun', p. 3, AWM 67, item 3/317.
112. Robson, interview with Long, 9 Aug 44, AWM 67, item 2/57, p. 3; Long, *Greece, Crete and Syria*, p. 355.
113. Robson, interview, p. 3.
114. Porter, 'The Attack on Merdjayoun', pp. 3–4.
115. Grey, *Australian Brass*, p. 86.
116. Dougherty, 'The Final Stage of the Battle of Tobruch', p. 1, WD 2/4 Bn, Jan 1941, AWM 52, item 8/3/4.
117. Savige, 'Chapter VIII: The Battle of Bardia', p. 15.
118. Citations found in CARO personal dossiers: NX383, Burrows, F. A.; VX21, Campbell, I. R.; NX231, Chilton, F. O.; NX378, Crawford, J. W.; NX148, Dougherty, I. N.; NX3, Eather, K. W., NX6, England, V. T.; VX47819, Evans, B.; VX25, Godfrey, A. H. L.; VX20315, King, R.; WX3346, Lloyd, J. E.; QX6049, Martin, J. E. G.; SX2889, Moten, M. J.; VX133, Porter, S. H. W. C.; NX49, Stevenson, J. R.; VX32, Walker, T. G.; NX396, Windeyer, W. J. V.; and QX6291, Withy, C. B.
119. Raymond Anderson, CO 2/32nd Battalion, died of wounds sustained when the vehicle he was travelling in at Tobruk struck a mine. The other CO casualties in 1941 were: F. A. Burrows, 2/13 Bn, T. S. Louch, 2/11 Bn, S. H. W. C. Porter, 2/31 Bn, H. Wrigley, 2/6 Bn and J. R. Stevenson, 2/3 Bn, wounded; and I. R. Campbell, 2/1 Bn, R. F. Marlan, 2/15 Bn, R. L. Sandover, 2/11 Bn and T. G. Walker, 2/7 Bn captured.
120. DSO citation, NX3, Eather, K. W., CARO Dossier NX3, Eather, K. W.
121. Fearnside and Clift, *Dougherty*, pp. 56, 83.
122. Markley, 'CO (Lt Col WALKER) endeavours to join men in attack on enemy 26 May 1941', WD 2/7 Bn, Jul 1941, AWM 52, item 8/3/7.
123. Savige, 'The Brigadier Reports', in Bolger and Littlewood, The *Fiery Phoenix*, p. 104.
124. Robson, interview with Long, p. 7.

125. Smith, 'Greece – Crete – Syria', p. 10, AWM PR 85/223, item 1.
126. Smith, notes on Greece, p. 10.
127. Allen, interview, p. 9.
128. Cullen, interview, 27 Feb 2001.
129. Wick, *Purple Over Green*, p. 190.
130. *White over Green*, p. 45; Chappel to Vasey, 2 Jun 1941 in Fearnside and Clift, *Dougherty*, p. 87.
131. Hurst, 'My Army Days', unpublished MS, p. 21, AWM PR, item MSS 1656.
132. Bolger and Littlewood, *The Fiery Phoenix*, p. 7.
133. S. G. Savige, 'Chapter VII. Before Bardia' (comments on draft of *To Benghazi*), p. 9; AWM 67, item 3/348, part 2.
134. Blamey to Sturdee, 26 Jun 1941, p. 2, AWM 3DRL 6643, item 1/11.
135. See for example Marshall, diary, AWM 54, item 255/4/12, pp. 14–15; 'Operation Order 26'; WD 2/7 Bn, May 1941, AWM 52, item 8/3/7.
136. R. Savige cited in Bolger and Littlewood, *The Fiery Phoenix*, p. 139; Hurst, 'My Army Days', p. 79.
137. Bolger and Littlewood, *The Fiery Phoenix*, p. 97.
138. Long, *Greece, Crete and Syria*, p. 307; Savige in Bolger and Littlewood, *The Fiery Phoenix*, p. 105.
139. Bolger and Littlewood, *The Fiery Phoenix*, p. 98.
140. WD 2/7 Bn, 31 May 1941.
141. Hay, interview, 31 May 2003.
142. Allen, interview, pp. 9–10.
143. Hurst, 'My Army Days', pp. 79, 129.
144. Chilton, interview.
145. Godfrey, diary, 21 and 22 Jan 1941, AWM 54, item 523/7/30.
146. A. W. Potts to D. Potts, 13 May 1940 and 11 Jul 1941 cited in Edgar, *Warrior of Kokoda*, pp. 67, 97–8.
147. Chilton, interview.
148. Boileau, letter, no date, May 1941, p. 23, NAUK WO 201, item 2752. Blamey similarly reported in August 1941 that 'great strength and resilience both of body and mind' was 'essential in Divisional commanders', which, by extension, could also be applied to brigade and battalion commanders: T. A. Blamey, AUSTFORCE GOC 158, 5 Aug 1941, AWM 3DRL 6643, item 1/24.
149. Disher cited in Walker, *Middle East and Far East*, p. 392.
150. Edgar, *Warrior of Kokoda*, p. 74.
151. Godfrey, diary, 26 Jan 1941.
152. Course report, British Army Senior Officers School, Mitchell, J. W., DSO citation, Mitchell, J. W., Bar to DSO citation, Mitchell, J. W., CARO dossier VX40, Mitchell, J. W.
153. G. A. Vasey to J. Vasey, 17 Mar 1941, NLA MS3782, item 12; Stevens, 'A personal story', p. 20; J. Magill, pers. comm. undated, 2003; Griffiths-Marsh, *The Sixpenny Soldier*, pp. 97, 140.
154. Grey, *Australian Brass*, p. 80; J. W. Mitchell to S. G. Savige, 7 Oct 1940, AWM 3DRL 2529, item 17.

155. The 2/8th acquitted itself particularly well during the attack on Tobruk on 21–22 January where it encountered the heaviest resistance of any Australian battalion there, including large numbers of tanks, both dug in and mobile. See Long, *To Benghazi*, pp. 226–9.
156. E. R. Wilmoth to 'Greg', 7 Mar 41, AWM PR, item 86/370.
157. WD 2/8 Bn, 31 Jan 1941, AWM 52, item 8/3/8; WD HQ 19 Bde, 31 Jan 1941, AWM 52 item 8/2/19. Jeff Grey (*Australian Brass*, p. 89) suggests that Mitchell mounted the night attack having been shamed into it by an identical order that was dispatched by Robertson to both him and Dougherty. The 2/8th Battalion's war diary, however, indicates that the battalion had secured the escarpment before Robertson's instructions arrived. Robertson's own account of the operations suggests Mitchell gained the escarpment on his own initiative (H. C. H. Robertson, '19th Australian Infantry Brigade Diary of Events Libyan Campaign', p. 15, AWM 54, item 521/2/2).
158. Robertson, '19th Australian Infantry Brigade Diary of Events Libyan Campaign', pp. 18–19.
159. 'Defence', METS Senior Wing précis, p. 5, AWM 3DRL 6599, item 3.
160. During the subsequent battle at least one company was required to withdraw to better fire positions further up the slope (Bentley, *The Second Eighth*, p. 59).
161. Long, *Greece, Crete and Syria*, p. 58. Mitchell may have indeed visited all of the companies along his front, but the only indication of this trip forward is contained in an after-action report prepared by the OC of C Company, which was the left flank company on the morning of 11 April ('Action of "C" Coy 2/8 Bn in Greece', p. 2, WD 2/8 Bn, Apr 1941; Magill correspondence, p. 1).
162. 'Report by Capt. N. F. Ransom, Adj 2/8 Bn, AIF Events between 1200 hrs and 2200 hrs 12 Apr 41', WD 2/8 Bn, Apr 1941.
163. 'Report by Capt. N. F. Ransom, Adj 2/8 Bn, AIF'.
164. WD HQ 19 Bde, 13 Apr 1941.
165. WD 2/8 Bn, 13 Apr 1941; G. A. Vasey, handwritten note, WD HQ 19 Bde, 13 Apr 1941.
166. Like those of so many of the Australian units involved in the campaigns in Greece and Crete, the archival record of the 2/8th's activities is patchy. The battalion's records for Greece up to and including the Vevi battle were destroyed during the withdrawal from Vevi. Other battalion records were lost when the *Costa Rica* was sunk during the evacuation from Greece. The records that remain extant were reconstructed in Palestine on the battalion's return there.
167. Bentley, *The Second Eighth*, p. 67.
168. Savige, 'The Campaign in Greece. Notes on chapter 16' (comments on draft of *Greece, Crete and Syria*), p. 3, AWM 67, item 3/348, part 2.
169. Vasey, handwritten note, WD HQ 19 Bde, 13 Apr 1941.
170. McAllester, *Men of the 2/14th Battalion*, p. 2; McAllester and Trigellis-Smith, *Largely a Gamble*, p. 181; WD 2/14 Bn, 29 Jun 1941, AWM 52, item 8/3/14.

171. Stevens, 'A personal story', p. 59.
172. A. W. Potts to D. Potts in Edgar, *Warrior of Kokoda*, p. 93.
173. Williamson, interview, 9 Feb 2001.
174. Allen, interview, 6 Feb 2001; J. Anning, interview, 15 Feb 2001.
175. Daly, interview.
176. Peterson, interview, 10 Feb 2001.
177. Rowell to Blamey, 17 May 1941, p. 2, AWM 3DRL 6643, item 1/24. Brigadier Stevens similarly reflected on his decision to remove Cannon: 'When I offered him command of this Battalion, on its formation, I realised that, because of his World War I wound and his age, he could probably raise and train a new battalion, take it into its fight, and then need relief. My early forecast had been correct, and George was not now physically fit enough to lead a battalion, in another fight. And so I parted with a man who had done a splendid job in his modest way' (Stevens, 'A personal story', p. 59).
178. OBE citation, Cook, T. P., CARO dossier VX44, Cook, T. P.
179. Blamey, GOC 74, 2 Jun 1941, AWM 3DRL 6643, item 1/19.
180. G. A. Vasey to J. Vasey, 24 Jun 41, NLA MS3782, item 12; V. A. H. Sturdee to T. A. Blamey, 17 Jun 1941, p. 3, AWM 3DRL 6643, item 1/6; Russell, *The Second Fourteenth Battalion*, p. 99.
181. CBE citation, Cook, T. P., CARO dossier VX44, Cook, T. P.
182. CARO dossier 4/2, MacDonald, A. B.
183. Stevens, 'A personal story', p. 59.
184. Guthrie in Edgar, *Warrior of Kokoda*, p. 86; DSO citation, Potts, A. W., CARO dossier WX1574, Potts, A. W.
185. Godfrey, diary, 26–27 Jan 1941; Bar to DSO citation, Godfrey, A. H. L., CARO dossier VX25, Godfrey, A. H. L.; Jackson, 'Autobiography', p. 219; Johnston and Stanley, *Alamein*, p. 36.
186. Auchinleck to Dill, BM/GO/214, 13 Oct 41, Rylands University Library, University of Manchester, Auchinleck Papers, item 377, pp. 2–3.
187. Matthews, diary, Dec 1941–Jan 1942, AWM PR89/79.

4 Desert epilogue: El Alamein, 1942

1. Both the Finschhafen and Balikpapan landings, of 1943 and 1945 respectively, could be considered full divisional operations. In neither case, however, did the divisions involved employ a full complement of supporting arms nor operate as part of a larger force.
2. Anning, interview, 15 Feb 2001.
3. See, for instance: Barr, *Pendulum of War*, pp. 261–6; Johnston and Stanley, *Alamein*, pp. 144–6; French, *Raising Churchill's Army*, pp. 249–50.
4. Serle, *The Second Twenty-Fourth*, p. 197.
5. 'Wireless sets held by 9 Aust Div units', AWM 72, item 29.
6. AIF WE 11/12F/3, 'An Infantry Battalion', 15 Dec 1941, p. 6, AWM 54, 327/6/9.
7. Ibid., pp. 3, 6, 11.
8. The 2/48th Battalion's anti-tank platoon, for example, was initially equipped with four 37mm A/T guns (Glenn, *Tobruk to Tarakan*, p. 125).

9. For example, the 2/23rd Battalion was forced from its gains on 16 July due to a lack of artillery, medium machine guns and anti-tank guns: 'Report on action – 2/23 Aust Inf Bn, 16 Jul 42', WD 2/23 Bn, Jul 1942, AWM 52, item 8/3/23.
10. 'Captured and other weapons held by 9 Aust Div units on 23 Oct 42', AWM 72, item 29.
11. '9 Aus Div General Staff Instn No 58', p. 1, '9 Aust Div General Staff Instruction No. 66: Formation of MG Pls Inf Bns', WD 9 Div (GS), Aug 1942, item 1/5/20; WD 2/48 Bn, 16 Aug 1942.
12. Appendix A to '9 Aust Div Report on Operations El Alamein – 23 October 42–5 November 42', 'Detailed Conclusions', p. 4, AWM 54, item 527/6/1A.
13. Horner, *The Gunners*, p. 325; Maughan, *Tobruk and El Alamein*, p. 622.
14. '9 Aust Div General Staff Instruction No. 61: Notes on the tactical handling of A Tk guns', p. 2, WD 9 Div (GS), Aug 1942; WD 2/48 Bn, 19 Aug 1942, AWM 52, item 8/3/36.
15. Appendix A to '9 Aust Div Report on Operations El Alamein – 23 October 42–5 November 42', 'Detailed Conclusions', p. 8.
16. Lt Col A. Spowers, CO 2/24 Bn, and Lt Col D. A. Whitehead, CO 2/32 Bn.
17. Grace cited in Austin, *Let Enemies Beware*, p. 150.
18. Broadbent, interview, October 2000. Don Jackson, brigade major of the 24th Brigade, also commented on the 9th Division's lack of experience in mobile offensive operations in early 1942 (Jackson, 'Autobiography', p. 224).
19. Simpson, CO's report for March, WD 2/17 Bn, Mar 1942, AWM 52, item 8/3/17.
20. Broadbent, interview.
21. Glenn, *Tobruk to Tarakan*.
22. Gill to mother, 7 May 1942, AWM 3DRL 7945, item 1.
23. Hammer, diary, Jan–Jun 1942, AWM 52, item 2642.
24. Serle, interview, 7 Feb 2001.
25. Evans, CO's summary, WD 2/23 Bn, Sep 1942, AWM 52, item 8/3/23.
26. See, for example, '2/48 Aust Inf Bn Trg Instn No 7', 20 Aug 1942, '2/48 Aust Inf Trg Instruction No 8', 23 Aug 1942, 'Notes on Trg – Mobile Force', 23 Aug 1942, '2/48 Aust Inf Bn Trg Instn No 9' 8 Sep 1942, 'Trg – Night attacks and fighting patrols', 9 Sep 1942, 'Training', 16 Sep 1942, WD 2/48 Bn, Aug–Sep 1942, AWM 52, 8/3/36.
27. WD 2/48 Bn, 31 Aug 1942, 7 Sep 1942.
28. 'Standing Orders for 2/23 Aust Inf Bn Gp – Mob Activities', p. 1, WD 2/23 Bn, Aug 1942, AWM 52, item 8/3/23.
29. Evans, 'Foreword', in 'Standing Orders for 2/23 Aust Inf Bn Gp – Mob Activities', p. 1, WD 2/23 Bn, Aug 1942.
30. 'Training. Syllabus of extra trg for NCOs to be carried out in addition to normal coy trg', WD 2/48 Bn, Aug 1942.
31. WD 2/48 Bn, 17 Oct 1942.
32. Broadbent, interview; Fearnside (ed.), *Bayonets Abroad*, p. 225; Share (ed.), *Mud and Blood*, p. 208.

33. Evans, COs report, WD 2/23 Bn, Sep 1942.
34. 2/48 WD, 7, 10 Jul 1942.
35. 'Report on operations: 9th Australian Division, the battle of El Alamein, 23 Oct–5 Nov 42', p. 6, AWM 54, item 527/6/1A.
36. Fearnside (ed.), *Bayonets Abroad*, p. 261.
37. Montgomery, untitled memorandum, cited in Jackson, 'Autobiography', p. 273; French, *Raising Churchill's Army*, p. 247.
38. 'Conference of all comds 9 Aust Div 10.10.1942', pp. 5–6, AWM 3DRL 2632, item 6/16.
39. Fearnside (ed.), *Bayonets Abroad*, p. 242.
40. Evans cited in Share (ed.), *Mud and Blood*, p. 212. One 2/23rd soldier, Private 'Pay-day' Cleary, who had been posted into the battalion from the 2/5th, took Evans to task for this remark and was rewarded with an apology from his CO.
41. Weir cited in Serle (ed.), *The Second Twenty-Fourth*, p. 197.
42. Johnston and Stanley, *Alamein*, p. 156.
43. Clothier, diary, 22 Oct 1942, AWM PR, item 00588.
44. Watkins, 'As I Remember It', unpublished MS, AWM PR, item MSS 1587, p. 61.
45. Masel, *The Second 28th*, p. 95.
46. Kehoe, 'That Minefield', in Masel, *The Second 28th*, p. 111.
47. Masel, *The Second 28th*, p. 115.
48. Ibid., p. 93.
49. Morshead to Blamey, 28 Jul 1942, AWM 3DRL 2632, item 6/14.
50. Masel, *The Second 28th*, pp. 67–68.
51. 'Conference of all comds 9 Aust Div 10.10.1942', p. 5.
52. Hammer, 'Night Attack – Alamein – Oct 23 to Nov 1, 1942', p. 5, WD 2/48 Bn, Nov 1942.
53. Ibid.
54. '2/17 Aust Inf Bn OO No 4', 20 Oct 1942, p. 7, WD 2/17 Bn, Oct 1942, AWM 52, 8/3/17; '2/13 Aust Inf Bn O. O. No. 27', 21 Oct 1942, p. 9, WD 2/13 Bn, Oct 1942, AWM 52, item 8/3/13; Appendix F to '2/15 Aust Inf Bn Brief Narrative of Ops 23 Oct–5 Nov 42', pp. 4–5, WD 2/15 Bn, Nov 1942, AWM 52, item 8/3/15; Hammer, 'Night Attack – Alamein – Oct 23 to Nov 1, 1942', p. 5; 'Report on signal comms during operations 30 Oct/5 Nov '42', p. 1, WD 2/32 Bn, Nov 1942, AWM 52, item 8/3/32; R. Cowie, interview, 15 Feb 2001.
55. Williamson, interview, 9 Feb 2001; Hammer, 'Attack', p. 1, WD 2/48 Bn, Dec 1942, AWM 52, item 8/3/36.
56. Fearnside (ed.), *Bayonets Abroad*, p. 271.
57. WD 2/48 Bn, 25 Oct 1942.
58. WD 2/17 Bn, 23 Oct 1942, AWM 52, item 8/3/17; 'Report on ops 23 Oct–4 Nov 42', p. 4, WD 2/23 Bn, Nov 1942.
59. 'Preliminary report on operations for period 23 Oct–3 Nov 42', p. 3, WD 2/48 Bn, Nov 1942.
60. Hammer, 'Re-organisation – Alamein – October 23 to November 1', p. 2, WD 2/48 Bn, Dec 1942.
61. Masel, *The Second 28th*, p. 131.

62. Hammer, 'Re-organisation – Alamein – October 23 to November 1', p. 3. The consolidated report on the lessons of the October battle prepared by HQ Middle East similarly commented: 'The value of siting Headquarters of every level well forward was constantly borne out throughout the operations. It allowed of easy physical contact between commanders or rapid receipt of information and it encouraged the troops': 'Lessons from Operations, Oct and Nov 1942', p. 25.
63. Hammer, 'Re-organisation – Alamein – October 23 to November 1', p. 5.
64. CO casualties at El Alamein 23 October–5 November 1942: J. W. Balfe, 2/32 Bn, WIA – gunshot; G. E. Colvin, 2/13 Bn, WIA – shrapnel; H. H. Hammer, 2/48 Bn, WIA – gunshot; C. K. M. Magno, 2/15 Bn, KIA – shrapnel; R. W. G. Ogle, 2/15 Bn, WIA – mine; R. W. N. Turner, 2/13 Bn Bn, DOW – shrapnel; W. J Wain, 2/43 Bn, WIA – concussion/combat exhaustion; C. G. Weir, 2/24 Bn, WIA, booby trap.
65. Johnston and Stanley, *Alamein*, pp. 202–3.
66. Share (ed.), *Mud and Blood*, p. 223.
67. Johnston and Stanley, *Alamein*, pp. 202–3.
68. Ibid., pp. 216–22; Glenn, *Tobruk to Tarakan*, pp. 162–70; Serle, *The Second Twenty Fourth*, pp. 214–22.
69. Serle, interview; WD 2/48 Bn, 30 Oct 1942; WD 2/24 Bn, 30 Oct 1942; Johnston and Stanley, *Alamein*, pp. 220–1.
70. Serle, *The Second Twenty-Fourth*, p. 219.
71. 'Lessons from Operations, Oct and Nov 1942', p. 25, NAUK WO201, item 2825.
72. Williamson, interview.
73. French, *Raising Churchill's Army*, p. 39.
74. Wesbrook, 'The Potential for Military Disintegration', in Sarkesian (ed.), *Combat Effectiveness*, p. 246.
75. See Johnston and Stanley, *Alamein*, pp. 212–32.
76. Fearnside (ed.), *Bayonets Abroad*, p. 242.
77. Colvin's acting command was confirmed after the battle.
78. Fearnside (ed.), *Bayonets Abroad*, pp. 274–5.
79. Morshead to Blamey, 28 Oct 1942, AWM 3DRL 2632, item 6/14.
80. WD 2/48 Bn, 27 Oct 1942. Johnston and Stanley, *Alamein*, pp. 190–1. The official history makes no mention of this incident, simply noting: 'The relief of the 26th Brigade by the 20th Brigade on the night of the 27th–28th was made difficult by strong enemy counter-attacks on both the 2/24th and 2/48th Battalions at the time set for their reliefs' (Maughan, *Tobruk and Alamein*, p. 700). Johnston and Stanley similarly fail to acknowledge the role of the battalion's own weapons at close range during the lull in the artillery fire.
81. The XXX Corps report on the October battle indicates that often the main impact of artillery during the battle was on the morale of the Axis troops. On one instance it states: 'Although the enemy may not have been killed in large numbers [by concentrated artillery barrages], his morale was undoubtedly affected' ('Extracts from draft report by 30 Corps dated 21 November 42', Appendix A to 'Report on operations: 9th Australian Division, the battle of El Alamein, 23 Oct–5 Nov 42', pp. 6–7; 'Report on

operations: 9th Australian Division, the battle of El Alamein, 23 Oct–5 Nov 42', p. 48).
82. Hammer, 'Attack', p. 2; WD 2/48 Bn, 25 Oct 1942; 'Report on operations: 9th Australian Division, the battle of El Alamein, 23 Oct–5 Nov 42', p. 27.
83. Fearnside (ed.), *Bayonets Abroad*, p. 261.
84. Johnston and Stanley, *Alamein*, pp. 224–6.
85. Jackson, 'Autobiography', p. 222.
86. WD 2/13 Bn, 29 Oct 1942.
87. Whitehead, 'Report on operations of 2/32 Aust Inf Bn 16/17 Jul 42', p. 4, 'Report on operations of 2/32 Aust Inf Bn 21/22 Jul 42', p. 4, WD 2/32 Bn, Jul 1942, AWM 52, item 8/3/32; B. Evans, 'Report on action – 2/23 Aust Inf Bn 16 Jul 42', WD 2/23 Bn, Jul 1942.
88. WD 2/48 Bn, 30 Oct 1942.
89. For example see the experiences of Bernard Evans, CO 2/23rd Battalion on 16 July (Johnston and Stanley, *Alamein*, pp. 79–80; Evans, 'Report on action – 2/23 Aust Inf Bn, 16 Jul 42'), Charles Weir, CO 2/24th Battalion on 22 July (Maughan, *Tobruk and El Alamein*, pp. 580–1; Serle, *The Second Twenty-Fourth*, pp. 180–3) and Colin Grace, acting CO 2/15th Battalion on 1 September (Austin, *Let Enemies Beware*, pp. 135–46).
90. CARO dossier VX23, Ferguson, M. A.

5 Victims of circumstance: Battalion command in the 8th Division

1. Long, notebook, AWM 67, item 3/109, facing p. 7.
2. Maxwell, interview with Long, Dec 1946, AWM 67, item 2/109, p. 6.
3. Taylor, interview with Long, 10 Feb 1947, AWM 67, item 3/109, p. 68.
4. Maxwell, interview, p. 6.
5. It was originally proposed that the 9th Division would be brought up to strength by raising an additional brigade in Australia, but after lobbying from Blamey in the Middle East, the 24th Brigade joined the 9th Division. Blamey argued that the two brigades in the UK had been trained to a high degree of efficiency during the invasion scare and that waiting for a completely new brigade would delay the operational employment of the new division.
6. Pratten, 'The "old man"', p. 193.
7. 'Add to Brigadier Taylor's Notes', p. 1, AWM 73, item 7.
8. Henning, *Doomed Battalion*, pp. 1–5.
9. Ibid., pp. 7–8.
10. Geoffrey Youl had served in the British Army as an artillery officer during the Great War and did not begin serving in the militia until 1936.
11. Campbell, 'A Model for Battalion Command', p. 40.
12. Oakes, 'Singapore Story', p. 17.
13. Major General Gordon Bennett was appointed to the command of the 8th Division as a result of the death of three senior government ministers and the experienced and widely respected CGS, Sir Brudenell White, in a plane crash near Canberra on 13 August 1940. Vernon Sturdee, with his wide experience, was promoted to replace White, and another regular soldier, Major General John Northcott, was nominated to take command of the 8th Division. Northcott, however, was the deputy CGS, and it was subsequently

decided that the AMF needed to retain his experience in this position while Sturdee got to grips with his new appointment.
14. Oakes, 'Singapore Story', p. 17.
15. Campbell, 'A Model for Battalion Command', p. 40.
16. Bennett, interview with Kirby and Wigmore, 30 Jan 1953, p. 2, AWM 73, item 7.
17. Taylor, interview, p. 69.
18. Oakes, 'Singapore Story', p. 5.
19. Ibid., p. 58.
20. Cahill, interview, 16 May 2002.
21. Oakey cited in Burfitt, *Against All Odds*, p. 20.
22. Taylor, diary, 5 Mar 1941, AWM 67, item 3/394, part 2. After an exercise in May 1941, Taylor noted: 'Commenced exercise with 19 Bn Layout of HQ good – still too much time taken up giving orders. Actually five hrs is used up in messages and orders, two hours in movement' (1 May 1941), and on the following day, 'Exercise with troops improved on yesterday – orders and organisation better. Orders not crisp enough – still tendency to discuss rather than order' (2 May 1941).
23. Bennett, diary, 19 Sep 1941, AWM PR, item MF0020.
24. Hodel, interview, 16 Feb 2001.
25. Hodel, interview.
26. Taylor, interview, p. 69.
27. Wall, *Singapore and Beyond*, pp. 3, 7, 13.
28. Bennett, diary, 22 May 1941.
29. WD 2/20 Bn, Feb–Jul 1941, AWM 52, item 8/3/20; Wall, *Singapore and Beyond*, pp. 14–15.
30. Jeater, 'The CO congratulates us', *Second Two Nought: Weekly Bulletin of the 2/20th Bn, AIF, Malaya*, 14 Jun 1941, p. 1, WD 2/20 Bn, Jun 1941, AWM 52, item 8/3/20.
31. WD 2/20 Bn, Apr–Aug 1941; Taylor, diary, 6–7 Jun 1941; Bennett, diary, 9 Jun 1941; CARO Dossier NX34992, Jeater, W. D.
32. *Second Two Nought*, 14 Jun 1941, p. 2.
33. Wall, *Singapore and Beyond*, p. 17.
34. Ibid., p. 14.
35. Ibid.
36. Taylor, diary, 7 Jun 1941.
37. Taylor, interview with Long, AWM 67, item 3/109.
38. Campbell, 'A Model for Battalion Command', p. 28.
39. 'We knew that if anyone went fast we had to go faster. We knew if we went first we would have to set such a gut busting pace that it would be difficult to match it' (Howells in Campbell, 'A Model for Battalion Command', p. 79).
40. Cahill, interview.
41. Cahill, interview.
42. Campbell, 'A Model for Battalion Command', p. 66.
43. Arneil, *Black Jack*, pp. 64, 67.
44. Campbell, 'A Model for Battalion Command', p. 63; Arneil, *Black Jack*, p. 64.

45. Heckendorf, interview with Nelson, 29 Jan 1990, AWM KMSA, item S763 (t), p. 9; 'Experimental taping made with idea of adding to 2/30th Bn History', p. 24, AWM PR, item 00686; Cahill, interview.
46. Campbell, 'A Model for Battalion Command', pp. 75–6; Arneil, *Black Jack*, pp. 71–2; Heckendorf, interview, p. 16.
47. Campbell, 'A Model for Battalion Command', pp. 75–6.
48. Not reporting a 'riot or disturbance' of troops under one's command to higher headquarters was in contravention of AMR&O (AMR&O, para. 1469).
49. Campbell, 'A Model for Battalion Command', p. 126; Arneil, *Black Jack*, p. 76.
50. Campbell, 'A Model for Battalion Command', pp. 68, 86.
51. Maxwell, interview, p. 18. 'Red flannel' is a reference to the red gorget patches and hat band worn by officers of the rank of colonel and above.
52. Bennett, diary, 15 Aug 1941.
53. Johnston, 'Experimental taping made with idea of adding to 2/30th Bn History', p. 23, AWM PR, item 00686.
54. AWM PR, item 00079, folder 15.
55. Johnston, 'Second day of special discussion group, 2/30th Battalion held at Stuart Peach's on 22nd August, 1983', p. 23, AWM PR, item 00686.
56. Ibid.
57. Campbell, 'A Model for Battalion Command', pp. 91–2.
58. Ibid., p. 28.
59. Galleghan, interview with Wigmore, 1 Feb 1950, p. 3, AWM 73, item 51.
60. Campbell, 'A Model for Battalion Command', p. 91.
61. Arneil, *Black Jack*, pp. 62, 72.
62. Galleghan, interview, p. 3.
63. Pond, interview with Wigmore, 2 Feb 1953, p. 2, AWM 73, item 7; Bennett, diary, 17, 23 Aug 1941.
64. See Lodge, 'Bennett, Henry Gordon', *Australian Dictionary of Biography*, Vol. 13: *1940–1980*, pp. 166–7; Lodge, *The Fall of General Gordon Bennett*; Warren, *Singapore 1942*, pp. 34–5, 37–8.
65. The antipathy between Bennett and Taylor is vividly illustrated in the diaries of the two men: AWM PR, item MF0020, and AWM 67, item 3/394, part 2, respectively.
66. Johnston, 'Second day of special discussion group', p. 16; Kappe and Morrison, interview with Kirby and Wigmore, 5 Feb 1953, p. 1, AWM 73, item 7.
67. Taylor, diary, 18 Jul, 3 Aug 1941.
68. Galleghan, 'History of the War in Malaya. Comments by Brigadier F. G. Galleghan on draft, chapter 5 (The Malayan Scene)', p. 3, AWM 67, item 3/140.
69. Wigmore, *The Japanese Thrust*, p. 86.
70. Jessup, interview with Kirby and Wigmore, 23 Jan 1953, p. 4, AWM 73, item 7; Pond, interview with Wigmore, 6 Feb 1953, AWM 73, item 7.
71. Maxwell, interview, p. 19.

72. Warren, *Singapore 1942*, p. 198; Farrell, *The Defence and Fall of Singapore 1940–1942*, pp. 305–6.
73. Peach, 'Second day of special discussion group', pp. 5–6; 'Narrative 3: Namazie Estate. 41. MP', P. 2; WD 2/26 Bn, Jan 1942, AWM 52, item 8/3/26.
74. Bennett, diary, 10 Apr 1941.
75. Ibid.
76. Bennett, diary, 6 Dec 1941. Jeater remained 'mental' during his time in captivity following the fall of Singapore. Lloyd Cahill, the 2/19th Battalion's MO, shared a hut with Jeater at Changi and described him as being 'mad as a meat axe' and a 'bloody nuisance' (Cahill, interview).
77. Taylor, interview with Long, 10 Feb 1947, p. 74.
78. Taylor, diary, 19 Sep 1941.
79. Bennett, diary, 15 Dec 1941; Christie (ed.), *A History of the 2/29 Battalion – 8th Division AIF*, p. 45.
80. Maxwell to Kirby, 4 Aug 1953, AWM 67, item 3/261.
81. Galleghan certainly believed there was a conspiracy to oust himself and Boyes. He wrote to Lionel Wigmore: 'Should a historian raise the question why Maxwell had two COs moved, one to hospital and one relieved of command at such a vital period' (Galleghan to Wigmore, 26 Mar 1954, p. 18, AWM 93, item 50/2/23/285).
82. Oakes remained on the east coast to oversee the handover of the 2/19th's positions to the relieving battalion of Johore when the battalion redeployed to Bakri.
83. 'Narrative 5', pp. 4–5, WD 2/26 Bn, AWM 52, item 8/3/26; Thyer, 'Operations of 8th Australian Division in Malaya 1942', p. 129, AWM 54, item 553/5/23; 'Memo for General Kirby in answer to letter from Mr Wigmore to Lt Col Oakes (CO 2/26 Bn AIF)', AWM 73, item 7; Oakes, 'Singapore Story', p. 174.
84. Anderson to Wigmore, 6 Nov 1954, p. 2, AWM 67, item 3/9, part 1.
85. Anderson in Newton et al., *The Grim Glory of the 2/19 Battalion AIF*, pp. 213, 305.
86. Newton et al., *The Grim Glory of the 2/19 Battalion AIF*, pp. 298–9, 302, 304–5.
87. Warren, *Singapore 1942*, p. 227.
88. Farrell, 'The Defence and Fall of Singapore', original manuscript, pp. 155–6; Maxwell, interview, p. 10; Taylor, diary, 24, 28 Feb 1941; Cahill, interview; Burnett, *Against All Odds*, p. 27; 'Add to Brig Taylor's Notes', p. 2; 'Draft letter to Brigadier J. K. Coffey', p. 1, NAUK WO106, item 2591; Warren, *Singapore 1942*, p. 199; Farrell, *The Defence and Fall of Singapore 1940–1942*, pp. 122, 248–9, 301–2, 303–5.
89. Warren, *Singapore 1942*, pp. 37, 46; Farrell, *The Defence and Fall of Singapore 1940–1942*, pp. 118–22.
90. 'Training Policy and Weakness of Training as it Affected the Malayan Campaign from Pioneer Pl. Point of View', pp. 1–3, WD GS 8 Div, 'Notes and Reports on the Malayan Campaign', AWM 52, item 1/5/17; 'The Carrier Pl. (in Malaya)', p. 1, WD GS 8 Div, 'Notes and Reports on the

Malayan Campaign'; 'Lessons Learned in Malaya – Rifle Coys', p. 5, WD GS 8 Div, 'Notes and Reports on the Malayan Campaign'; Report on Visit to 2/26th Battalion [title obscured], 'Date of Visit: 1900 hrs 14 Jan to 1000 hrs 10 Jan 41', p. 2, 'Notes on Bn Exercises, Wambool Area', pp. 1–8, WD HQ 27 Bde, May 1941, AWM 52, item 8/2/27.

91. Warren, *Singapore 1942*, pp. 156–7; McCure, interview with Nelson, May 1984, in Finkemeyer, *It Happened to Us*, p. 10. Robertson told the attached troop commander from the 2/4th Anti-Tank Regiment: 'I have orders from the General that I should be accompanied by a troop of anti-tank guns, but as far as I am concerned, you're not wanted. I don't want you to interfere with us in any way. I don't expect the Japanese to use tanks, so for my part, you can go home.'
92. McCure, interview, p. 13.
93. Eaton, 'Experimental taping made with idea of adding to 2/30th Bn History', p. 11.
94. 'Notes on Lessons – AIF ops Malaya – Some points in relation to arty', p. 2, WD GS 8 Div, 'Notes and Reports on the Malayan Campaign'.
95. 'Artillery Troop Syndicate', p. 5, 'Notes on Lessons – AIF ops MALAYA – Some points in relation to ARTY', pp. 2, 4, WD GS 8 Div, 'Notes and Reports on the Malayan Campaign'.
96. 'Notes on Lessons – AIF ops MALAYA – Some points in relation to ARTY', p. 2; O'Brien and Whitelocke, *Gunners in the Jungle*, pp. 65, 67.
97. 'Suggestions', p. 1, WD GS 8 Div, 'Notes and Reports on the Malayan Campaign'; 'Notes on Operations in Malaya', p. 5.
98. Kappe, 'The Artillery Problem in the Western Area', p. 2, 'The Malayan Campaign', AWM PR, MSS 1393.
99. 'Notes on Operations in Malaya', p. 5; 'The Carrier Platoon (in Malaya)', pp. 1, 4; WD GS 8 Div, 'Notes and Reports on the Malayan Campaign'.
100. Newton et al., *The Grim Glory of the 2/19 Battalion AIF*, pp. 316–18; 'The Landing', pp. 1–2, WD 2/19 Bn, Feb 1942, AWM 52, item 8/3/19; WD 2/18 Bn, Jan–Feb 1942, pp. 21–2, AWM 52, item 8/3/18.
101. Galleghan was not alone in this view. Recounting the lessons of the Malaya campaign, Maurice Ashkanasy, DAAG 8th Division, told Long carriers were 'noisy', 'vulnerable' and of no use (Ashkanasy, interview with Long, 5 Aug 44, AWM 67, item 1/5, p. 80; Galleghan, '2/30 Bn War Diary – Gemenchah Action', p. 1; AWM PR, item 3DRL, 2313; Galleghan, 'History of the War in Malaya. Comments on draft chapters [*sic*] 11 by Brigadier F. G. Galleghan', p. 4, AWM 67, item 3/140).
102. 'The Carrier Pl. (in Malaya)', p. 4.
103. WD 2/18 Bn, 10 February 1942; Evans in Burfitt, *Against all Odds*, pp. 73–4.
104. 'It might be noted by the historian that armoured cars were of more use in Malaya than Carriers but no credit is given to 2/30 for "acquiring" the armoured cars' (Galleghan, 'History of the War in Malaya. Comments on chapter 13 by Brigadier F.G. Galleghan', p. 4, AWM 67, item 3/140).
105. Campbell, 'A Model for Battalion Command', p. 118.

106. Galleghan, 'History of the War in Malaya. Comments on draft chapters [*sic*] 11 by Brigadier F. G. Galleghan', p. 4; Galleghan, 'Notes on lessons from AIF ops in Malaya', p. 3; Newton et al., *The Grim Glory of the 2/19 Battalion AIF*, pp. 253–5; Bowring, Gibson, Jessup and Walpole, 'Malayan Campaign', p. 5, WD GS 8 Div, 'Notes and Reports on the Malayan Campaign'; Galleghan, 'Notes on lessons from AIF ops in Malaya', p. 3; Bennett, 'Report by Major General H. Gordon Bennett, CB, CMG, DSO, VD, GOC AIF Malaya on Malayan Campaign 7th Dec 1941 to 15 Feb 1942', p. 31, AWM 54, item 553/5/16.
107. Farrell, pers. comm. 30 Mar 2005; Wigmore, *The Japanese Thrust*, p. 277; Thyer, 'Operations of 8th Australian in Malaya 1942', p. 70, AWM 54, item 553/5/23, part 1.
108. Bennett, 'Malayan Campaign', p. 30.
109. Galleghan, for instance, travelled more than a thousand miles conducting reconnaissance while the 2/30th waited to be committed to action in early January 1942 (Ramsay to wife, 7 Jan 42, AWM PR 00079, folder 15).
110. Tsuji, *Singapore: The Japanese Version*, pp. 192–4, 196, 199.
111. Arneil, *Black Jack*, pp. 92–9; Campbell, 'A Model for Battalion Command', p. 114; Farrell, *The Defence and Fall of Singapore 1940–1942*, pp. 303–5, 347–8; Wigmore, *The Japanese Thrust*, pp. 217–20, 227, 229, 234, 251–2, 262–4, 277–9, 331–3; 'Narrative 2, Simpang Rengam', 'Narrative 3, Namazie Estate 41. MP', 'Narrative 4. 30 1/4 MP Johore Bahru–Segamat Rd', WD 2/26 Bn, Jan 1942, AWM 52, item 8/3/26; 'Narrative 5.', p. 2, WD 2/26 Bn, Feb 1942, AWM 52, item 8/3/26.
112. Galleghan, interview with Kirby, no date, 1953, p. 1, AWM 73, item 7; Oakes to Wigmore, 19 Feb 1953, AWM 73, item 7.
113. Campbell, 'A Model for Battalion Command', p. 109; 'Sketch of Gemas Ambush Area', Appendix A, WD 2/30 Bn, Jan–Feb 1942, AWM 52, item 8/3/30.
114. Farrell, *The Defence and Fall of Singapore 1940–1942*, p. 245.
115. Wigmore, *The Japanese Thrust*, p. 217; Peach, 'Experimental taping made with idea of adding to 2/30th Bn History', p. 11.
116. Campbell, 'A Model for Battalion Command', pp. 113–14; Wigmore, *The Japanese Thrust*, p. 218.
117. Wigmore, *The Japanese Thrust*, pp. 231–2.
118. The reliability of wireless communications had been a matter of concern for the divisional signals since their arrival in Malaya and led to a series of tests in May 1941. Although the 108 sets used for internal battalion communications were not tested, the 101 set used for communication with brigade headquarters was found to be reliable even in rubber plantations. Atmospheric conditions in Malaya were also found to cause some difficulties, but the times of this interference could be quite accurately predicted. The portable 108 sets were more susceptible to both the climatic and atmospheric conditions in Malaya and required well-trained battalion signallers to employ them to their full potential (Fisk, 'Suitability 109 and 101 Sets . . . Malaya', NAA MP508/1, item 305/761/1500).

119. No firm evidence has ever been produced that DF equipment was used by the Japanese in Malaya, and reports compiled by signals officers insisted that this did not occur ('Suggestions', p. 2; 'Signals in the Malayan Campaign', p. 3).
120. 'Signals in the Malayan Campaign', p. 3, 'Suggestions', p. 2; Thyer to Wigmore, 22 Sep 1951, p. 3, AWM 93 item 50/2/23/480; Barker, *Signals*, p. 240; Cahill, interview; Jacobs and Bridgland (eds), *Through*, p. 90.
121. Thyer, 'Comments on chapter 9. Australians into Battle', p. 2, AWM 93, item 50/2/23/480.
122. Kappe, 'The Malayan Campaign', Ch. 20, p. 30, AWM PR, item MSS 1393; Thyer, 'Comments on chapter 9. Australians into Battle', p. 2. Similar remarks are to be found in: 'Notes On Operations in Malaya', p. 5; and Thyer to Wigmore, 22 Sep 1951, p. 3.
123. Barker, *Signals*, p. 241.
124. 'Company Commanders' Observations', p. 3, WD GS 8 Div, 'Notes and Reports on the Malayan Campaign'.
125. Ibid.
126. O'Brien and Whitelocke, *Gunners in the Jungle*, p. 66; Johnston, 'Experimental Taping made with idea of adding to 2/30th Bn History', p. 6; Galleghan interview with Wigmore, 1 Feb 1950, AWM 73, item 51, p. 2; Arneil, *Black Jack*, pp. 81–9.
127. WD 2/18 Bn, 26–27 Jan 1942, AWM 52, item 8/3/18.
128. Reid, 'Road Block', p. 2, AWM PR, item MSS 1629.
129. Oakes to Wigmore, 22 March 1954, AWM 67, item 3/291; Cahill, interview; Anderson, 'Chapter 12. The Battle of Muar' (comments on draft of *The Japanese Thrust*), p. 2; AWM67, item 3/9.
130. Wall, *Singapore and Beyond*, p. 86.
131. Campbell, 'Second Day of Special Discussion Group, 2/30th Battalion', p. 3.
132. Warren, *Singapore 1942*, p. 173.
133. WD 2/29 Bn, 18 Jan 1942, AWM 52, item 8/3/29.
134. '2/29 Bn Comments by CO and additions to unit war diary for period subsequent to 25 Jan 42', pp. 9–14, WD 2/29 Bn, Feb 1942.
135. WD 2/18 Bn, Jan–Feb 1942, AWM 52, item 8/3/18, pp. 15–16.
136. Galleghan, 'Notes on Lessons from AIF Ops in Malaya', p. 3.
137. Anderson to Wigmore, 6 Nov 1954, p. 3, AWM 67, item 3/9.
138. Johnston and Eaton, 'Experimental taping made with idea of adding to 2/30th Bn History', p. 6.
139. Manston in Campbell, 'A Model for Battalion Command', p. 126; Maxwell, interview with Kirby and Wigmore, 26 Jan 1953, AWM 73, item 7; McLeod in Campbell, 'A Model for Battalion Command', p. 127.
140. Brown in Campbell, 'A Model for Battalion Command', p. 148.
141. Maxwell, interview with Long, Dec 1946, pp. 37–8.
142. Maxwell, interview with Kirby and Wigmore, 26 Jan 1953, p. 4, AWM 73, item 7. In an interview in 1953, Brigadier Arthur Blackburn, who was captured on Java, recalled a conversation he had with an 8th Division brigadier while a prisoner of war: '[He] talked to him about the crossing of the straits to Singapore Island. Blackburn said that he couldn't understand

how the Japs had made the crossing, and questioned the brigadier closely. After some time the brigadier said: "Look here, Arthur, I'll tell you what happened. I knew it was hopeless so I drew my men back from the beaches and let the Japs through"' (Blackburn, interview with Wigmore, Adelaide, 21 Jan 1953, p. 1, AWM 73, item 7). The brigadier could only have been Maxwell or Taylor. The attitudes expressed in this conversation conform with those of Maxwell, but the actions described in it are more akin to those of Taylor's troops. When Maxwell's withdrew they were ordered to maintain an unbroken front, whereas Taylor's were ordered to fall back on defended localities. Maxwell's views are well documented, but in post-war interviews Taylor also expressed the hopelessness of the situation that faced his brigade, strung out along a front that was almost 13 kilometres long.

143. Summary of interviews with Ramsay and Maxwell by Kirby and Wigmore, 26 Jan 1953, pp. 2–3; AWM 73, item 7; Thyer, interview with Kirby and Wigmore, Adelaide, 19 Jan 1953, pp. 1–2, AWM 73, item 7; 'Report by GSO 1 – HQ Malaya Comd', NAUK WO106, item 273c.
144. Eaton, 'Chapter 15' (comments on draft of *The Japanese Thrust*), pp. 2–3, AWM PR, item 00079.
145. Summary of interviews with Ramsay and Maxwell by Kirby and Wigmore, 26 Jan 1953, pp. 2–3; Kappe, 'The Malayan Campaign', chap. 33, pp. 46, 48; Singapore narrative, p. 17, WD 2/30 Bn, Jan–Feb 1942, AWM 52, item 8/3/30.
146. Eaton, 'Chapter 16' (comments on draft of *The Japanese Thrust*), p. 2; Wigmore, *The Japanese Thrust*, pp. 354–5; Singapore narrative, p. 17, WD 2/30 Bn, Jan–Feb 1942.
147. WD 2/18 Bn, Jan 1942, p. 7; Farrell, *The Defence and Fall of Singapore*, pp. 379–80.
148. Sturdee to Long, 8 Feb 1953, p. 2, AWM 67, item 3/384.
149. Beaumont, *Gull Force*; Nelson, *POW*; Henning, *Doomed Battalion*.
150. Hearder, 'Careers in Captivity'.
151. Thyer, 'Operations of 8th Australian Division in Malaya 1942', p. 193.
152. Oakes, comments on chapter 13 of the draft of *The Japanese Thrust*, p. 66, AWM 67, item 3/291.
153. Thyer, 'Operations of 8th Australian Division in Malaya 1942', p. 193.

6 'No place for half-hearted measures': Australia and Papua, 1940–42

1. 'The will to win', p. 1, WD HQ Nth Comd, Mar 1942, AWM 52, item 1/7/1.
2. Proceedings of the Promotions and Selection Board, NAA A2653, 1940 Vol. 4.
3. HQ Eastern Command selection and appointment files: NAA SP1048/6, C2/1/60, C2/1/177, C2/1/181, C2/1/209.
4. See Macfarlan, *Etched in Green*, p. 60.
5. Speed, *Esprit de Corps*, p. 278.
6. Mackay, 'Western Command', p. 2, AWM 3DRL 6850, item 121.
7. Ibid., p. 4.

8. Report on #26 COs and 2iCs Course, C&S School, Duntroon: Stewart, A. W. J., CARO Dossier, V41021, Stewart, A. W. J.
9. Report on #26 COs and 2iCs Course, C&S School, Duntroon: Moss, R. G., CARO Dossier, V43048, Moss, R. G.
10. Confidential Report on Officers: Moss, R. G., 30 Aug 1940, NAA MP 508/1, 251/704/53.
11. Report on #26 COs and 2iCs Course, C&S School, Duntroon: Sadler, R. M., CARO Dossier, V64336, Sadler, R. M.
12. McCarthy, *South-West Pacific Area*, p. 30.
13. Mackay, 'Preparedness of home forces for active operations', p. 1, AWM 3DRL 6853, item 127.
14. Mackay, 'Visit to units in Southern Command', p. 1, AWM 3DRL 6850, item 127.
15. For example, Savige wrote that his 'accustomed habits of good discipline and turnout' were offended by the low standards he encountered when taking command of the 3rd Division in June 1942 (Savige in Russell, *There Goes a Man*, p. 253).
16. WD HQ First Aust Army (GOC), 17 Jun 1942, AWM 52, item 1/3/1.
17. Dougherty, interview with Long, 18 Aug 1945, AWM 67, item 2/86, p. 39; Fearnside and Clift, *Dougherty*, p. 102; 'Training Instruction No 21', WD HQ 23 Bde, Apr 1942, AWM 52, item 8/2/23.
18. Milcommand Moresby, G.2887 of 4 Jun 1942, NAA MP729/6, 42/401/142.
19. Long, diary, 5 July 1942, AWM 67, item 1/8, pp. 34–5.
20. Morris, diary, 25 May 1941, AWM 3DRL 3402.
21. Ibid., 25 May 1941 and 16 August 1941.
22. Mackay, 'Message from GOC-in-C, Home Forces, to All Officers', p. 1, AWM 3DRL 6850, item 121.
23. WD HQ First Aust Army (GOC), 21 July 1942, AWM 52, item 1/3/1.
24. Long, diary, 30 Jun 1942, AWM 67, item 1/8, pp. 13–14.
25. Louch, 'Bde Commander's notes for the month', WD HQ 29 Bde, Jan 1942, AWM 52, item 8/2/29.
26. Stevens, 'Notes on Trg', 19 May 1942, p. 1, WD HQ 4 Div GS, May 1942, AWM 52, item 1/5/8.
27. Ibid.
28. Dougherty, interview with Long, 18 Aug 1945, AWM 67, item 2/86, pp. 39–40.
29. 'Confidential Report on Officers – Hale, F. W. H', 25 May 1940, NAA MP 508/1, item 251/704/31.
30. Pooler, correspondence, 7 February 2000, p. 1.
31. Mathews, *Militia Battalion at War*, p. 8.
32. '23 Aust Inf Bde Training Instruction No. 17', 6 November 1941, p. 1, WD HQ 23 Bde, November 1941, AWM 52, item 8/2/23.
33. Maitland, 'Memo re trg for month', July 1942, WD 43 Bn, June 1942, AWM 52, item 8/3/82.
34. Isaachsen, interview; Brune, *A Bastard of a Place*, pp. 560–2; Austin, *To Kokoda and Beyond*, p. 62; Cranston, *Always Faithful*, p. 171; Paull, *Retreat from Kokoda*, p. 14.

35. Mackay, 'Some Functions of Command', p. 2, AWM 3DRL 6850.
36. Austin, *To Kokoda and Beyond*, p. 12.
37. Blair, *Young Man's War*, p. 32.
38. Ibid., p. 43.
39. Horner, *Crisis of Command*, p. 54.
40. Dougherty, interview with Long, 17 Aug 1945, AWM 67, item 2/86, p. 41; Dougherty, correspondence with Herring, 5 Aug 1942, p. 2, WD HQ 23 Bde, Jul–Aug 1942, AWM 52, item 8/2/23.
41. Dougherty, correspondence with Herring, 5 Aug 1942, p. 1.
42. Long, diary, 9 July 1942, AWM 67, item 1/8.
43. 'Disposal of over-age and unfit personnel', 27 Apr 1942, WD First Aust Army (AG Br), Apr–May 1942, AWM 52, 1/3/10; 'Disposal of over-age and unfit personnel', 29 Apr 1942, WD HQ 5 Div (GS Br), Apr 1942, AWM 52, 1/5/11.
44. It was not long before the 3rd Battalion's CO, Albert Paul, was replaced, however. On the first day of the battalion's operations on the Kokoda Track Paul was evacuated after having been unable to complete the first major climb, Imita Ridge (Kennedy, *Port Moresby to Gona Beach*, p. 34).
45. Brig A. A. Brackpool, Comd 31 Bde, Mar 42–18 Feb 1943; Brig F. Hosking, Comd 10 Bde, Apr 1942–Sep 42, Comd 15 Bde, Oct 42–Jun 43; Brig G. F. Langley, Comd 2 Bde, Apr 1942–Jan 1944.
46. O. A. Kessels, CO 9/49 Bn, Aug 1939–Jul 1940, CO 9 Bn, Jul 1940–Jan 1942, CO 49 Bn, Jan 1942–Jul 1943; A. Meldrum, CO 61 Bn, May 1941–Aug 1943; E. S. Miles, CO 29 Bn Jun 1941–Jan 1944; K. H. Ward, CO 20/19 Bn, Sep 1939–Jun 1941, CO 20 Bn, Jul 1941–Sep 1941, CO 53 Bn, Sep 1941–Aug 1942.
47. Seventeen from the Middle East, one from Malaya and one from the South-West Pacific.
48. Like most aspects of officers' careers, the courses they attended were not consistently recorded in their personnel files. It can be determined, however, that of the 38 COs appointed during the period of the purge at least three had attended command schools run by the regional commands, five the LHQ Regimental Commanders School, one the Senior Officers Wing of the LHQ Tactical School, six the Senior Officers Wing of METS, one the AMF Command and Staff School, and one had graduated from the British Army's Senior Officers' School in 1918. In addition, at least five had completed the company commanders course at METS.
49. Crang, *The British Army and the People's War 1939–1945*, p. 47.
50. Army Council Paper 61 (1941), NAUK WO163, item 50 cited in Crang, *The British Army and the People's War 1939–1945*, pp. 47–8.
51. Executive Committee of the Army Council Paper 6 (1943), NAUK WO163, item 90 cited in Crang, *The British Army and the People's War 1939–1945*, p. 49.
52. Russell, *There Goes a Man*, p. 250.
53. Wilmoth, interview, 27 September 2000.
54. Clifford Frith was among a group of 9th Division officers recalled to Australia for duty with the militia.

55. Dougherty, 'Comments on Volume 5, Chapter 14 (draft) of Official War History (Gona)', p. 6, AWM 67, item 3/109, part 4.
56. WD First Aust Army (GOC), 18 May 1942, AWM 52, item 1/3/1, Apr–Oct 1942. This incident and Purser's dismissal receive no coverage in the 36th Battalion's history. See S. and L. Brigg, *The 36th Australian Infantry Battalion 1939–1945*, p. 16.
57. Dougherty, correspondence with Herring, 5 Aug 1942, p. 2.
58. Rowan in Blair, *A Young Man's War*, p. 60. See also Bilney, *14/32 Australian Infantry Battalion AIF 1940–1945*, p. 39, and Austin, *Bold, Steady, Faithful*, p. 167.
59. For example: 'The effort asked of all ranks has been considerable but it is only in keeping with the seriousness of the threat to this country, and the possibility of our being called upon to fight any day now without any further tr[aining]. Every minute is precious and the burden of responsibility must rest on off[ice]rs and NCOs who will lead their men into battle, trained only to the degree to which their efforts have raised them' ('2 Inf Bde Gp Trg Instn', 10 May 1942, WD HQ 2 Bde, May 1942, AWM 52, item 8/2/2).
60. Coyle in Austin, *Bold, Steady, Faithful*, p. 167.
61. Wilmoth, interview.
62. See Macfarlan, *Etched in Green*, p. 61.
63. Austin, *Bold, Steady, Faithful*, p. 167.
64. Charlott, *The Unofficial History of the 29/46th Australian Infantry Battalion*, p. 6.
65. Coyle in Austin, *Bold, Steady, Faithful*, p. 167.
66. Pearce in ibid., p. 187.
67. Wilmoth, 'Wandering recollections of John Wilmoth aged 83 years', p. 83.
68. Wilmoth, interview.
69. WD HQ 12 Div GS, Jun–Dec 1942.
70. WD HQ 4 Div GS, Jul–Dec 1942.
71. 'Combat efficiency report on formations of Allied Land Forces as at 20 Oct 42', sheet 2, AWM 54, item 305/2/1.
72. 'Combat efficiency report on formations of Allied Land Forces as at 20 Nov 42', sheet 3, AWM 54, item 305/2/1.
73. C. B. Withy, CO 2/25 Bn, Oct 1940–Oct 1942.
74. Rhoden, interview, Oct 2000.
75. Austin, *To Kokoda and Beyond*, pp. 56, 128; Russell, *The Second Fourteenth Battalion*, pp. 120, 161.
76. Each 16th Brigade battalion carried two of each weapon, while those of the 25th only carried one ('Report on operations in New Guinea. 16 Aust Inf Bde', pp. 1–6, AWM 54, item 577/7/29; '25 Australian Infantry Brigade: Some notes on the New Guinea Campaign, Sept to Dec 1942', part 11, vii–viii; AWM 54, item 577/7/31). This section has been informed by: Rhoden, interview; Cullen, interview, 27 Feb 2001; Sublet, interview, 12 Sep 2000; Cooper, interview, 8 Feb 2001; 'Notes on recently expressed concepts of tactics' (30 Bde), p. 8, AWM 54, item 923/1/6; Burns, *The Brown and Blue Diamond at War*, pp. 109, 111.
77. Allchin, *Purple and Blue*, p. 92.

78. Rhoden, interview; Cullen, interview; Sublet, interview; Porter, 'Notes on Japanese tactics etc, and lessons learnt Kokoda–Alola Area', p. 1, AWM 54, item 577/7/28.
79. Paull, *Retreat from Kokoda*, p. 198; '25 Australian Infantry Brigade notes on the New Guinea Campaign, Sept to Dec 1942', AWM 54, item 577/7/31; Isaachsen, interview.
80. Rhoden, interview.
81. '25 Australian Infantry Brigade notes on the New Guinea Campaign, Sept to Dec 1942'; 'Notes on recently expressed concepts of tactics' (30 Bde), p. 8.
82. Laffin, *Forever Forward*, p. 86.
83. Cooper, interview; J. Burns, *The Brown and Blue Diamond at War*, p. 147.
84. Brune, *A Bastard of a Place*, pp. 452–3.
85. Cooper served as a company commander with the 2/10th Battalion during the siege of Tobruk.
86. Brune, *A Bastard of a Place*, p. 327.
87. Brune, *The Spell Broken*, pp. 60–1, 66, 67, 72.
88. '25 Australian Infantry Brigade notes on the New Guinea Campaign'.
89. '21 Australian Infantry Brigade report on operations in the Owen Stanley Range, 15 August–20 September 1942', p. 62, AWM 54, item 577/7/3.
90. Rhoden, interview.
91. Cullen, interview.
92. Ibid.; '25 Australian Infantry Brigade notes on the New Guinea Campaign'; Sublet, *Kokoda to the Sea*, p. 119.
93. '25 Australian Infantry Brigade notes on the New Guinea Campaign'; 'Report on operations in New Guinea. 16 Aust Inf Bde', pp. 5–6, AWM 54, item 577/7/29.
94. Honner, 'Notes on and lessons from recent operations in Gona and Sanananda areas', p. 1, AWM 67, item 3/170, part 2.
95. Parry-Okeden, notebook, pp. 77–8, AWM PR, item 00321.
96. 'Report on part played by 30 Aust Inf Bde HQ in Owen Stanley Ranges', p. 5, AWM 54, item 923/1/6.
97. The strategic and operational politics of the Papuan campaign have been extensively explored by David Horner. See Horner, *Crisis of Command*, and *Blamey*, pp. 316–84.
98. Burns, *The Brown and Blue Diamond at War*, p. 147.
99. Isaachsen, interview with Brune, Jun 1989, cited in Brune, *The Spell Broken*, p. 232; Sublet, *Kokoda to the Sea*, p. 168.
100. Cooper, interview; Cullen, interview; Sublet, interview; Dunbar to Sweeting, 19 Mar 1957, p. 2, AWM 67, item 3/112; Brune, *A Bastard of a Place*, pp. 464, 500, 534–5; Sublet, *Kokoda to the Sea*, pp. 159, 175; Eather, *Desert Sands, Jungle Lands*, p. 118.
101. Dougherty, 'Comments on Volume 5, Chapter 14 (Draft) of Official War History (Gona)', p. 5.
102. Each stretcher required up to eight men to carry it and a similar number to relieve them.

103. Burns, *The Brown and Blue Diamond at War*, p. 131.
104. Paull, *Retreat from Kokoda*, p. 236.
105. Brune, *The Spell Broken*, pp. 59–61, 65, 82, 84; *A Bastard of a Place*, pp. 320–1; McCarthy, *South-West Pacific Area*, pp. 166–7.
106. Sublet, *Kokoda to the Sea*, p. 105; McCarthy, *South-West Pacific Area*, p. 287; Cullen, interview.
107. McCarthy, *South-West Pacific Area*, p. 290; Sublet, *Kokoda to the Sea*, p. 106; Cullen, interview.
108. Cullen, interview.
109. McCarthy, *South-West Pacific Area*, pp. 302–3; WD 2/3 Bn, 27–28 Oct 1942, AWM 52, item 8/3/3.
110. Cummings, interview with Baker and Knight in Brune, *A Bastard of a Place*, p. 500.
111. Spencer, *In the Footsteps of Ghosts*, pp. 124–5.
112. Cullen, interview; Cooper, interview.
113. Cooper, interview.
114. Cullen, interview.
115. Cooper, interview.
116. Honner, interview with Brune, no date, cited in Brune, *A Bastard of a Place*, p. 468.
117. Cooper, interview.
118. Eather, *Desert Sands, Jungle Lands*, p. 118; Sublet, *Kokoda to the Sea*, p. 159.
119. Dougherty, 'Comments on Volume 5, Chapter 14 (Draft) of Official War History (Gona)', p. 15.
120. Eather, *Desert Sands, Jungle Lands*, p. 118.
121. Honner, 'The 39th at Gona', pp. 1–2, AWM 67, item 3/170, part 3.
122. Dunbar to Sweeting, 19 Mar 1957, pp. 1–2; Cullen, interview; Stevens, 'A personal story', p. 79.
123. Honner in Austin, *To Kokoda and Beyond*, p. 157.
124. Honner, interview with Brune, no date, cited in Brune, *A Bastard of a Place*, p. 469.
125. Honner, 'The 39th at Gona', p. 3; Honner to Sublet, 2 Feb 1993, cited in Sublet, *Kokoda to the Sea*, p. 169.
126. Honner, 'The 39th at Gona', p. 4; McCarthy, *South-West Pacific Area*, pp. 440–1; Brune, *Those Ragged Bloody Heroes*, pp. 256–8.
127. Sublet, interview.
128. Ibid.
129. McCarthy, *South-West Pacific Area*, p. 302. Gavin Long, after meeting Hutchison at Ravenshoe in July 1943, described him as 'slow and singled-minded' (Long, diary, 19 Jul 1943, AWM 67, item 1/2).
130. Stevens, 'A personal story', p. 79.
131. Cullen, interview; Rhoden, interview.
132. Sublet, interview.
133. Dougherty, 'Comments on Volume 5, Chapter 14 (Draft) of Official War History (Gona), p. 7; Brune, *Those Ragged Bloody Heroes*, pp. 199–206; McCarthy, *South-West Pacific Area*, pp. 334–5 (fn); Horner, *Blamey*, pp. 352–3, 367; Paull, *Retreat from Kokoda*, pp. 257–8.

134. Parry-Okeden, notebook, p. 73.
135. Lovett, interview with Brune, 5 Oct 1998, cited in Brune, *We Band of Brothers*, p. 146.
136. Allchin, *Purple and Blue*, pp. 292–3; Parry-Okeden, notebook, p. 45.
137. Parry-Okeden, notebook, p. 18.
138. Scott, in Brune, *A Bastard of a Place*, p. 455.
139. Lovett, interview with Brune, 5 October 1998, cited in Brune, *We Band of Brothers*, pp. 141, 150; McCarthy, *South-West Pacific Area*, p. 141.
140. Merritt, in Austin, *To Kokoda and Beyond*, p. 154.
141. G. A. Vasey to J. Vasey, 10 Jul 41, pp. 5–6, NLA MS 3782, folder 12.
142. Gwillim, correspondence with Brune, Jul 1987, cited in Brune, *We Band of Brothers*, p. 154.
143. Ibid.; Russell, *The Second Fourteenth Battalion*, p. 114.
144. WD 39 Bn, 17 Aug 1942, AWM 52, item 8/3/78; Honner, interview with Long, 27 Sep 1944, AWM 67, item 2/50.
145. Brune, *A Bastard of a Place*, pp. 464, 538; *Those Ragged Bloody Heroes*, pp. 265–6; *Gona's Gone*, pp. 93–4. This argument is also made in Sublet, *Kokoda to the Sea*, p. 175 (fn).
146. Dunbar to Sweeting, 19 Mar 1957, p. 2.
147. Brune, *A Bastard of a Place*, p. 538.
148. Sublet, correspondence with Brune, 10 Jul 1987, cited in Brune, *We Band of Brothers*, p. 156; Dougherty subsequently wrote in the adverse report that resulted in Caro's dismissal that he was 'likely to waver with decisions and has NOT the strength of personality to create a highly aggressive attitude and high standard of discipline in the tps under his comd' (Dougherty to Vasey, 8 Dec 1942, p. 1, AWM 3DRL 6643, item 3/60).
149. Laffin, *Forever Forward*, p. 94.
150. WD 2/27 Bn, 29 Nov 1942, AWM 52, item 8/3/27; Dougherty, 'Comments on Volume 5, Chapter 14 (draft) of Official War History (Gona)', p. 13; McCarthy, *South-West Pacific Area*, p. 428. Cooper, in an interview with Brune, maintains he did not wait and Brune states that the 2/27th reached the start line on time (Brune, *Those Ragged Bloody Heroes*, p. 234). Both of these assertions are contradicted by the 2/27th war diary.
151. Dougherty, 'Some background for Gona', p. 5, AWM 67, item 3/109, part 4.
152. Honner, interview with Brune, cited in *We Band of Brothers*, p. 56.
153. Cooper, interview.
154. Cudden, correspondence.
155. Rhoden, interview.

7 'There is no mystery in jungle fighting': The New Guinea offensives

1. The 17th Brigade report on its operations in the Salamaua hinterland exemplified this attitude: 'Our troops gained a moral ascendency over the Japanese in the early fighting at Wau . . . and this was retained throughout. They were better trained and equipped than the Jap. They had complete mastery of their weapons, were skilfully led by their junior leaders, were keen, imaginative and quick "on the draw", their jungle craft had been

learned in a hard school of bitter experiences and had risen to a high pitch of efficiency. In combat with the Jap our men were ruthless and merciless. The Jap knew this and feared them' ('Lessons learnt from the operations', p. 2, Appendix 14 to 'Report on the operations of 17 Aust Inf Bde Gp in the Mubo–Salamaua Area from 23 Apr to 24 Aug 43', AWM 54, item 587/7/21, part 2; hereafter '17 Bde Lessons – Mubo–Salamaua').
2. Dexter, *The New Guinea Offensives*.
3. C. C. F. Bourne, 2/12 Bn, WIA, Shaggy Ridge, 21 Jan 1944; R. H. Honner, 2/14 Bn, WIA, Ramu Valley, 4 Oct 1943; T. H. Scott, 2/32 Bn, WIA, Lae, 14 Sep 1943; R. E. Wall, 2/23 Bn, KIA, Lae, 4 Sep 1943. A full analysis of CO casualties appears at appendix 1.
4. In 1945 the LHQ Tactical School was merged with the School of Military Intelligence to become the Land Headquarters School of Tactics and Intelligence.
5. 'Directorate of Military Training. Account of activities: 3 Sep 39–15 Aug 45', p. 3, AWM 54, item 937/1/2.
6. Ibid., pp. 3, 11.
7. Ibid., p. 11; 'Memo to students in explanation of series "Tac Principles and Lessons"', p. 1, AWM 54, item 881/1/4, part 1.
8. 'Directorate of Military Training. Account of activities: 3 Sep 39–15 Aug 45', p. 11.
9. 'LHQ Tactical School. List of précis', AWM 54, item 881/1/4, part 1.
10. For instance, Colin Norman of the 2/28th Battalion attended the Senior Officers Course at METS, the Middle East Combined Operations Training Centre, the Joint Operations Overseas Training School and the Senior Officers wing at the LHQ Tactical School, as well as courses covering the employment of anti-tank guns and the duties of a motorised contact officer. By the time he attended the LHQ Tactical School, Norman had served in the 9th Division's two campaigns in the Middle East and commanded the 2/28th Battalion throughout the Huon Peninsula operations (CARO dossier, WX3421, Norman, C. H. B.).
11. 'LHQ Tactical School. Some points about the course', p. 1, AWM 54, item 881/1/4, part 1.
12. AWM 54, item 225/1/14.
13. CGS 130774, 'LHQ Tactical School', 20 Sep 1944, pp. 1–2, AWM 193, item 292.
14. 'Memo to students in explanation of series "Tac Principles and Lessons"', p. 1.
15. 'LHQ Tactical School. Some points about the course', p. 2.
16. 'LHQ Tactical School – Grading of students', 30 Nov 1942, NAA MP 742/1, 240/1/1325; 'LHQ Tactical School. Some points about the course', p. 4.
17. Various course reports for A Wing, LHQ Tactical School contained in CARO dossiers; Anderson (BGS Adv LHQ), to Comdt LHQ Tactical School, 21 Aug 43, NAA MP 742/1, 240/1/1325.
18. In his diary for 7 September 1944, William Parry-Okeden of the 30th Battalion reflected on the friendships he made while attending the LHQ

Tactical School, men whom he described as 'grand chaps and amongst the cream of the AIF' (Parry-Okeden, diary, 7 Sep 1944, AWM PR00321).
19. Rhoden, interview.
20. 'LHQ Tactical School Standing Orders', pp. 3, 5.
21. CGS 130774, 'LHQ Tactical School', 20 Sep 1944, p. 1, AWM 193, item 292.
22. T. R. W. Cotton, 2/33rd Bn; R. H. Marson, 2/25 Bn; and E. M. Robson, 2/31st Bn.
23. P. W. Crosky, 4 Bn; D. F. Rae, 35th Bn.
24. A. J. Anderson, 24 Bn; F. H. M. Armstrong, 35 Bn; E. Barnes, 19 Bn; B. J. Callinan, 26 Bn; J. de M. Carstairs, 22 Bn; G. S. Cox, 2/4 Bn; H. L. E. Dunkley, 7 Bn; F. J. Embrey, 37/52 Bn; C. H. Green, 2/11 Bn; J. G. Hendry, 2/28th; D. A. C. Jackson, 2/6 Bn; M. R. Jeanes, 2/43 Bn; J. L. A. Kelly, 31/51 Bn; W. M. Mayberry, 58/59 Bn; H. H. McDonald, 15 Bn; D. D. Pitt, 37/52nd Bn; P. M. Shanahan, 55/53 Bn; and H. G. Sweet, 11 Bn.
25. P. G. C. Webster, 57/60 Bn.
26. C. J. A. Coombes, 47 Bn; N. W. P. Farrell, 2/4 Bn; W. G. Fry, 47 Bn; and L. de L. Miell, 19 Bn.
27. T. J. Daly, 2/10 Bn; L. J. Loughran, 8 Bn; and P. K. Parbury, 2/7 Bn.
28. J. N. Abbott, 26 Bn; R. I. Ainslie, 2/48 Bn; B. N. Berry, 41 Bn; W. B. Caldwell, 14/32 Bn; P. A. Cullen, 2/1 Bn; A. B. Gillespie, 2/24 Bn; H. G. Guinn, 2/7 Bn; W. S. Howden, 2/8 Bn; I. Hutchison, 2/3rd Bn; D. J. H. Lovell, 55/53 Bn; G. R. Matthews, 9 Bn; C. H. B. Norman, 2/28 Bn; W. N. Parry-Okeden, 30 Bn; A. Pope, 27 Bn; P. E. Rhoden, 2/14 Bn; R. L. Sandover, 2/11 Bn; and F. H. Sublet, 2/16 Bn.
29. 'Note on the Australian Military Forces', 2 Aug 1945, p. 2, NAUK WO203, item 4125.
30. Rhoden, interview.
31. Palazzo, *The Australian Army*, p. 153.
32. 'Organization for Jungle Warfare', AWM 54, 721/2/11.
33. 'The Jungle Div and Jungle Corps', p. 2; 'Organization for Jungle Warfare', p. 2.
34. 'The Jungle Div and Jungle Corps', p. 2; AMF WE II/12C/2, 'An Infantry Battalion (Tropical Scale)', AWM 54, 327/6/9.
35. 'Tank Attack Guns in Infantry Battalions (Tropical Scale)', p. 1, AWM 54, 905/23/11.
36. AMF WE II/12C/3, 'An Infantry Battalion (Tropical Scale)', p. 4; Amendment No. 7 to AMF WE II/12C/3, 'An Infantry Battalion (Tropical Scale)'.
37. 'The Jungle Div and the Jungle Corps', pp. 2, 3.
38. Norman, 'Notes on operations Red Beach–Lae, 15 Sep–16 Sep 43', p. 2, AWM 54, 589/7/36. Norman's full name was Colin Hugh Boyd Norman, but he seems to have preferred his middle name. He signed his reports 'H. B. Norman' and was known by both the officers and men of the 2/28th Battalion as 'Hugh'.
39. 'Report on operations of 9 Australian Division, Lae–Finschhafen–Huon Peninsula, 4 Sep 1943–15 Jan 1944', p. 77. This is a common theme of

post-operations reports and other reflections on the New Guinea offensives. See, for example: Savige, 'Volume VI. Chapter VII. Komatium and Mount Tambu' (notes on draft of *The New Guinea Offensives*), p. 1, AWM 67, item 3/348, part 3) and 'Report on operations – New Guinea. Ramu Valley–Shaggy Ridge. Operations by 18 Aust Inf Bde 1 Jan –6 Feb 44', p. 12, AWM 54, item 595/6/3 (hereafter '18 Bde Report – Ramu Valley–Shaggy Ridge Operations').

40. Starr, comments on Chapter 3 of *The New Guinea Offensives*, p. 2, AWM 172, item 53; 'Doublet. Lessons from operations – own tactics 30 Jun–6 Aug 43', p. 6, AWM 54, item 587/7/18, part 2; Dexter, *The New Guinea Offensives*, pp. 129, 170; Mathews, *Militia Battalion at War*, p. 39.
41. Savige, 'Volume VI. Chapter VII. Komiatum and Mount Tambu', p. 1, AWM 67, item 3/343, part 2; 'Report on operations of 3 Aust. Div. in Salamaua area from 22 April 43 to 25 Aug 43', p. 137, AWM 54, item 587/7/12, part 2 (hereafter '3 Div Report – Salamaua'); 'Report on operations of 9 Australian Division, Lae–Finschhafen–Huon Peninsula, 4 Sep 1943–15 Jan 1944', p. 79 (hereafter '9 Div Report – Lae–Finschhafen–Huon Peninsula'); '17 Bde Lessons – Mubo–Salamaua', pp. 4, 5.
42. '18 Bde Report – Ramu Valley–Shaggy Ridge', p. 11. See also '17 Bde Report – Mubo–Salamaua', p. 2.
43. Savige, 'Volume VI. Chapter VII. Komiatum and Mount Tambu', p. 2; 29 Aust Inf Bde Lessons from Operations – Salamaua Campaign', p. 1, WD HQ 29 Bde, Sep–Oct 1943, AWM 52, item 8/2/29.
44. Norman to Long, 6 April 1956, p. 3, AWM 93, item 50/2/23/578.
45. '18 Bde Report – Ramu Valley–Shaggy Ridge', p. 9.
46. Dexter, *The New Guinea Offensives*, p. 747.
47. Graeme-Evans, *Of Storms and Rainbows*, p. 368.
48. '18 Bde Report – Ramu Valley–Shaggy Ridge', p. 9. The 2/12th's CO, Charles Bourne, was wounded by shrapnel from a Japanese mountain gun during this attack.
49. Long, diary, 26 Jul 1943, pp. 50–1, AWM 67, item 1/2.
50. Lieutenant General Herring discussed this aspect of operations in a letter to the official historian, David Dexter: Herring to D. St A. Dexter, 21 Jan 1952, pp. 1–2, AWM 67, item 3/167, part 3.
51. '3 Div Report – Salamaua', p. 137.
52. '9 Div Report – Lae–Finschhafen–Huon Peninsula', p. 81; Savige, 'Volume VI. Chapter V. The Capture of Mubo', p. 16; Coates, *Bravery Above Blunder*, p. 184.
53. Savige, 'Volume VI. Chapter V. The Capture of Mubo', p. 16; '9 Div Report – Lae–Finschhafen–Huon Peninsula', p. 77.
54. '15 Aust Inf Bde Report on Operations, Part IV: Lessons from operations Ramu Valley–Orgoruna–Mindjim Valley–Bogadjim–Madang 1 Jan 44 to 30 Apr 44', p. 11, AWM 54, item 595/6/7, part 1 (hereafter '15 Bde Report – Ramu Valley–Madang'); '9 Div Report – Lae–Finschhafen–Huon Peninsula', p. 77; '17 Bde Lessons – Mubo–Salamaua', p. 6.
55. '17 Bde Lessons – Mubo–Salamaua', p. 4.

56. '9 Div Report – Lae–Finschhafen–Huon Peninsula', p. 78.
57. Norman, 'Report on operations of 2/28 Aust. Inf. Bn 5–16 Sept '43', p. 1, AWM 54, item 589/7/36; 'Notes on operations Red Beach–Lae, 15 Sep–16 Sep 43', p. 1, AWM 54, item 589/7/36.
58. '9 Div Report – Lae–Finschhafen–Huon Peninsula', p. 80; '29 Aust Inf Bde Training Instruction period 2 Oct to 24 Oct 43', p. 2, WD HQ 29 Bde, Sep–Oct 1943, AWM 52, item 8/2/29.
59. '17 Bde Lessons – Mubo–Salamaua', p. 5. See also 3 Div Report – Salamaua, p. 136.
60. '9 Div Report – Lae–Finschhafen–Huon Peninsula', p. 80.
61. C. H. Grace (2/15 Bn), 'Chapter 17' (comments on Dexter, *The New Guinea Offensives*), para. p. 75, AWM 93, item 50/2/23/170.
62. Brigadier Hammer recalled ordering the 57/60th Battalion to make a night move off the Prothero feature on Shaggy Ridge in order to maintain contact with the withdrawing Japanese. Hammer noted the 'misgivings' of the 57/60th's CO Robert Marston but recorded that the move was a success and that the rapid regaining of contact was 'notable' and 'worthy of recording' (Hammer, 'Ramu' (comments on draft of Dexter, *The New Guinea Offensives*), p. 2, AWM 93, item 50/2/23/440).
63. '9 Div Report – Lae–Finschhafen–Huon Peninsula', p. 82.
64. There are numerous examples of this occurring. In one instance, the Victoria Cross winning efforts of Sergeant Tom 'Diver' Derrick at Sattelberg were precipitated by an order from the 2/48th Battalion's CO, Robert Ainslie, for B Company to break off its attack and withdraw because there was no chance of it succeeding before nightfall (Coates, *Bravery Above Blunder*, pp. 225–6; Wigmore et al., *They Dared Mightily*, p. 144). In another, Colin Grace halted the 2/15th Battalion's advance towards a suspected Japanese headquarters in Nanda late in the afternoon of 31 December 1943 because he considered it too late in the day to attack (Grace, 'Chapter 27' (comments on draft of Dexter, *The New Guinea Offensives*), p. 1, AWM 93, item 50/2/23/170).
65. '9 Div Report – Lae–Finschhafen–Huon Peninsula', p. 78.
66. '18 Bde Report – Ramu Valley–Shaggy Rudge', p. 11.
67. First Australian Army School of Infantry – Mobile Tactical Wing, précis No. 9, 'Control in Battle', p. 2, AWM 54, item 881/11/2.
68. '15 Bde Report – Ramu Valley–Madang', p. 13.
69. Bunbury, '24th Aust. Inf. Battalion (AIF) Kooyong Regiment', *Red and White Diamond*, March 1989, p. 4.
70. '15 Bde Report – Ramu Valley–Madang', p. 14.
71. '18 Bde Report – Ramu Valley–Shaggy Ridge', p. 9.
72. '3 Div Report – Salamaua', pp. 144, 145.
73. An insight into Dexter's character is provided by his description of Lababia Ridge in a letter to his brother. Asked to comment on remarks made by Henry Guinn of the 2/7th Battalion that the ridge was a 'terrific feature' with the 'appearance of some huge pre-historic monster that had come to rest on Earth' (D. St A. Dexter to W. R. Dexter, 29 Sep 1953, p. 1, AWM 67, item 50/2/23/99), Dexter replied: ' . . . it was certainly high; but actually

just another hill. Once you were used to her you could run up her with a loaded pack' (W. R. Dexter to D. St A. Dexter, 5 Oct 1953, p. 1, AWM 67, item 50/2/23/99).
74. Moten, interview with D. St A. Dexter, 11 April 1951, AWM 172, item 9; W. R. Dexter 'Chapter III, Jungle Stalemate' (comments on draft of Dexter, *The New Guinea Offensives*), p. 2, AWM 93, item 50/2/23/99.
75. WD 2/6 Bn, 21 Jun 1943, AWM 52, item 8/3/6.
76. Savige, 'Notes on Volume V. Chapter 3' (comments on draft of Dexter, *The New Guinea Offensives*).
77. DSO citation, VX5172, Dexter, W. R., CARO Dossier VX5172, Dexter, W. R.
78. Coates, *Bravery Above Blunder*, pp. 191, 219.
79. Glenn, *Tobruk to Tarakan*, pp. 271, 273.
80. Graeme-Evans, *Of Storms and Rainbows*, p. 371.
81. '18 Bde Report – Ramu Valley–Shaggy Ridge', p. 9.
82. '15 Aust Inf Bde. Lessons extracted from report by 57/60 Aust Inf Bn', p. 3, AWM 54, item 595/6/7; Dexter, *The New Guinea Offensives*, p. 776.
83. Moten, interview with D. St A. Dexter, 11 April 1951, AWM 172, item 9.
84. 'Handful of Australians routs Japanese on Tambu', p. 1, AWM 3DRL 5043, item 23.
85. Dexter, *The New Guinea Offensives*, p. 163.
86. Savige, 'Notes on Volume V, Chapter 6. Closing In' (comments on draft of Dexter, *The New Guinea Offensives*), p. 20, AWM 67, item 3/348, part 2. Savige commented: 'I felt . . . that Conroy was developing too many frontal attacks, which, in turn, obtained heavy casualties without any recompense in progress.' He also noted an interview with Major Vernon Walters, OC A Company, 2/5th Battalion, in which Walters stated he had lost confidence in Conroy and was openly bitter at some of the frontal attacks that were ordered. The Tambu operations were the 2/5th's last of any consequence during the New Guinea offensives, and Conroy remained in command when the battalion returned to Australia. It was subsequently recommended that, as 'a skilled industrial chemist', he be returned to his civilian occupation. He was seconded as an instructor at the LHQ Tactical School for a short period before being discharged in October 1944 (CARO dossier, SX8886, Conroy, T. M.).
87. 'Doublet. Lessons from operations – own tactics 30 Jun–6 Aug 43', p. 1.
88. '15 Bde Report – Ramu Valley–Madang', p. 7.
89. Ibid., p. 4.
90. Norman, 'Notes on operations Red Beach–Lae, 15 Sep–16 Sep 43', p. 1, AWM 54, 589/7/36.
91. 'Doublet. Lessons from operations – own tactics 30 Jun–6 Aug 43', p. 3.
92. Ibid., p. 5. See also 'Attack and Defence of Old Vickers Position', p. 1, AWM 54, item 587/7/18, part 1.
93. HQ Kanga Force G19/18, 20 Mar 1943, 'VX 46 Major (T/Lt-Col) P. D. Starr', AWM 3DRL 5043, item 38b.
94. DADO 074, 28 Feb 1943, AWM 3DRL 5043, item 38b.

95. 0316, 15 Mar 1943, AWM 3DRL 5043, item 38a.
96. Hammer, interview with D. St A. Dexter, 26 May 1951, p. 1, AWM 93, item 50/2/23/440.
97. '15 Bde Report – Ramu Valley–Madang', p. 19.
98. 'Lessons extracted from report by 24 Aust Inf Bn', p. 4, AWM 54, item 595/6/7.
99. '15 Bde Report – Ramu Valley–Madang', p. 19.
100. Whitehead, 'Chapter 22' (comments on draft of Dexter, *The New Guinea Offensives*), pp. 1–2, AWM 93, item 50/2/23/476.
101. 'Signals', Appendix F to '7 Aust Division Report on Operation Outlook', p. 3, AWM 54, item 589/7/2, part 2.
102. Wade, 'Report by Lieut-Col R. E. Wade. Observer with 25 Aust Inf Bde Gp during operations 6/16 leading to the capture of Lae', p. 6, AWM 54, item 589/7/33; Norman, 'Report on operations of 2/28 Aust. Inf. Bn 5–16 Sept '43', p. 4; Whithead, 'Chapter 14', pp. 1–2, 'Chapter 22', p. 1 (comments on draft of Dexter, *The New Guinea Offensives*), AWM 93, item 50/2/23/476.
103. '15 Bde Report – Ramu Valley–Madang', p. 19.
104. Graham-Sutton, 'Observer report by Major H. Graham-Sutton on Cape Gloucester Operations', p. 13, AWM 54, item 609/7/14.
105. DSO citation, NX12217, Colvin, G. E., CARO Dossier NX12217, Colvin, G. E.
106. Windeyer, interview with D. St A. Dexter, 11 May 1952, p. 1, AWM 93, item 50/2/23/250; Coates, *Bravery Above Blunder*, p. 114; Simpson, interview with D. St A. Dexter, 10 September 1951, p. 3, AWM 93, item 50/2/23/474.
107. Windeyer, 'Comments on draft Chapter 12 – Assault on Lae and Nadzab' (comments on draft of Dexter, *The New Guinea Offensives*), p. 2, AWM 93, item 50/2/23/250.
108. Dexter, *The New Guinea Offensives*, p. 347.
109. Windeyer, 'Chapter 14 – The Fall of Lae' (comments on draft of Dexter, *The New Guinea Offensives*), p. 1, AWM 93, item 50/2/23/250.
110. Unknown 'contemporary' in Coates, *Bravery Above Blunder*, p. 153.
111. Windeyer, 'Comments on draft Chapter 12 – Assault on Lae and Nadzab', p. 2.
112. Simpson, interview, p. 3.
113. Long, diary, 11 Nov 1943, p. 85, AWM 67, item 1/3.
114. DSO citation, NX457, Grace, C. H., CARO Dossier NX457, Grace, C. H.
115. Coates, *Bravery Above Blunder*, p. 153.
116. Stuart in Austin, *Let Enemies Beware*, p. 293.
117. Austin, *Let Enemies Beware*, pp. 142–5; Cowie, interview, 15 Feb 2001. Despite being 2iC, Grace was not appointed CO after operation Bulimba, but was superseded by Keith Magno, from the 2/17th.
118. Coates, *Bravery Above Blunder*, p. 153.
119. Windeyer, interview, p. 3.
120. Windeyer, 'Comments on draft Chapter 12 – Assault on Lae and Nadzab', pp. 2–3.

121. Newton in 2/17 Battalion History Committee, *'What We Have . . . We Hold!'*, p. 158.
122. Windeyer, 'Comments on draft Chapter 12 – Assault on Lae and Nadzab', pp. 2–3.
123. Broadbent in 2/17 Battalion History Committee, *'What We Have . . . We Hold!'*, p. 85.
124. Broadbent, interview; Broadbent in 2/17 Battalion History Committee, *'What We Have . . . We Hold!'*, p. 85; Windeyer, 'Comments on draft Chapter 12 – Assault on Lae and Nadzab', pp. 2–3; Coates, *Bravery Above Blunder*, p. 192.
125. Fearnside, *Bayonets Abroad*, p. 178.
126. Simpson had been known as the 'Red Fox' when he was the 2iC of the 2/13th Battalion. Its battalion history simply introduces him as such (Fearnside, *Bayonets Abroad*, p. 6). Neither Coates nor Johnston and Stanley offer any explanation either. Not even Simpson himself throws much light on the issue. Commenting on Chapter 22 of the draft of *The New Guinea Offensives*, he explained: 'My "nick-name" was "The Red Fox". During an exercise on [sic] Syria in early 1942 this sobriquet was really confirmed – although I was so known in the 2/13 Bn. A fox was seen to run across the battalion front and a signaller of one of the forward companies sent a message to its neighbouring company to the effect that a fox was moving across the battalion front. Simultaneously I moved across to see how the exercise was progressing . . . I asked the late Lt. Col. Turner (2/13 Bn) the reason for this "nick-name". He replied, "You crafty old . . . [sic], you always turn up in the most unexpected places"' (N. W. Simpson, 'Towards Sattelberg' (comments on draft of *The New Guinea Offensives*), p. 4, AWM 93, item 50/2/23/474).
127. Ibid.
128. 2/17 Battalion History Committee, *'What We Have . . . We Hold!'*, p. 268.
129. Norman to Long, 6 Apr 1956 (comments on draft of *The New Guinea Offensives*), p. 3, AWM 93, item 50/2/23/578.
130. Coates, *Bravery Above Blunder*, p. 219.
131. Unknown 2/48th soldier, interview with G. M. Long, 21 Nov 1943, AWM 67, item 2/23, p. 43.
132. '9 Div Report – Lae–Finschhafen–Huon Peninsula', p. 79.
133. Windeyer, 'Comments on Chapter 27 – "To Sio"', p. 1, AWM 93, item 50/2/23/250. This method of operations was maintained throughout the Australian pursuit of the retreating Japanese all the way to Madang. The 8th Brigade's report on its operations noted: 'From the commencement of operations it was understood that casualties should be kept to a minimum and artillery support was always to be available to cover the advance' ('Operations. 8th Australian Infantry Brigade Group', p. 1, WD HQ 8 Bde, AWM 54, item 8/2/8, Jan 1944).
134. Grace, 'Chapter 27' (comments on draft of *The New Guinea Offensives*), p. 1, AWM 93, 50/2/23/170.
135. Ibid.

136. 'Report on employment of tanks in operations of 9 Aust Div in area north of Finschhafen, New Guinea', p. 4, AWM 54, item 519/6/65.
137. Coates, *Bravery Above Blunder*, p. 222.
138. Stuart in Austin, *Let Enemies Beware!*, p. 293.
139. Norman, 'The greatest problem I've had to face', p. 4, AWM 93, item 50/2/23/578.
140. Ibid., p. 3.
141. Ibid., p. 1.
142. Coates, *Bravery Above Blunder*, p. 66.
143. 'Report of Operations – 8 Aust Inf Bde Gp. Period 10 Jan '44–11 Feb '44', p. 3, WD HQ 8 Bde, Jan 1944, AWM 52, item 8/2/8.
144. Ibid., p. 4.
145. Cameron, interview with D. St A. Dexter, n.d., p. 2, AWM 93, item 50/2/23/494.
146. See also 'Report on Operations 5 Aust. Div. Operations from Sio to Saidor, 20 Jan –29 Feb 44', pp. 1–2, AWM 54, item 519/6/48.
147. Cameron, interview, p. 2.
148. 'Doublet. Lessons from operations – own tactics 30 Jun–6 Aug 43', p. 10.
149. Savige, 'Notes on Volume V. Chapter 3', p. 14.
150. Stevens, 'A personal story', pp. 29, 33–4; Burns, The *Brown and Blue Diamond at War*, p. 3.
151. Savige, interview with D. St A. Dexter, 10 Apr 1951, AWM 172, item 7.
152. Moten, interview; 'Stud Book' of Officers of 17 Australian Infantry Brigade AIF, AWM 3DRL 5043, item 100; Moten, interview with D. St A. Dexter, 11 April 1951, AWM 172, item 9.
153. Unknown 2/5 Bn officer in Trigellis-Smith, *All the King's Enemies*, p. 215.
154. Savige, 'Notes on Volume V. Chapter 3', p. 14.
155. Hammer, 'Notes on Ops Missim Area by 15 Aust Inf Bde', 8 Jul 1943, p. 1, WD HQ 15 Bde, Appendices Jul 1943, AWM 52, item 8/2/15.
156. Mathews, *Militia Battalion at War*, p. 39.
157. WD HQ 15 Bde, Jun–Aug 1943, AWM 52, item 8/2/15.
158. Refshauge, interview with Long, 3 Sep 1945, AWM 67, item 2/98, p. 1.
159. Hammer, interview with D. St A. Dexter, 26 May 1951, p. 1, AWM 93, item 50/2/23/440. The transcript of this interview (a direct copy of Dexter's original notes), conducted at Hammer's club in Melbourne, is a remarkable document. It features such gems as 'Interview sitting now. Previous at bar but Dexter fell a . . . over tip' and 'Dexter and Hammer now get off backsides and head for bar'.
160. Starr to Long, 23 Jan 1957, p. 3, AWM 172, item 53.
161. Warfe to D. St A. Dexter, 3 Mar 1954, AWM 93, 50/2/23/489.
162. 15 Bde Report – Ramu Valley–Madang', p. 2.
163. Pooler, pers. comm., 7 Feb 2000, p. 2.
164. Hammer to Savige, 13 Aug 1943, WD HQ 15 Bde, Aug 1943, Appendix 53, AWM 52, item 8/2/15.
165. Hammer, 'Chapter 11' (comments on draft of *The New Guinea Offensives*), p. 8, AWM 93, item 50/2/23/440.
166. Savige, 'Notes on Volume V. Chapter 3', pp. 15–16.

167. Hammer, 'Chapter 11', p. 7.
168. W. R. Dexter, 'Chapter III. Jungle Stalemate', pp. 2–5, AWM 93, item 50/2/23/99.
169. CARO dossier VX33, Guinn, H. G. Savige later recalled: 'Guinn, when I sent him to administer command 15th Bde, was recognised to be the champion walker over the mountains within the Bde. Shortly afterwards, I saw him moving out aided by a stick marking a few hundred yards between rests' (Savige, 'Notes on Volume V. Chapter 3', p. 15, AWM 67, item 3/342, part 2).
170. Hammer to Savige, 'VX33 Lt-Col H. G. Guinn', 5 Aug 43, WD HQ 15 Bde, Aug 1943, AWM 52, item 8/2/15.
171. Moten, interview.
172. Moten to Guinn, 4 Nov 1944, pp. 1–2, AWM 3DRL 5043, item 101; Stevens to Moten, 3 Nov 1944, AWM 3DRL 5043, item 101.
173. Moten to Stevens, 3 Nov 1944.
174. HQ 6 Aust Div A11538, 'VX33 Lt-Col H. G. Guinn, DSO, ED LHQ Tactical School – Formerly CO 2/7 Aust Inf Bn – Complaint', 28 Feb 1945, AWM 3DRL 5043, item 101.
175. Moten to Stevens, 3 Nov 1944, p. 1.
176. HQ Kanga Force, G19/19, 20 Mar 1943, AWM 3DRL 5043, item 36b.
177. Refshauge, interview, p. 2.
178. McCarthy, *South-West Pacific Area*, p. 556.
179. 'Lt Geo Warfe could not get over the wall with his carrier, so removed his Bren and "hosed" the trenches, moving steadily along with his legs widely stretched as he bestrode it. The Italians, firing through the shell-torn cover, could neither hit the intrepid Warfe's balls or body. Out of ammunition, the future legendary commando hurled in his grenades, then taking charge, called on his CO, Godfrey, and his Company Commander, Smith, to get off their bellies and come over to hurl in their grenades as he lifted the cover, which they meekly did' (Smith, '2/6th Aust. Inf. Bn AIF', p. 32, AWM PR 85/223).
180. Christensen (ed.), *That's the Way It Was*, p. 70.
181. Mathews, *Militia Battalion at War*, p. 84.
182. Hammer, 'Chapter 11', p. 7.
183. Mathews, *Militia Battalion at War*, p. 84.
184. Pooler, pers. comm., p. 2; Mathews, *Militia Battalion at War*, p. 150.
185. Pooler, pers. comm., p. 2.
186. Mathews, *Militia Battalion at War*, p. 150.
187. Ibid., p. 89. 'I remember once . . . I was in a holding patrol with Wharfie when a patrol of Japs came down the track. Instead of waiting for them to come level with us he rushed out firing his trusty .45 and missed the lot. We got one in the mad rush and when chasing them came under heavy fire and had to withdraw' (Pooler, pers. comm., p. 3).
188. 'VX185 Major G. F. Smith – 2/6 Aust Inf Bn', 1 Oct 1942, AWM 3DRL 5043, item 38b.
189. Hay, interview, 31 May 2003.
190. Wrigley, 'Confidential report on Major G. F. Smith. 2/6 Aust Inf Bn', 15 Jan 1942, AWM 3DRL 5043, item 38b.

191. Savige, 'Notes on Volume V. Chapter 3', p. 24. Falconer was later found guilty of being AWL and was dismissed from the AMF.
192. Christensen, *As It Happened*, p. 50.
193. Refshauge, interview, p. 1.
194. Obituary, G. F. Smith, *Herald Sun*, 9 February 1995, p. 59.
195. Hammer, interview, p. 1.
196. DSO citation, VX185, Smith, G. F., CARO dossier, VX185, Smith, G. F.
197. Comment by unnamed 22 Bn soldier in MacFarlan, *Etched in Green*, p. 131.
198. Field, diary, 4 Dec 1943, AWM 3DRL 6937, item 5.
199. Field, 'Notes on draft of Australian Official History, Vol. VII, Chapter 6', p. 1, AWM 3DRL 6937, item 40.
200. The influence of Savige's patronage in the appointment of former 17th Brigade officers to battalion commands is also apparent in Howard Dunkley's appointment to command of the 7th Battalion in October 1944. Dunkley wrote in a letter home: 'Am part of Gen. Savige's team so am fortunate in that at least. In fact I fancy I see his hand in my getting a command over so many crowned soldiers of appalling seniority' (Dunley to 'Lola', 21 Nov 1944, AWM PR 84/035). Dunkley visited Savige's headquarters in March 1945 and was presented with a specially bound and signed copy of his 'new tactical book' (Dunkley to 'Lola', 8 Mar 1945, AWM PR 84/035).
201. Field, diary, 1944, AWM 3DRL 6937, item 6.
202. Chilton, interview, 5 October 2000.
203. Field, diary, Apr 1943, 6 Aug 1943.
204. Stevens, interview with Long, 4 Oct 1945, AWM 67, item 2/105, p. 12.
205. Bishop, *Word from John*, p. 226. It is a little ironic that Blamey introduced an edited collection of Bishop's letters, which included his musings on the end of his command.
206. Jackson, 'Autobiography', p. 333, AWM PR, item MSS 1193.
207. Ibid., p. 353.
208. Herring, 'Chapter 10' (comments on draft of Dexter, *The New Guinea Offensives*), p. 8, AWM 67, item 3/167, part 2; Herring interview with D. St A. Dexter, 6 Apr 1951, p. 2, AWM 67, item 3/167, part 3.
209. Macfarlane, interview, 24 Oct 2003.
210. Lewin interview with Coates, no date, cited in *Bravery Above Blunder*, p. 302; Cail, interview, 24 Oct 2003.
211. Unnamed 2/48 Bn soldiers, interview with Long, 20 November 1943, AWM 67, item 2/23, pp. 50–1.
212. Stanley, *Tarakan*, p. 34.
213. MC citation, VX15117, Joshua, R., CARO dossier VX15117, Joshua, R.
214. Evans, interview with D. St A. Dexter, 13 December 1951, p. 2, AWM 93, item 50/2/23/475.
215. Jackson, 'Autobiography', p. 222; Evans, interview, p. 1; Simpson, interview, p. 2.
216. Williamson, interview, 9 Feb 2001; Coates, *Bravery Above Blunder*, p. 186.

217. Evans, interview, p. 2; Porter, interview with D. St A. Dexter, 12 Dec 1951, AWM 172, item 5.
218. Lush, *Campaigning*, cited in Coates, *Bravery Above Blunder*, p. 187; Porter, interview.
219. Coombe, interview with Coates, cited in *Bravery Above Blunder*, p. 186.
220. Coombe, interview.
221. Coates, *Bravery Above Blunder*, p. 239.
222. Ibid., pp. 187–8.
223. Porter, interview.
224. Coates, *Bravery Above Blunder*, p. 305. Simpson's impact on the 2/43rd is clearly visible in the battalion history. Joshua's departure rates only a few token comments, whereas in the following pages several glowing references are made to Simpson and the pride and confidence he engendered in the battalion (Combe, Ligertwood, Gilchrist, *The Second 43rd Australian Infantry Battalion*, pp. 185, 189, 190, 201).
225. Hay, interview; Moten, interview.
226. Hay, interview.

8 'Experienced, toughened, competent': 1945

1. Stanley, 'An OBOE concerto: Reflections on the Borneo landings, 1945' in Wahlert (ed.), *Australian Army Amphibious Operations in the South-West Pacific: 1942–45*, p. 32.
2. In mid-1944 Gavin Long noted the comments of several AMF senior officers that a militia division on a par to those of the AIF could be formed from the 4th, 15th and 29th Brigades (Long, notebook, p. 19, AWM 67, item 2/36). In particular, Major General Alan Ramsay commented that the 4th and 15th Brigades were as proficient as any AIF brigade when it came to tropical warfare, and better than some (Ramsay, interview with Long, 30 Jul 1944, AWM 67, item 1/5, p. 66). The most laudatory assessment of militia brigade comes from Brigadier 'Tack' Hammer about the 15th: 'I would gladly have fought with the [15th] Bde at Alamein or in the taking of Tokyo with the experience and efficiency it had gained by the end of April [1944] in the Ramu Valley' (Hammer, 'Queries – Chapter 11', p. 6, AWM 93, item 50/2/23/440).
3. Cullen, interview; Hay, interview; Rhoden, interview; Chilton to Long, 23 Oct 1957, p. 2, AWM 93, item 50/2/23/322.
4. Unidentified Australian officer cited by Wood, 'Preface', in Hay, *Nothing Over Us*, p. viii.
5. Cullen, interview.
6. The 'lessons learned' sections of post-operations reports for the operations of 1945 are remarkably similar regarding infantry tactics. For example, the 2/32nd Battalion's report on its operations in British North Borneo reported: 'There were no outstanding lessons learned, as stated before the operation was more of an exercise than an operation, normal procedure and drills were carried out and proved efficient and satisfactory' ('2/32 Australian Infantry Battalion Report on Operations: Weston', p. 5,

AWM 54, item 619/7/58). See also '2/10 Aust Inf Bn Report on Operations – Oboe Two', sect. 3, para. 11, AWM 52, item 8/3/10 and 'Part II – Lessons from Operations [Aitape–Wewak]', para. 1, AWM 54, 603/7/27, part 1.
7. Long, notebook, p. 10, AWM 67, item 2/74.
8. Field, diary, 8 Aug 1944, AWM 3DRL 6937, item 6.
9. Pratten, 'The "old man"', p. 317.
10. J. de M. Carstairs (29), 22 Bn, and W. M. Mayberry (29), 58/59 Bn.
11. P. M. Shanahan (27), 55/53 Bn, and H. G. Sweet (29), 11th Bn.
12. AMF Record of Service Book (AAP83), NX121, Green, C. H., AWM PR, item 00466.
13. Chilton to O. Green, 9 Sep 1980, AWM PR 00466, item 3.
14. Long, notebook, p. 10, AWM 67, item 2/74.
15. Peterson, interview, 10 Feb 2001.
16. Corfield, *Hold Hard, Cobbers*, pp. 192–4.
17. Butler, 'The youngest CO: Charles Green', in Argent, Butler and Shelton, *The Fight Leaders*, pp. 47, 48.
18. In July 1945, 82 per cent of the 2/11th's personnel hailed from Western Australia (Williams, 'New Guinea, the Aitape–Wewak Campaign' in Johnson (ed.), *The History of the 2/11th (City of Perth) Australian Infantry Battalion, 1939–1945*, p. 156).
19. Butler, 'The youngest CO', p. 47.
20. Burley to O. Green, AWM PR, item 00466, folder 11; Mawhinney, 'Charles Hercules Green: Through the eyes of T. H. M.', AWM PR, item 00466, folder 11; Butler, 'The youngest CO', p. 47.
21. Chilton to O. Green, 9 Sep 1980.
22. C. H. Green to O. Green, 20 Apr 1945, AWM PR 00466, item 1.
23. Butler, 'The youngest CO', p. 48.
24. Stevens, 3 Jun 1946, AWM PR 00466, item 3.
25. Peterson, interview, 10 Feb 2001.
26. McElwain, 'Commanding officers of the infantry battalions of 2nd New Zealand Division', in Harper and Hayward (eds), *Born to Lead?* p. 182.
27. Long, diary, 7 Jul 1944, AWM 67, item 1/5.
28. Crang, *The British Army and the People's War*, p. 52; unknown corps commander to Mackley, 13 Aug 1944, LHCMA, Papers of Lt. Col. H. P. Mackley, OBE; figure based on details in McElwain, 'Commanding officers of the infantry battalions of 2nd New Zealand Division', pp. 193–7; 'Notes on infantry battalion commanders 1939–1945', p. 1; M. Dorosh, 'Treatise on battalion commanders', http://www.battlefront.com/community/archive/index.php/t-32644.html, accessed 19 October 2008.
29. 'The problem of officers' ages for command', 25 Nov 1946, NAUK WO 32, item 13253 cited in Crang, *The British Army and the People's War*, p. 54.
30. Hay, *Nothing Over Us*, p. 410.
31. Stanley, *Tarakan*, p. 35.
32. Hay, *Nothing Over Us*, p. 410.
33. '2/10 Aust Inf Bn Report on Operations – Oboe Two', sect. 3, para. 11.
34. Daly, interview.

35. '15th Australian Infantry Brigade Report on Operations, South Bougainville, from Puriata River to Mivo River 13 Apr 45 to 15 Aug 45', p. 10, AWM 54, item 613/7/12.
36. Pratten, 'The "old man"', p. 322.
37. Ibid., p. 323.
38. Brougham, diary, 9 May 1945 in Hay, *Nothing Over Us*, p. 440.
39. J. R. Broadbent, 2/17 Bn; G. E. Colvin, 2/13 Bn; T. W. Cotton, 2/33 Bn; T. J. Daly, 2/10 Bn; W. N. Farrell, 2/4 Bn; C. H. Grace, 2/15 Bn; I. Hutchison, 2/3 Bn; H. Mc. L. Binks, R. H. Marson, 2/25 Bn; C. H. Norman, 2/28 Bn, P. E. Rhoden, 2/14 Bn; E. M. Robson, 2/31 Bn; F. H. Sublet, 2/16th Bn; 2/11 Bn; and F. G. Wood, 2/6 Bn.
40. Compiled from McAllester, *Men of the 2/1 Battalion*, pp. 376–525; World War Two Nominal Roll, http://www.ww2roll.gov.au, accessed 23 Jan 2004.
41. 2/17 Battalion History Committee, *'What We Have . . . We Hold!'*; and *Gradation List of Officers of the Australian Military Forces: Active List*, Vols I and II, 18th January 1945 (hereafter 'AMF Gradation List').
42. Data compiled from AMF Gradation List, Vol. II, 18th January 1945; Givney (ed.), *The First at War*, pp. 80–1, 373–4; World War Two Nominal Roll, http://www.ww2roll.gov.au, accessed 23 Jan 2004.
43. Long, *The Final Campaigns*, pp. 76–7.
44. Data compiled from Mathews, *Militia Battalion at War*, p. 157; and AMF Gradation List, Vols I and II, 18th January 1945.
45. Data compiled from 'Field Return of Officers', 7 Apr 1945, WD 7 Bn, AWM 52, 8/3/44; AMF Gradation List, Vols I and II, 18th January 1945; and World War Two Nominal Roll, http://www.ww2roll.gov.au, accessed 23 Jan 2004.
46. Daly, interview, in Pratten and Harper (eds), *Still the Same*, p. 20.
47. Daly, interview, 26 Oct 2000.
48. Ibid.
49. Daly, interview, in Pratten and Harper (eds), *Still the Same*, p. 28.
50. Daly, interview, 26 Oct 2000.
51. Daly, 'Official War History – Vol VII Comments on Chapter 21 – The Seizure of Balikpapan', p. 1, AWM 93, item 50/2/23/417.
52. Daly, interview.
53. Efficiency Report, SX1436, Daly, T. J., 31 January 1946, CARO dossier, SX1436, Daly, T. J.
54. Rhoden, interview.
55. Daly in Pratten and Harper (eds), *Still the Same*, p. 28.
56. WD GS 6 Div, 30 Oct 1944, AWM 52, item 1/5/12.
57. Stevens, interview with Long, Nov 1944, AWM 67, item 2/105, p. 10.
58. Picken, interview with Long, 20 Aug 1945, AWM 67, item 2/94.
59. Daly, interview with Long, 18 Aug 1945, AWM 67, item 2/86.
60. Eather to parents, 29 July 1945 in Eather, *Desert Sands, Jungle Lands*, p. 161. Eather's comments are supported by the difficulties faced by the 2/23rd Battalion following the loss of three of its experienced company commanders during the first week on Tarakan, 'thrusting unaccustomed responsibility upon young lieutenants' (Stanley, *Tarakan*, pp. 92, 152).

61. Matthews, diary, 25 July 1943, AWM PR 89/79.
62. Ibid., 29 Apr 1945.
63. Field, diary, 18 Jan 1945, AWM 3DRL 6937, item 6; Savige, 'Notes on Volume VII. Chapter 4. The Bougainville campaign takes shape', p. 3, AWM 67, item 3/348, part 2. A lack of stability had been a feature of Monaghan's brigade headquarters since the New Guinea offensives of 1943. Tom Daly, GSO1 of the 5th Division at the time, observed: 'Monaghan tended to churn officers through his HQ like sausages. I lost track of his Brigade Staff long before the end of the campaign' (Daly, 'Notes on Chapter XI – Capture of Salamaua', p. 1, AWM 93, item 50/2/23/417).
64. Savige, 'Notes on Volume VII. Chapter 4. The Bougainville campaign takes shape', p. 3.
65. Field, diary, 18 Jan 1945.
66. Report on METS Senior Officers Course, WX19, Egan, E. W., Adverse report, WX19, Egan, E. W., CARO dossier WX19, Egan, E.W.
67. Jackson, 'Autobiography', p. 329.
68. See AWM 67, items 50/2/23/388, 50/2/23/578.
69. Rhoden, interview. Fred Chilton echoed Rhoden's comments stating that the Balikpapan campaign was the only one in which, 'with very minor exceptions', the 7th Division was provided with 'everything needed in the way of equipment and support' (Chilton to Long, 23 Oct 1957, p. 1, AWM 93, item 50/2/23/322).
70. A 'sniping gun' is a single artillery piece used in a direct fire role against a point target, such as a bunker.
71. 'Report on Operation Oboe Two', p. 40, AWM 54, item 621/7/1.
72. Ibid., p. 30.
73. '24 Australian Infantry Brigade Report on Operations Oboe Six', p. 27, AWM 54, item 619/7/67.
74. Examples are discussed in: 'Report on Operation Oboe Two', p. 41; '15th Australian Infantry Brigade Report on Operations, South Bougainville, from Puriata River to Mivo River 13 Apr 45 to 15 Aug 45', pp. 5–6, 8–9, AWM 54, item 617/3/12.
75. Russell, *The Second Fourteenth Battalion*, p. 285
76. Rhoden, interview.
77. Account of Manggar action derived from: WD 2/14 Bn, 4–9 July 1945, AWM 52, item 8/3/14; and Russell, *The Second Fourteenth Battalion*, pp. 282–95.
78. Russell, *The Second Fourteenth Battalion*, p. 294.
79. '19 Aust Inf Bde Report on Operations Aitape–Wewak Campaign 16 December 1944–29 July 1945', p. 12, AWM 54, item 603/7/21, part 2.
80. '15th Australian Infantry Brigade Report on Operations, South Bougainville, from Puriata River to Mivo River 13 Apr 45 to 15 Aug 45', p. 10.'
81. Dougherty in Long, *The Final Campaigns*, p. 531.
82. Stanley, *Tarakan*, pp. 83, 139; Ainslie, interview with Nicholson, 10 Aug 1945, pp. 2–3, AWM 54, item 617/7/17.
83. Stanley, *Tarakan*, p. 139.
84. Matthews, diary, 14 Dec 1944, AWM PR 89/79.

85. Ibid., 16 & 18 Dec 1944, 3 Feb 1945.
86. '19 Aust Inf Bde Report on Operations Aitape–Wewak Campaign 16 December 1944–29 July 1945', p. 34.
87. 'Report on Operation Oboe Two', pp. 40–1.
88. '2/10 Aust Inf Bn Report on Operations – Oboe Two', sect. 3, para. 6, 10.
89. Hay, interview; Matthews, diary, vols 12 & 13, November 1944–February 1945; Cullen, interview; Long, notebook 68, p. 34, AWM 67, item 2/68.
90. Dwyer, *Facts. Fallout! Fantasy?* pp. 108–9.
91. Dunkley to 'Lola', 2 May, 5 Aug 1945, AWM PR, item 84/035.
92. Dwyer, *Facts. Fallout! Fantasy?* p. 110.
93. Matthews, diary, 23 & 29 Nov 1944.
94. '15th Australian Infantry Brigade Report on Operations, South Bougainville, from Puriata River to Mivo River 13 Apr 45 to 15 Aug 45', p. 6.
95. Howard, 'Leadership in the British Army in the Second World War: Some personal observations', in Sheffield (ed.), *Leadership and Command*, pp. 117–18.
96. 1st Australian Field Censorship Company field censorship reports 1944–45, AWM 54, item 175/3/4; J. H. Ewers, diary, 8 May 1945, AWM PR, item 89/190. 'Observations of Acting Minister for the Army [J. M. Fraser] on Operations in New Guinea, New Britain and the Solomon Islands', AWM 3DRL 6643, item 2/136.2; Long, notebook 74, pp. 22, 25–6, notebook 78, p. 10, AWM 67, items 2/74, 2/78; Hammer, 'Notes on Bougainville', p. 1, AWM 67, item 3/156; 'Medical Report for March 1945', p. 2, WD 2/3 Bn, AWM 52, item 8/3/3.
97. Blamey, instruction to Hammer, 9 Jun 1945 in Long, *The Final Campaigns*, p. 193. Long records Major General Bridgeford making similar remarks to his battalions: 'Bridgeford had told 2 bns of his div that the op on Bougainville was to be taken slowly, and our losses were going to be slow. Everything was going to be very carefully considered. There was no hurry' (Long, notebook, pp. 35–6, AWM 67, item 2/64).
98. Stanley, *Tarakan*, p. 102.
99. Stanley, 'An Oboe concerto', p. 139.
100. Cullen, 'Address by NX163 LT COL P. A. Cullen DSO to 2/1 Aust Inf Bn at 0800hrs Saturday 20 Jan 45', WD 2/1 Bn, AWM 52, item 8/3/1.
101. Cullen, interview.
102. Cullen, 'Address . . . to 2/1 Aust Inf Bn'.
103. '15 Australian Infantry Brigade, Report on Operations South Bougainville', p. 5.
104. Stevens to Long, 21 Sep 1955, p. 2, AWM 67, item 3/378.
105. Christian, interview, 30 May 2003.
106. Stevens, interview with Long, 4 Oct 1945, p. 8.
107. Parry-Okeden, diary, 9 Jan 1945, AWM PR, item 00321.
108. Ibid., 20 Jan 1945.
109. Parry-Okeden, diary, 9 Feb 1945. After receiving the orders to withdraw Parry-Okeden flew to brigade headquarters in Madang where Brigadier Fergusson congratulated him on his efforts. The two officers then prepared

an appreciation on the situation in the Sepik–Ramu area, justifying offensive action, which was distributed to the GOC First Australian Army, Lieutenant General Sturdee, and the heads of all three services (ibid., 11 Feb 1945).
110. Ibid., 15 Feb 1945.
111. Ibid., 16 May, 27 Aug 1945.
112. Cullen, interview.
113. Stevens, interview with Long, p. 8.
114. Hastings, *Overlord*, p. 165.
115. McElwain, 'Commanding officers of the infantry battalions of 2nd New Zealand Division', p. 185.
116. Major George Taylor observed the actions of his CO at Normandy: 'Slightly flushed of face, the CO calmly worked out the time factor on a piece of paper. At least two hours, was his reply. The brigadier adding another half hour for good measure, fixed zero hour at 20.30 hours. On these occasions, time is always the relentless enemy of the battalion commander . . . Hampered by the vagueness of local information, his decision has to be reached often when moving at speed in a carrier or scout car. He must marry in the armour with the infantry and devise the covering fire plan of the artillery, mortars and machine guns. Then he must ensure that his troops undaunted by enemy fire, are at the battle start line by the appointed hour. Heavier becomes the weight of responsibility on his shoulders as the hands of his watch seem to turn faster and faster' (Taylor, *Infantry Colonel*, p. 38).
117. Hastings, *Overlord*, p. 165.
118. Long, *The Final Campaigns*, pp. 385, 237, 451, 545.
119. Long, notebook 79, pp. 55–6, AWM 67, item 2/79.
120. Hastings, *Overlord*, p. 165.
121. Long, notebook 78, pp. 7–8, 10, AWM 67, item 2/78.
122. 'Medical Report for March 1945', p. 4, WD 2/2 Bn, AWM 52, item 8/3/2.
123. Cullen, interview.
124. The bulk of this account of the 9th Battalion mutiny is drawn from Matthews' remarkably frank diary, and all quotations are his (Matthews, diary, 30 January–3 March 1945). An additional source was the battalion's war diary for the same period – AWM 52, 8/3/46 – which makes no mention of the troubles within the battalion.
125. Long, *The Final Campaigns*, p. 147; Field, diary, 31 Jan 1945.
126. The 9th Battalion spent the period 27 February–11 March resting on the beach at Mawaraka, during which time Matthews was upbraided by General Savige, through Brigadier Field, for the poor dress standards and apparent idleness of his troops: 'If he should visit here he would see most of us with nothing on. It appears I must take care to ensure that even though resting strict discipline must be maintained, organized sports, lectures, etc. to be arranged to keep them occupied. What the men need they are now getting, *REST* and I don't intend to alter things unless forced to as everyone is extremely happy and I have no fears as to discipline' (Matthews, diary, 3 Mar 1945).

127. Field, 'VX5172 Lt Col W. R. Dexter 61 Aust Inf Bn', p. 2, AWM 3DRL, item 6937, part 32. Long, notebook 71, 11 Feb 1945, AWM 67, item 2/71. Dexter, 'The Battalion – My home, unpublished MS, pp. 164–5, AWM PR01182, item 4.
128. Dexter, interview with Long, 10 February 1945, AWM 67, item 2/68.
129. Dexter to J. Field, 19 Mar 1945, p. 1, AWM 3DRL, item 6937, part 32.
130. Ellis, *The Sharp End*, pp. 249–50.
131. Johnston, *At the Frontline*, pp. 61–2.
132. Field, 'VX5172 Lt Col W. R. Dexter 61 Aust Inf Bn', p. 2.
133. Hay, interview.
134. Field, 'VX5172 Lt Col W. R. Dexter 61 Aust Inf Bn', p. 2.
135. Picken, interview with Long, 20 Aug 1945, AWM 67, item 2/94.
136. Matthews, diary, 27 Apr 1945.
137. J. N. Abbott, 26 Bn; F. J. Embrey, 37/52 Bn; D. J. H. Lovell, 55/53 Bn; and L. de L. Miell, 19 Bn.
138. CARO dossier NX59, Abbott, J. N, CARO dossier NX53, Lovell, D. J. H.
139. DSO citation DSO (GO 40/41), NX59, Abbott, J. N., CARO dossier NX 59, Abbott, J. N.
140. Long, notebook 69, p. 93, AWM 67, item 2/93.
141. Entry dated 18 Feb 1946 in CARO dossier NX256, Hendry, J. G.
142. Jackson, 'Autobiography', p. 353.

Conclusion

1. Rhoden, interview.
2. Kippenberger, *Infantry Brigadier*, pp. 1–4.
3. 'Notes on infantry battalion commanders 1939–1945', http://www.canadiansoldiers.com/Bncomdnotes.htm; Blaxland, 'Strategic cousins: Australia, Canada and their use of expeditionary forces from the Boer War to the War on Terror', pp. 249–50; Hayes, 'A search for balance: Officer selection in the wartime Canadian Army, 1939–1945', p. 6; McElwain, 'Commanding officers of the infantry battalions of 2nd New Zealand Division', in Harper and Hayward, *Born to Lead*, pp. 179–80, 193–7.
4. Hingston, 'The development of gifted amateurs – New Zealanders at war in the Western Desert', paper for Defence Studies program, University of New England, 1996, pp. 3, 10.
5. Ellis, *At the Sharp End*, p. 227; Gavin in Meloy, 'Gen. Ridgway's Personnel Management Policy'.
6. Taylor, *Infantry Colonel*, p. 48.
7. McElwain, 'Commanding officers of the infantry battalions of 2nd New Zealand Division', p. 190.
8. Howard, 'Leadership in the British Army in the Second World War: Some personal observations' in Sheffield (ed.), *Leadership and Command*, p. 118; Carver in 'Thoughts on command in battle', *British Army Review*, No. 69, December, 1981, p. 5.
9. Hayes, 'A search for balance', pp. 27–8, 31.
10. 'Notes on infantry battalion commanders', p. 1.
11. Lindsay in 'Thoughts on command in battle', p. 4.

12. G. E. Colvin, CO 66 Bn, Oct 1945–Dec 1947; R. H. Marson, CO 65 Bn, Oct 1945–Aug 1948.
13. C. H. Green, CO 3 RAR, Sep–Oct 1950 (DOW 1 Nov 1950); I. Hutchison, CO 1 RAR, Nov 1951–Nov 1952.
14. Cotton, interview with Long, 11 Aug 1944, AWM 67, item 256.
15. McCarthy, *The Once and Future Army*, p. 28. The experience of the CMF in the immediate post-war period is well explored in McCarthy's book.
16. The other CGS who had commanded a battalion during the Second World War was Alwyn Ragnar Garrett, CO 2/31st Battalion, Sep 1940–Feb 1941.
17. Daly, interview.
18. Chilton, interview.
19. Daly, interview.
20. Rhoden, interview.

Appendix 1

1. Brennan, 'Amateur warriors'.
2. With only a few exceptions, the war diaries for militia battalions held in AWM 52 do not begin until the period December 1941–January 1942.
3. The organisation and activities of 2nd Echelon can be found in 'History of Central Army Records Office, War of 1939–1945', AWM 54, item 755/5/7. At the time of research the service records were still in the custody of CARO, then a part of the Soldier Career Management Agency (SCMA), but they have been subsequently transferred to the National Archives of Australia where they form two file series: B883 (AIF service records) and B884 (militia service records).
4. *Census of the Commonwealth of Australia, 30th June*, 1933. (Hereafter Commonwealth Census, 1933.)
5. Johnson, *At the Front Line*, p. 212. The major omission of this census was the 8th Division, which was in Japanese captivity when the census was conducted.
6. 'Army Census 1942–43. Summary of tabulation of a 10 per cent sample', NAA MP729/6, item 58/401/485.
7. Commonwealth Census, 1933, p. 101.
8. Two British-born COs – Raymond Sandover (CO 2/11 Bn, Apr 1941–May 1943) and William Mayberry (CO 58/59 Bn, Jan 1945–Oct 1945) – had emigrated to Australia as adults; Sandover's service record indicates that he was still serving in the Territorial Army in 1933. Charles Anderson (CO 2/19 Bn, Aug 1941–Feb 1942) had emigrated from South Africa in 1934.
9. See Henning, *Doomed Battalion*, pp. 1–14.
10. Originally numbered, respectively, as the 70th, 71st and 72nd Battalions.
11. For example, the 2/10th Battalion was raised in South Australia in 1939 almost entirely from personnel recruited in that state. By 1945, only 40 per cent of its personnel were from South Australia (Allchin, *Purple and Blue*, p. 364).
12. E. P. Foster, CO 19 Bn, Jan 1943–Apr 1944.

13. A. 'Jiggy' Spowers, CO 24/39 Bn, Jan 1940–May 1940, CO 2/24 Bn, Jul 1940–Jul 1942.
14. F. G. 'Black Jack' Galleghan, CO 17 Bn, Mar 1937–Oct1940, CO 2/30 Bn, Oct 1940–Feb 1942.
15. C. J. A. Coombes, CO 47 Bn, Jan–Apr 1945.
16. J. Moyes, CO 41 Bn, Dec 1939–Jan 1942.
17. Although the 'Commerce' categories of the 1933 Census included a diverse range of individual occupations, including greengrocers, bag, sack and jute dealers, tobacconists and florists, the COs in this group were drawn from a relatively narrow strata running through this group and were employed as company representatives, salesmen and sales executives. Because these men's employment did not depend on professional skills and generally did not involve the supervision of staff, but was still of a reasonably high social standing, it was decided to group them together in a separate category.
18. Figures derived from McElwain, 'Commanding officers of 2nd New Zealand Division', in Harper & Hayward, *Born to Lead?*, pp. 193–7.
19. Approximates based on 'Australia. Schooling – Persons receiving instruction according to age', Commonwealth Census 1933, p. 1152.
20. Cullen, interview, 27 Feb 2001.
21. Chilton, interview, 25 Oct 2000.
22. Rhoden, interview, Oct 2000.
23. It is informative to note that lawyers also represented the most numerous single professional group among the battalion commanders of the Canadian army in the First World War (see Brennan, 'Amateur warriors', p. 7).
24. J. R. Broadbent, CO 2/17 Bn, Feb 1944–Feb 1946; F. O. Chilton, CO 2/2 Bn, Dec 1940–Nov 1941; G. D. T. Cooper, CO 2/27 Bn, Jan 1942–Dec 1942; P. A. Cullen, CO 2/1 Bn, June 1942–Aug 1945; T. J. Daly, CO 2/10 Bn, Oct 1944–Dec 1945; O. C. Isaachsen, CO 36 Bn, Sep 1942–Aug 1945; P. E. Rhoden, CO 2/14 Bn, Oct 1943–Oct 1945; F. H. Sublet, CO 2/16 Bn, Apr 1943–Oct 1945 and J. A. Wilmoth, CO 7 Bn, Apr 1942–Nov 1943.
25. Chilton, interview.
26. Cullen, interview.
27. Charles Moorhouse recalled in discussions with Michael Cathcart that the Melbourne University Rifles 'was generally believed to be more likely to open the way to a commission than other corps' (Cathcart, *Defending the National Tuckshop*, p. 51).
28. For an insight into the ethos of the university regiments see Lilley, *Sydney University Regiment*, p. 4.
29. This has persisted into the current day; senior SUR alumni, for instance, are collectively known as the 'Rum Corps' (see McCarthy, *The Once and Future Army*, pp. 206, 215–16).
30. Chilton, interview.
31. Different forms were used to document the service history of ORs and officers in the pre-war militia. ORs and officers files were also stored and handled differently in units; the chief clerk was responsible for ORs' records while the adjutant administered those of officers. It does not appear to have been the practice for the two files to have been amalgamated when an OR

was commissioned. Hence it is often difficult to determine the extent, if any, of an officer's service as an OR, particularly as the AMF's officers lists do not detail dates of enlistment. Pre-war officers record of service forms did have a section in which to list service in the ranks, but it was often haphazardly completed or not filled out at all.

32. E. Barnes (pte, 2/9 Bn, Nov 1939), CO 19 Bn, Mar–Aug 1945; J. D. M. Carstairs (pte, 2/7 Bn, Nov 1939), CO 22 Bn, Apr 1944–Aug 1945; T. R. W. Cotton (pte, 2/11 Bn, Nov 1939), CO 2/33 Bn, May 1943–Dec 1945; G. S. Cox (sgt, 127 Bty, AFA), CO 2/4 Bn, Mar–Dec 1945; W. R. Dexter (sgt, 23/21 Bn, Sep 1939), CO 61 Bn, Apr 1944–May 1945; H. L. E. Dunkley (pte, 2/6 Bn, Nov 1939), CO 7 Bn, Oct 1944–Nov 1945; A. J. Lee (pte 43/48 Bn, Sep 1939); CO 2/9 Bn, Aug 1944–Aug 1945; W. M. Mayberry (cpl, 2/5 Bn, Nov 1939), CO 58/59 Bn, Jan–Oct 1945; W. S. Melville (sgt, SUR, Sep 1939), CO 11 Bn, Nov 1943–Aug 1945; C. H. B. Norman (cpl, 25 LHR, Sep 1939), CO 2/28 Bn, Jul 1943–Jul 1945.
33. A. W. Buttrose, CO 2/33 Bn, Apr 1942–Apr 1943, CO 2/5 Bn, Jul 1944–Nov 1945; T. M. Conroy, CO 2/32 Bn, Jun 1941–Feb 1942, CO 2/5 Bn, Mar 1943–Jul 1944; N. W. Simpson, CO 2/17 Bn, Mar 1942–Feb 1944, CO 2/43 Bn, Feb 1944–Mar 1945; F. G. Wood, CO 2/5 Bn, Nov 1941–Jan 1942, CO 2/6 Bn, Jan 42–Jul 1945; and H. Wrigley, CO 2/5, Dec 1940–Jan 1941, CO 2/6, Mar 1941–Jan 1942.
34. J. H. Byrne, CO 2/31 Bn, March-May 1943, CO 42 Bn, Sep 1944–Oct 1945; J. A. Corby, CO 2/33 Bn, May 1941–Apr 1942, CO 34 Bn, Sep–Dec 1943; R. Joshua, CO 2/33 Bn, Jul 1943–Feb 1944, CO 13/33 Bn, Feb 1944–Nov 1945; and P. D. S. Starr, CO 5 Bn, Jan 1942–Mar 1943, CO 58/59 Bn, Apr 1943–Sep 1943.
35. Stanley, *Tarakan*, p. 128.
36. Walker, *Middle East and Far East*, p. 228; Johnston, *At the Front Line*, p. 50; Ross, 'Psychotic casualties in New Guinea', p. 1, AWM 54, 804/1/4, part 1; Stanley, *Tarakan*, p. 128.
37. Appendix C to SA/99/5/Med, 21 Jun 44, 'Psychiatric Summary week ending Sunday 18 Jun 44', NAUK, WD Medical Branch, HQ Second Army, WO 177, item 321.
38. Figures compiled from McElwain, 'Commanding officers of the infantry battalions of 2nd New Zealand Division', pp. 193–7, and Cody, 28 *Maori Battalion*.

Bibliography

INTERVIEWS

Personal collection
Addison, K., Stonyfell, 8 Feb 2001
Allen, B. C., Glenelg, 6 Feb 2001
Anning, J., Coorparoo, 15 Feb 2001
Barker, G., Chermside West, 13 Feb 2001
Bellow, R., Jolimont, 24 Oct 2003
Bletchly, D., The Gap, 16 Feb 2001
Broadbent, J. R., Sydney, Oct 2000
Cahill, L., Pymble, 16 May 2002
Cail, W., Jolimont, 24 Oct 2003
Chappel, B., Aldgate, 7 Feb 2001
Chilton, F. O., Avalon, 25 Oct 2000
Christian, D. W., Blackburn South, 30 May 2003
Clark, B., Jolimont, 24 Oct 2003
Cooper, G. D. T., Heathpool, 8 Feb 2001
Cowie, R., Cooperoo, 15 Feb 2001
Cullen, P. A., Marulan, 27 Feb 2001
Daly, T. J., Bellevue Hill, 26 Oct 2000
Gilbert, G., Holland Park West, 15 Feb 2001
Hamilton, H. M., Hampton East, Jul 2001
Hay, D. A., South Yarra, 31 May 2003
Hodel, F., Brisbane, 16 Feb 2001
Homewood, A., Manly (Qld), 13 Feb 2001
Isaachson, O. C., Adelaide, 6 Feb 2001
Jones, E., Campbell, 5 Mar 2004
Lamb, J., Jolimont, 24 Oct 2003
Lloyd, J. K., Jul 1994
Londos, G., Ferny Hill, 14 Feb 2001
Macfarlane, A., Jolimont, 24 Oct 2004
McAllester, J., Brighton, 1 Jun 2003
Peterson, L. J., Myrtlebank, 10 Feb 2001
Pratten, R. E., Balwyn, 1992
Rhoden, P. E., Malvern, Oct 2001
Serle, F., Clearview, 7 Feb 2001

Sublet, F. H., Mount Eliza, 12 Sep 2000
Symington, G., Kensington Gardens, 10 Feb 2001
Williamson, H., Novar Gardens, 9 Feb 2001
Wilmoth, J., Brighton, 27 Sep 2000
Yeatman, J., Myrtlebank, 9 Feb 2001

Keith Murdoch Sound Archives, Australian War Memorial
S763   E. Heckendorf

CORRESPONDENCE
Boorman, C., 16 Dec 2002, 3 Jan 2003
Bowden, J., 17 Jan 2004
Brumby, S., 8 Apr 2003
Carey, P. R., 20 Jun 2000
Christian, D. W., 8 Apr 2003
Cundell, R., 30, 31 Dec 2003; 3, 4, 5, 21 Jan, 10 Feb 2004
Farrell, B. P., 30 Mar 2005
Johnson, K. T., 11 Jan, 9 Apr, 3 May, 2001
Lazzarini, P. F., 21 Feb 2003
Magill, J., undated 2003
McAllester, J., 19 May 2003
Pooler, J., 7 Feb 2000
Savige, S., 28 Feb 2001

PRIVATE PAPERS
Hutchinson, T., diary

ARCHIVAL SOURCES
Australian War Memorial
AWM 1   Pre-Federation and Commonwealth Records
AWM 49   Inter-war Army Records
AWM 52   Australian Military Forces Formation and Unit Diaries, 1939–45
AWM 54   Written Records, 1939–45 War
AWM 62   Southern Command Registry Files
AWM 67   Official History, 1939–45 War: Records of Gavin Long, General Editor
AWM 72   Official History, 1939–45 War, Series 1 (Army), Vol. 3: Records of Barton Maughan
AWM 73   Official History, 1939–45 War, Series 1 (Army), Vol. 4: Records of Lionel Wigmore
AWM 93   Australian War Memorial Registry Files
AWM 172   Official History, 1939–45 War, Series 1 (Army), Vol. 6: Records of David Dexter
AWM 193   Eastern Command 'G' Branch registry files
MFF 0020   Papers of Maj Gen H. G. Bennett
PR 00079   Papers of Col G. E. Ramsay

PR 00321   Papers Lt Col W. N. Parry-Okeden
PR 00446   Papers of O. Green
PR 00588   Papers of Sgt L. Clothier
PR 00686   Papers of Sgt E. E. Heckendorf
PR 84/035   Papers of Lt Col H. L. E. Dunkley
PR 85/42   Papers of Brig H. B. Taylor
PR 85/223   Papers of Lt Col G. F. Smith
PR 86/002   Papers of Maj I. F. Macrae
PR 86/070   Papers of F. P. Gardner
PR 86/370   Papers of Lt E. R. Wilmoth
PR 89/079   Papers of Col G. R. Matthews
PR 89/099   Papers of Lt Col W. W. Legatt
PR 89/190   Papers of Sgt J. H. Ewers
3DRL 1006   Papers of Maj G. F. Smith
3DRL 2313   Papers of Brig F. G. Galleghan
3DRL 2529   Papers of Lt Gen Stan Savige
3DRL 2632   Papers of Lt Gen L. J. Morshead
3DRL 2642   Papers of Maj Gen H. H. Hammer
3DRL 2691   Papers of Brig. A. L. Varley
3DRL 3402   Papers of Maj Gen B. M. Morris
3DRL 3561   Papers of Maj Gen J. E. S. Stevens
3DRL 4142   Papers of Maj Gen A. S. Allen
3DRL 5043   Papers of Brig M. J. Moten
3DRL 6224   Papers of FM T. A. Blamey
3DRL 6599   Papers of Lt Col P. D. S Starr
3DRL 6643   Papers of FM T. A. Blamey
3DRL 6850   Papers of Lt Gen Sir Iven Mackay
3DRL 6937   Papers of Brig J. Field
3DRL 7945   Papers of Lt G. T. Gill

Liddell Hart Centre for Military Archives
Papers of Lt Col H. P. Mackley, OBE

National Archives of Australia
A2653   Volumes of Military Board Proceedings
A5954   Shedden Collection
A6828   Chief of the General Staff's Periodical Letters to the Chief of the Imperial General Staff
A9787   Council of Defence Minutes and Agenda Papers
B883   Second Australian Imperial Force Personnel Dossiers, 1939–47
B884   Citizen's Military Forces Personnel Dossiers, 1939–47
B1535   Department of the Army Correspondence Files
MP 508/1   Army Headquarters General Correspondence Files
MP 729/6   Department of the Army Secret Correspondence Files
MP 742/1   Department of the Army General and Civil Staff Correspondence Files and Army Personnel Files

SP459/1 2nd Military District/Headquarters Eastern Command Correspondence Files
SP1048/6 Headquarters Eastern Command, General Correspondence, 'C' (Confidential) Series

National Library of Australia
MS 3782 Papers of George Alan Vasey

National Archives of the United Kingdom
WO106 War Office, Directorate of Military Operations and Military Intelligence, and predecessors: Correspondence and Papers
WO163 War Office and Ministry of Defence and predecessors: War Office Council, later War Office Consultative Council, Army Council, Army Board and their various committees: Minutes and Papers
WO177 War Office: Army Medical Services: War Diaries, Second World War
WO201 War Office: Middle East Forces; Military Headquarters Papers, Second World War
WO202 War Office: British Military Missions in Liaison with Allied Forces; Military Headquarters Papers, Second World War
WO203 War Office: South East Asia Command: Military Headquarters Papers, Second World War
WO216 War Office: Office of the Chief of the Imperial General Staff: Papers
WO236 Papers of General Sir George Erskine

Rylands University Library, University of Manchester
Auchinleck Papers

COMMONWEALTH GOVERNMENT REPORTS AND PUBLICATIONS

Census of the Commonwealth of Australia, 30th June, 1933, Commonwealth Government Printer, Canberra, 1936–40
Official Year Book of the Commonwealth of Australia, No. 32 – 1939, L. F. Johnston, Commonwealth Government Printer, Canberra, 1939

MILITARY PAMPHLETS AND DOCTRINAL PUBLICATIONS

Australian Military Forces and Australian Army
AIF (ME) Orders and AIF (ME) Lists with Indices, Volumes 1–3, Australian Imperial Force (Middle East), 1940–43
The Army List of the Australian Military Forces, Part 1: *Active List*, H. J. Green, Government Printer, Melbourne, 1935
The Army List of the Australian Military Forces, Part 1: *Active List*, H. J. Green, Government Printer, Melbourne, 1939
The Army List of the Australian Military Forces, Part 2: *The Australian Imperial Force*, Vol. 4 – *Gradation List of Officers*, H. E. Daw, Government Printer, Melbourne, 1941

The Army List of the Australian Military Forces, Part 1: *The Active List*, Vol. 1, Command and Staff List of Officers, Brown Prior Anderson Limited, Government Printer, Melbourne, 1942

Australian Army Orders, General Orders, Appointment Orders, Posting Orders and Gazette Notices, 1939–45, H. E. Daw, Government Printer, Melbourne, 1939–45

Australian Imperial Force, Gradation Lists of Officers, and Lists of Honours and Awards for Service in the Field: July 1918, Albert J. Mullet, Government Printer, Melbourne, 1918

Australian Imperial Force, Staff and Regimental Lists of Officers: August 1918, Old Westminster Press, London, 1918

Australian Imperial Force (Middle East), Staff and Regimental Lists of Officers, No. 1–6, Australian Imperial Force (Middle East), 1941–43

Australian Military Forces Command, Staff and Extra-Regimental Appointments List, No. 1–8, LHQ Press, Brisbane, 1944–45

'The Basis of Human Behaviour [&] the Military Leader', Australian Army First Appointment Course Leadership Précis 3–1, 1996

'Behaviour of the Individual Soldier', Australian Army First Appointment Course Leadership Précis 3–2, 1996

The Defence Act 1903–41 and Regulations and Orders for the Australian Military Forces and Senior Cadets, 1927, H. E. Daw, Government Printer, Melbourne, 1941

'Discipline, Morale and Esprit de Corps', Australian Army First Appointment Course Leadership Précis 3–6, 1996

Durant, J. M. A., *Leadership: Some Aspects of the Psychology of Leadership, More Particularly as affecting the Regimental Officer and the Australian Soldier*, A. H. Tucker, Government Printer, Brisbane, 1941

Field Service Regulations, Vol. 1: *Organization and Administration*, Wilke & Company (by authority), Melbourne, 1940

The Fundamentals of Land Warfare, Australian Army, Georges Heights, 1993

The Fundamentals of Land Warfare, Australian Army, Sydney, 1998

Gradation List of Officers of the Australian Military Forces: Active List, Vol. 1, Wilke & Company, Melbourne, 1945

Hints to Officers on Command, Discipline and Care of the Men, AMF, Melbourne, 1916

Infantry Section Leading (Modified for Australia), 1939, AMF, Melbourne, 1939

Instructions for Medical Officers (Australia), 1942, AMF, Melbourne, 1942

Instructions on Medical Boarding and Reclassification, 1944, F. J. Hilton & Company, Melbourne, 1944

Leadership Theory and Practice, Australian Army, Sydney, 1973

Military Board Instructions, 1938, AMF, Melbourne, 1938

Military Board Instructions and General Routine Orders, 1939–1945, H. E. Daw, Government Printer, Melbourne, 1939–45

Regimental Lists and Manning Tables of Officers on the Active List, Vol. 1, No. 1–2, Stevenson, Government Printer, Adelaide, 1944–45

British Army and War Office
Army Doctrine Publication, Vol. 2, Command, British Army, 1995
Field Service Regulations, Vol. 2, Operations-General, 1935, HMSO, London, 1939
The Infantry (Rifle) Battalion, Part 1: Organization and Characteristics – The Carrier Platoon, War Office, London, 1940
Infantry Section Leading, 1938, HMSO, London, 1940
Infantry Training. Part I: The Infantry Battalion, 1943, HMSO, London, 1943
Infantry Training: Training and War, 1937, HMSO, London, 1937
Regulations for the Territorial Army (including the Territorial Army Reserve) and For County Associations, 1936, War Office, London, 1936
Signal Training (All Arms), 1938, HMSO, London, 1940

New Zealand Army
Infantry in Battle, 1950, New Zealand Army, Wellington, 1949

ARTICLES, PAPERS AND THESES
Amin, A. H., 'Orders and obedience: An in depth analysis', *Pakistan Army Journal*, Vol. 23, No. 1, March 1991, pp. 16–22
Baldwin, S., 'The leadership formula', *Proceedings*, Vol. 118, July 1992, pp. 82–6
'The Basis of Expansion for War' (unsigned article), *Australian Army Journal*, No. 12, May 1950, pp. 5–9
Berger, P., 'Leadership', *Australian Defence Force Journal*, No. 61, November–December 1986
Brennan, P. H., 'Amateur warriors: Infantry battalion commanders in the Canadian Expeditionary Force', in author's collection.
Bridge, C., Harper, G. J., Spence, I., 'The Australian way of war: Some thoughts', draft in author's collection
Budge, K., 'Tribute to a C. O.', *The Red and White Diamond*, December 1987, p. 11
Campbell, I. T., 'A model for battalion command: Training and leadership in the 2nd AIF: A case study of Brigadier F. G. Galleghan', thesis, MA (Hons), University College, UNSW at ADFA, 1991
Crawford, D., 'Col. Warfe's strong point: Organisation', *Sun News Pictorial*, 5 August 1969
Dandridge, W. L., 'My philosophy of leadership', *Australian Defence Force Journal*, No. 58, May/June 1986, pp. 57–8
Dwyer, T., 'Sergeants are snakes', *Sydney Morning Herald*, 2 Dec 1944
Echevarria, A. J. II, 'Auftragstaktik: In its proper perspective', *Military Review*, October 1986, pp. 50–6
Evans, M., 'The Australian way in warfare', Australian Army – General Staff Division discussion paper, 3 Oct 1995, copy in author's collection
—— 'The tyranny of dissonance: Australia's strategic culture and way of war 1901–2005', Land Warfare Studies Centre, Study Paper No. 306, 2005.
Funnell, R. G., 'Leadership: Theory and practice', *Australian Defence Force Journal*, No. 35, July/August 1982, pp. 5–21

Graves, T. C., 'Choosing the best: Battalion command and the role of experience', monograph, School of Advanced Military Studies, United States Army Command and General Staff College, 1999
Hayes, G., 'A search for balance: Officer selection in the wartime Canadian Army, 1939–45', in author's collection
—— 'The Canadians in Sicily: Sixty years on', in author's collection
Hearder, R. S., 'Careers in captivity: Australian prisoner-of-war medical officers in Japanese captivity during World War II', PhD thesis, University of Melbourne, 2004
Henderson, F. P., 'Commandership: The art of command', *Marine Corps Gazette*, January 1992, pp. 41–3
Hingston, D., 'The development of gifted amateurs – New Zealanders at war in the Western Desert', paper for Defence Studies program, University of New England, 1996
Horner, D. M., 'Staff Corps versus militia: The Australian experience in World War II', *Australian Defence Force Journal*, No. 26, January/February 1981, pp. 13–26
—— 'Towards a philosophy of Australian command', in author's collection
Illing, H. C., 'Leadership and the commanding officer', *British Army Review*, No. 50, August 1975, pp. 43–6
Keating, G. M., 'The right man for the right job: An assessment of Lieutenant General S. G. Savige as a senior commander', honours thesis, BA, University of New South Wales, 1995
Kirkland, F. L., 'Combat leadership styles: Empowerment versus authoritarianism', *Parameters*, Vol. 20, No. 4, December 1990, pp. 61–72
Moremon, J., 'Most deadly jungle fighters? Australian infantry in Malaya and Papua: 1941–43', BA (Hons) thesis, University of New England, 1992
Neumann, C., 'Australia's citizen soldiers, 1919–39: A study of organization, command, recruiting training and equipment', MA thesis, UNSW at RMC Duntroon, 1978
Nicholson, M. J., Thoughts on Followership – Part I', *British Army Review*, No. 99, December 1991, pp. 11–17
—— 'Thoughts on Followership – Part II', *British Army Review*, No. 100, April 1992, pp. 37–41
Palazzo, A., 'The way forward: 1918 and the implications for the future', in P. Dennis & J. Grey (eds), *1918: Defining Victory, Proceedings of the Chief of Army's History Conference held at the National Convention Centre*, Australian Army History Unit, Canberra 1999
People, J., 'The Australian militia 1930–39', *Australian Defence Force Journal*, No. 33, March–April 1982, pp. 44–8
Perry, W., 'Is there an Australian style of command?', *Australian Defence Force Journal*, No. 55, November–December 1985, pp. 4–14
Porter, S. H. W. C., 'Why do it, colonel? Why not get someone else to do it?', *Australian Army Journal*, No. 57, February 1954, pp. 5–9
Pratten, G. M., '"Under rather discouraging circumstances": The CMF in Melbourne's eastern suburbs, 1921–39', BA (Hons) thesis, University of Melbourne, 1994

—— '"The old man": Australian battalion commanders in the Second World War', PhD thesis, Deakin University, 2006

Robson, L. L., 'The origin and character of the First AIF, 1914–18: Some statistical evidence', *Historical Studies*, No. 61, 1973, pp. 737–49

Rocke, M. D., 'Trust: The cornerstone of leadership', *Military Review*, August 1992, pp. 30–40

Ryan, A., '"Putting your young men in the mud": Change, continuity and the Australian infantry battalion', Land Warfare Studies Centre, Working Paper No. 124, 2003

Saint, C. E., 'Commanders still must go see', *Army*, Vol. 41, No. 6, June 1991, pp. 18–26

Sorley, L., 'Creighton Abrams and levels of leadership', *Military Review*, August 1992, pp. 3–10

'Spartacus', 'Command and leadership: A rational view', *British Army Review*, No. 1066, April 1994, pp. 24–7

Stanley, P., '"The men who did the fighting are now all busy writing": Australian post-mortems on defeat in Malaya and Singapore, 1942–45', paper presented at National University of Singapore, February 2002

—— 'The Green Hole: Exploring our neglect of the New Guinea campaigns of 1943–44', *Sabretache*, Vol. 34, April–June 1993, pp. 3–9

Steele, W. M., 'Army leaders: How you build them; how you grow them', *Military Review*, August 1992

Sutton, R., 'Lieutenant Colonel Sir John Overall CBE MC (and bar) FRAI FAPI AMTPI (1913–2001)', author's collection

Symons, J. N., 'Leadership in context', *British Army Review*, No. 100, April 1992, pp. 10–13

'Thoughts on Command in Battle' (unsigned article), *British Army Review*, No. 69, December 1981

Thyer, J. H., 'Leadership: An address delivered to the officers, warrant officers and senior non-commissioned officers in the Changi Prison Camp on 18th June 1942', author's collection

Travers, B. H., 'The Staff Corps–CMF conflict as seen by a young AIF officer 1940–41', paper presented to Australian War Memorial History Conference, 6–10 July 1987, author's collection

Wass de Czege, H., 'A comprehensive view of leadership', *Military Review*, August 1992, pp. 21–9

Wilmoth, J. R., 'Greece – Kalamata', author's collection

Wilson, D. E., 'Theory X, theory Y won't take the hill', *Army*, July 1992, pp. 35–8

INTERNET RESOURCES

Ambur, O., 'Reconsidering the higher-order legitimacy of French and Raven's bases of social power in the information age', http://users.erols.com/ambur/French&Raven.htm, accessed 20 Dec 2001

Dorosh, M., 'Treatise on Battalion Commanders', http://www.battlefront.com/community/archive/index.php/t-32644.html, accessed 19 October 2008

'Key leadership theories in the "Stimulating Leaders" Report', Manufacturing Foundation of the United Kingdom, http://www.manufacturingfoundation.org.uk/upload/leadership_theories.pdf, accessed 24 Apr 2005

'Leadership', Human Service Department, Flathead Valley Community College, http://www.fvcc.edu/academics/dept_pages/human.services/leadership.htm, accessed 12 Jan 2005

'Maslow's Hierarchy of Needs', http://web.utk.edu/~gwynne/maslow.htm, accessed 28 Dec 2001

Meloy, G. S., 'Gen. Ridgway's Personnel Management Policy', *Army Magazine*, November 2002, http://www3.ausa.org/webpub/DeptArmyMagazine.nsf/byid/CCRN-6CCS6A, accessed 19 Oct 2008

Sherman, L. W., Schmuck, R. & P., 'Kurt Lewin's contribution to the theory and practice of education in the United States', http://www.users.muohio.edu/shermalw/LewinConference_PaperV3.doc, p. 14, accessed 20 Apr 2005

'Vroom's taxonomy of leadership style', http://www.gsu.edu/~dscthw/413-813/groups/vroom.htm, accessed 27 Dec 2001

'World War 2 Nominal Roll', Department of Veterans Affairs (Australia), http://www.ww2roll.gov.au, accessed various dates 2004–05

UNPUBLISHED MANUSCRIPTS

Chilton, F. O., 'A Life in the Twentieth Century'
Dexter, W. R., 'The Battalion – My Home' (AWM PR 01192)
Field, J., 'Warriors for the Working Day' (AWM MSS785)
Hurst, P. J., 'My Army Days' (AWM MSS1656)
Jackson, D. R., 'Autobiography' (AWM MSS1193)
Kappe, C. H., 'The Malayan Campaign' (AWM MSS1393)
Oakes, R. F., 'Singapore Story' (AWM MS 00776)
Reid, M., 'Road Block' (AWM MSS1629)
Stevens, J. E. S., 'A personal story of the service, as a citizen soldier, of Major-General Sir Jack Stevens, KBE, CB, DSO, ED' (AWM 3DRL 3561)
Watkins, L. W., 'As I Remember It' (AWM MSS1587)
Wilmoth, J. A., 'Wandering Recollections of John Wilmoth aged 83 Years'

NEWSPAPERS AND PERIODICALS

Men May Smoke: Official Journal of 2/18th Battalion, 8th Division, AIF, Vol. 1, No. 1, Dec 1940

Second Two Nought: Weekly Bulletin of the 2/20th Bn, AIF, Malaya, 2/20th Battalion, 14 Jun 1941

The Red and White Diamond, 24th Battalion (AIF) Association, Dec 1987

OFFICIAL HISTORIES

Official History of Australia in the War of 1914–18

Bean, C. E. W., *The Australian Imperial Force in France, 1916*, University of Queensland Press, Brisbane, 1982

Official History of Australia in the War of 1939–45
Dexter, D. St A., *The New Guinea Offensives*, Australian War Memorial, Canberra, 1961
Long, G. M., *To Benghazi*, Australian War Memorial, Canberra, 1952
—— *Greece, Crete and Syria*, Australian War Memorial, Canberra, 1953
—— *The Final Campaigns*, Australian War Memorial, Canberra, 1963
Maughan, D. W. B., *Tobruk and El Alamein*, Australian War Memorial, Canberra, 1966
McCarthy, D., *South-West Pacific Area – First Year: Kokoda to Wau*, Australian War Memorial, Canberra, 1959
Walker, A. S., *Middle East and Far East*, Australian War Memorial, Canberra, 1953
Wigmore, L. G., *The Japanese Thrust*, Australian War Memorial, Canberra, 1957

Official History of Australia's Involvement in Southeast Asian Conflicts 1948–75
Dennis, P. & Grey, J., *Emergency and Confrontation: Australian Military Operations in Malaya and Borneo 1950–66*, Allen & Unwin, Sydney, 1996

Official History of the Canadian Army in the Second World War
Stacey, C. P., *The Canadian Army 1939–1945: An Official Historical Summary*, The King's Printer, Ottawa, 1948
—— *Six Years of War: The Army in Canada, Britain and the Pacific*, The Queen's Printer and Controller of Stationery, Ottawa, 1955
—— *The Victory Campaign: The operations in North-west Europe 1944–1945*, The Queen's Printer and Controller of Stationery, Ottawa, 1960

Official History of New Zealand in the Second World War 1939–45
Cody, J. F., *28 Maori Battalion*, War History Branch – Department of Internal Affairs, Wellington, 1956
Henderson, J., *22 Battalion*, War History Branch – Department of Internal Affairs, Wellington, 1958
Norton, F. D., *26 Battalion*, War History Branch – Department of Internal Affairs, Wellington, 1952
Ross, A., *23 Battalion*, War History Branch – Department of Internal Affairs, Wellington, 1959
Stevens, W. G., *Problems of 2 NZEF*, War History Branch – Department of Internal Affairs, Wellington, 1958

History of the Second World War: United Kingdom Military Series
Ellis, L. F., *Victory in the West*, Vol. 1: *The Battle of Normandy*, Her Majesty's Stationery Office, London, 1962
Playfair, I. S. O. & Molony, C. J. C., *The Mediterranean and Middle East*, Vol. 4: *The Destruction of the Axis Forces in Africa*, Her Majesty's Stationery Office, London, 1966

United States Army in World War II
Greenfield, K. R., Palmer, R. R. & Wiley, B. I., *The Organization of Ground Combat Troops*, Historical Division – Department of the Army, Washington, DC, 1947
Palmer, R. R., Wiley, B. I. & Keast, W. R., *The Procurement and Training of Ground Combat Troops*, Historical Division – Department of the Army, Washington, DC, 1948

BOOKS
Allchin, F., *Purple and Blue: The History of the 2/10th Battalion, AIF (The Adelaide Rifles) 1939–1945*, 2/10th Battalion Association, Adelaide, 1958
Argent, A., Butler, D. M. & Shelton, J. J., *The Fight Leaders: A Study of Australian Battlefield Leadership – Green, Ferguson and Hassett*, Australian Military History Publications, Loftus, NSW, 2002
Arneil, S., *Black Jack: The Life and Times of Brigadier Sir Frederick Galleghan*, Macmillan, Melbourne, 1983
Austin, R., *Bold, Steady, Faithful: The History of the 6th Battalion, The Royal Melbourne Regiment 1854–1993*, 6th Battalion Association, Melbourne, 1993
—— *Let Enemies Beware 'Caveant Hostes': The History of the 2/15th Battalion 1940–1945*, Slouch Hat Publications, McCrae, Vic., 1995
Austin, V., *To Kokoda and Beyond: The Story of the 39th Battalion 1941–1943*, Melbourne University Press, Carlton, 1988
Australian Dictionary of Biography, Vol. 13: 1940–80, A–De, Melbourne University Press, Carlton, Vic., 1993
Baker, K., *Paul Cullen – Citizen and Soldier: Life and Times of Major General Paul Cullen*, Rosenberg Publishing, Sydney, 2005
Barker, T., *Signals: A History of the Royal Australian Corps of Signals 1788–1947*, Royal Australian Signal Corps Committee, Canberra, 1988
Barr, N., *Pendulum of War: The Three Battles of El Alamein*, Jonathan Cape, London, 2004
Barrett, J., *We were There: Australian Soldiers of World War II*, Viking, Melbourne, 1987
Barter, M., *Far Above Battle: The Experience and Memory of Australian Soldiers in War, 1939–1945*, Allen & Unwin, Sydney, 1994
Beaumont, J., *Gull Force: Survival and Leadership in Captivity, 1941–1945*, Allen & Unwin, Sydney, 1988
Beaumont, J. (ed.), *Australian Defence: Sources and Statistics*, Oxford University Press, Melbourne, 2001
Benson, S. E., *The Story of the 42 Aust. Inf. Bn*, 42nd Australian Infantry Battalion Association, Sydney, 1952
Bentley, A., *The Second Eighth: A History of the 2/8th Australian Infantry Battalion*, Second Eighth Battalion Association, Melbourne, 1984
Bidwell, S. & Graham, D., *Fire-Power: British Army Weapons and Theories of War 1904–1945*, George Allen & Unwin, London, 1982

Bilney, K., *14/32 Australian Infantry Battalion AIF 1940–1945*, 14/32nd Australian Infantry Battalion Association, Melbourne, 1994
Bishop, J. A., *Word from John: An Australian Soldier's Letter to His Friends*, Cassell, Melbourne, 1944
Blair, D. J., *Dinkum Diggers: An Australian Battalion at War*, Melbourne University Press, Carlton, 2001
Blair, R. C., *A Young Man's War: A History of the 37th/52nd Australian Infantry Battalion in World War Two*, 37/52 Australian Infantry Battalion Association, Melbourne, 1992
Bolger, W. P. & Littlewood, J. G., *The Fiery Phoenix: The Story of the 2/7 Australian Infantry Battalion 1939–1946*, 2/7 Battalion Association, Parkdale, Vic., 1983
Bradley, P., *On Shaggy Ridge: The Australian Seventh Division in the Ramu Valley: From Kaiapit to the Finisterres*, Oxford University Press, Melbourne, 2004
Brigg, S. & L., *The 36th Australian Infantry Battalion 1939–1945: The Story of an Australian Infantry Battalion and Its Part in the War against Japan*, 36th Battalion Association, Sydney, 1967
Brown, D. E. & Matthews, L. J. (eds), *The Challenge of Military Leadership*, Pergamon-Brassey's, Washington, 1989
Brune, P., *A Bastard of a Place: The Australians in Papua*, Allen & Unwin, Sydney, 2003
—— *Gona's Gone! The Battle for the Beach-head 1942*, Allen & Unwin, Sydney, 1994
—— *The Spell Broken: Exploding the Myth of Japanese Invincibility*, Allen & Unwin, Sydney, 1997
—— *Those Ragged Bloody Heroes: From the Kokoda Trail to Gona Beach*, Allen & Unwin, Sydney, 1992
—— *We Band of Brothers: A Biography of Ralph Honner, Soldier and Statesman*, Allen & Unwin, Sydney, 2000
Buck, J. H. & Korb, L. J., *Military Leadership*, Sage Publications, Beverly Hills, CA, 1981
Budden, F. M., *That Mob: The Story of the 55/53rd Australian Infantry Battalion, AIF*, F. M. Budden, Sydney, 1973
Burfitt, J., *Against All Odds: The History of the 2/18th Battalion AIF*, 2/18th Battalion (AIF) Association, Sydney, 1991
Burns, J., *The Brown and Blue Diamond at War: The Story of the 2/27th Battalion AIF*, 2/27th Battalion Ex-Servicemen's Association, Adelaide, 1960
Calvert, M., *Slim*, Pan Books, London, 1973
Cartwright, D. (ed.), *Studies in Social Power*, University of Michigan, Ann Arbor, 1959
Cathcart, M., *Defending the National Tuckshop: Australia's Secret Army Intrigue of 1931*, McPhee Gribble, Melbourne, 1988
Charlott, R. (ed.), *The Unofficial History of the 29/46th Australian Infantry Battalion AIF September 1939–September 1945*, Halstead Press, Melbourne, 1951
Charlton, P., *The Thirty-niners*, Macmillan, Melbourne, 1981

Christensen, G. (ed.), *That's the Way It Was: The History of the 24th Australian Infantry Battalion (AIF) 1939–1945*, 24th Battalion (AIF) Association, Melbourne, 1982
Christie, R. W. (ed.), *A History of the 2/29th Battalion – 8th Australian Division AIF*, 2/29 Battalion AIF Association, Sale, Vic., 1983
Clift, K. R., *War Dance: A Story of the 2/3 Aust. Inf. Battalion AIF*, P. M. Fowler and 2/3rd Battalion Association, Kingsgrove, NSW, 1980
Clift, K. R. & Fernside, G. H., *Dougherty: A Great Man Among Men*, Alpha Books, Sydney, 1979
Coates, J., *Bravery Above Blunder: The 9th Australian Division at Finschhafen, Sattelberg, and Sio*, Oxford University Press, Melbourne, 1999
Combe, G., Gilchrist, T., Ligertwood, F., *The Second 43rd Australian Infantry Battalion 1940–1946*, Second 43rd Battalion AIF Club, Adelaide, 1972
Coombes, D., *Morshead: Hero of Tobruk and El Alamein*, Oxford University Press, Melbourne, 2001
Copeland, N., *Psychology and the Soldier: The Art of Leadership*, George Allen & Unwin, London, 1944
Corfield, R. S., *Hold Hard, Cobbers: The Story of the 57th and 60th and 57/60th Australian Infantry Battalions*, Vol. 2: 1930–90, 57/60th Battalion (AIF) Association, Glenhuntly, 1991
Coulthard-Clark, C. D., *Duntroon: The Royal Military College of Australia, 1911–1986*, Allen & Unwin, Sydney, 1986
—— *Soldiers in Politics: The Impact of the Military on Australian Political Life and Institutions*, Allen & Unwin, Sydney, 1996
Crang, J. A., *The British Army and the People's War 1939–1945*, Manchester University Press, Manchester, 2000
Cranston, F., *Always Faithful: A History of the 49th Australian Infantry Battalion 1916–1982*, Boolarong Publications, Brisbane, 1983
Crotty, M., *Making the Australian Male: Middle-class Masculinity 1870–1920*, Melbourne University Press, Carlton, 2001
Dennis, P., *The Territorial Army 1906–1940*, Royal Historical Society, Woodbridge, UK, 1987
Dennis, P. & Grey, J. (eds), *1918: Defining Victory*, Australian Army History Unit, Canberra, 1999
Dennis, P., Grey, J., Morris, E., Prior, R. (eds), *The Oxford Companion to Australian Military History*, Oxford University Press, Melbourne, 1995
Devine, W., *The Story of a Battalion*, Melville & Mullen, Melbourne, 1919
Dunlop, W. A. S., *Hints to Young Officers in the Australian Military Forces*, Angus & Robertson, Sydney, 1940
—— *The Fighting Soldier*, Angus & Robertson, Sydney, 1940
Dunn, J. B., *Eagles Alighting: A History of 1 Australian Parachute Battalion*, 1 Australian Parachute Battalion Association, East Malvern, Vic., 1999
Dwyer, M., *Facts. Fallout! Fantasy!* Self-published, Melbourne, 1985
Eather, S., *Desert Sands, Jungle Lands: A Biography of Major General Ken Eather*, Allen & Unwin, Sydney, 2003
Edgar, B., *Warrior of Kokoda: A Biography of Brigadier Arnold Potts*, Allen & Unwin, Sydney, 1999

Ellis, J., *The Sharp End: The Fighting Man in World War II*, Pimlico Press, London, 1990
English, J. A., *On Infantry*, Praeger Publishers, Westport, CT, 1994
Farrell, B. P., *The Defence and Fall of Singapore 1940–1942*, Tempus Publishing, Stroud, UK, 2005
Fearnside, G. H. (ed.), *Bayonets Abroad: A History of the 2/13 Battalion, AIF in the Second World War*, John Burridge Military Antiques, Swanbourne, WA, 1993
Finkemeyer, C. E., *It Happened to Us: The Unique Experiences of 20 Members of the 4th Anti-Tank Regiment*, C. E. & D. J. Finkemeyer, Melbourne, 1994
French, D., *Raising Churchill's Army: The British Army and the War Against Germany 1919–1945*, Oxford University Press, Oxford, 2001
Fuller, J. F. C., *Generalship, Its Diseases and Their Cure: A Study of the Personal Factor in Command*, Faber & Faber, London, 1933
Gabriel, R. A., *Military Incompetence: Why the American Military Doesn't Win*, Hill & Wang, New York, 1985
Gerster, R., *Big-noting: The Heroic Theme in Australian War Writing*, Melbourne University Press, Carlton, Vic., 1987
Givney, E. C., *The First at War: The Story of the 2/1st Australian Infantry Battalion 1939–45, The City of Sydney Regiment*, Association of First Infantry Battalions, Earlwood, NSW, 1987
Glenn, J. G., *Tobruk to Tarakan: The Story of the 2/48th Battalion AIF*, Rigby, Adelaide, 1960
Gower, S. N., *Guns of the Regiment*, Australian War Memorial, Canberra, 1981
Graeme-Evans, A. L., *Of Storms and Rainbows: The Story of the Men of the 2/12th Battalion AIF*, Vol. 1, 2/12th Battalion Association, Hobart, 1989
—— *Of Storms and Rainbows: The Story of the Men of the 2/12th Battalion AIF*, Vol. 2, 2/12th Battalion Association, Hobart, 1991
Green, O., *The Name's Still Charlie*, University of Queensland Press, Brisbane, 1993
Grey, J., *A Military History of Australia*, Cambridge University Press, Melbourne, 1990
—— *Australian Brass: The Career of Lieutenant General Sir Horace Robertson*, Cambridge University Press, Melbourne, 1992
Griffiths-Marsh, R., *The Sixpenny Soldier*, Angus & Robertson, Sydney, 1990
Grossman, D. A., *On Killing: The Psychological Cost of Learning to Kill in War and Society*, Little, Brown, Boston, 1996
Harper, G. J., *Kippenberger: An Inspired New Zealand Commander*, Harper Collins, Auckland, 1997
Harper, G. J. & Hayward, J. (eds), *Born to Lead? Portraits of New Zealand Commanders*, Exisle Publishing, Auckland, 2003
Hastings, M., *Overlord: D-Day and the Battle for Normandy*, Pan Books, London, 1985
Hay, D. A., *Nothing Over Us: The Story of the 2/6th Australian Infantry Battalion*, Australian War Memorial, Canberra, 1984

Henning, P., *Doomed Battalion: Mateship and Leadership in War and Captivity*, Allen & Unwin, Sydney, 1995

History of the Activities of the Office of the Military Secretary during the War 1939–1945, Australian Military Forces, Melbourne, 1946

Hogg, I. V., *British and American Artillery of World War 2*, Arms & Armour Press, London, 1978

Holmes, R., *Firing Line*, Pimlico Press, London, 1985

Holmes, R. & Keegan, J., *Soldiers: A History of Men in Battle*, Hamish Hamilton, London, 1985

Hopkins, R. N. L., *Australian Armour: A History of the Royal Australian Armoured Corps 1927–1972*, Australian War Memorial, Canberra, 1978

Horner, D. M., *Armies and Nation Building: Past Experience – Future Prospects*, Strategic & Defence Studies Centre, Australian National University, Canberra, 1995

—— *Blamey: The Commander-in-Chief*, Allen & Unwin, Sydney, 1998

—— *Crisis of Command: Australian Generalship and the Japanese Threat 1941–1943*, Australian National University Press, Canberra, 1978

—— *The Commanders: Australian Military Leadership in the Twentieth Century*, George Allen & Unwin, Sydney, 1984

—— *The Gunners: A History of Australian Artillery*, Allen & Unwin, Sydney, 1995

Horner, D. M. (ed.), *High Command: Australia's Struggle for an Independent War Strategy 1939–1945*, Allen & Unwin, Sydney, 1982

Jacobs, J. W. & Bridgland, R. J. (eds), *Through: The Story of Signals 8 Australian Division and Signals AIF Malaya*, 8 Division Signals Association, Sydney, 1949

Johnson, K. T., *The 2/11th (City of Perth) Australian Infantry Battalion 1939–45*, John Burridge Military Antiques, Swanbourne, WA, 2000

Johnston, M., *At the Front Line: Experiences of Australian Soldiers in World War II*, Cambridge University Press, Melbourne, 1996

Johnston, M. & Stanley, P., *Alamein: The Australian Story*, Oxford University Press, Melbourne, 2002

Jungle Warfare: With the Australian Army in the South-West Pacific, Australian War Memorial, Canberra, 1944

Keegan, J., *The Mask of Command: A Study of Generalship*, Pimlico Press, London, 1999

Kennedy, C., *Port Moresby to Gona Beach: 3rd Australian Infantry Battalion 1942*, C. Kennedy, Turner, ACT, 1992

Kippenberger, H., *Infantry Brigadier*, Oxford University Press, London, 1949

Kuring, I., *Red Coats to Cams: A History of Australian Infantry, 1788–2001*, Australian Military History Publications, Loftus, NSW, 2004

Lack, J. (ed.), *No Lost Battalion: An Oral History of the 2/29th Battalion AIF*, Slouch Hat Publications, McCrae, Vic., 2005

Laffin, J., *Digger: The Story of the Australian Soldier*, Cassell, London, 1959

—— *Forever Forward: The Story of the 2/31st Infantry Battalion 2nd AIF, 1940–45*, 2/31st Australian Infantry Battalion Association, Newport, NSW, 1994

Lambert, E. F., *The Twenty Thousand Thieves*, Frederick Muller, London, 1952

Lilley, A. B., *Sydney University Regiment: A Description of the Insignia Worn from 1900 to 1973 by Military Units at the University of Sydney Together with Information on Honorary Colonels and Commanding Officers*, Military Historical Society of Australia, Sydney, 1974

Lodge, A. B., *The Fall of General Gordon Bennett*, Allen & Unwin, Sydney, 1986

—— *Lavarack: Rival General*, Allen & Unwin, Sydney, 1998

Louch, T. S., *From the CO's Notebook*, 2/11th Battalion Association, Perth, n.d.

MacFarlan, G., *Etched in Green: The History of the 22nd Australian Infantry Battalion 1939–1946*, 22nd Australian Infantry Battalion Association, Melbourne, 1961

Maclean, I., *A Guide to the Records of Gavin Long*, Australian War Memorial, Canberra, 1993

Magarry, W. R., *The Battalion Story: 2/26th Battalion, 8th Division – AIF*, W. R. Magarry, Jindalee, Qld, 1994

Marshall, A. J. (ed.), *Nulli Secundus Log*, 2/2nd Australian Infantry Battalion, AIF, Sydney, 1946

Masel, P., *The Second 28th: The Story of a Famous Battalion of the Ninth Australian Division*, 2/28th Battalion and 24th Anti-tank Company Association, Perth, 2000

Matthews, R., *Militia Battalion at War: The History of the 58/59th Australian Infantry Battalion in the Second World War*, 58/59th Battalion Association, Brunswick, 1998

McAllester, J. C., *Men of the 2/14 Battalion*, 2/14 Battalion Association, Melbourne, 1990

McCarthy, D., *The Once and Future Army: A History of the Citizen Military Forces 1947–1974*, Oxford University Press, Melbourne, 2003

Miller, D., *Commanding Officers*, John Murray, London, 2001

Moore, A., *The Secret Army and the Premier: Conservative Paramilitary Organisations in New South Wales 1930–32*, UNSW Press, Kensington, NSW, 1989

Myatt, F., *The British Infantry 1660–1945*, Blandford Press, Poole, UK, 1983

Newton, R. W. et al., *The Grim Glory of the 2/19 Battalion AIF*, 2/19 Battalion AIF Association, Sydney, 1975

O'Brien, G. & Whitelocke, C., *Gunners in the Jungle: A Story of the 2/15 Field Regiment Royal Australian Artillery, 8 Division, Australian Imperial Force*, 2/15 Field Regiment Association, Eastwood, NSW, 1983

O'Brien, M., *Conscripts and Regulars: With the Seventh Battalion in Vietnam*, Allen & Unwin, Sydney, 1995

Palazzo, A., *Defenders of Australia: The Third Australian Division, 1916–1991*, Australian Army History Unit and Australian Military History Publications, Canberra, 2002

—— *The Australian Army: A History of Its Organisation 1901–2001*, Oxford University Press, Melbourne, 2001

Paull, R., *Retreat from Kokoda*, Heinemann, Melbourne, 1958

Pedder, A. L., *The Seventh Battalion, 1936–1946*, 7th Battalion (1939–45) Association, Melbourne, 1989

Pedersen, P. A., *Monash as Military Commander*, Melbourne University Press, Carlton, Vic., 1985
Perry, F. W., *The Commonwealth Armies: Manpower and Organisation in Two World Wars*, Manchester University Press, Manchester, 1988
Pratten, G. M. & Harper, G. J. (eds), *Still the Same: Reflections on Active Service from Bardia to Baidoa*, Australian Army Doctrine Centre, Georges Heights, NSW, 1996
Ross, J., *The Myth of the Digger: The Australian Soldier in Two World Wars*, Hale & Iremonger, Sydney, 1985
Russell, W. B., *The Second Fourteenth Battalion: A History of an Australian Infantry Battalion in the Second World War*, Angus & Robertson, Sydney, 1948
Russell, W. B., *There Goes a Man: The Biography of Sir Stanley G. Savige*, Longmans, London, 1989
Sarkesian, S. C. (ed.), *Combat Effectiveness: Cohesion, Stress and the Volunteer Military*, Sage, Beverly Hills, CA, 1980
2/17 Battalion History Committee, *'What We Have – We Hold!' A History of the 2/17 Australian Infantry Battalion, 1940–1945*, Australian Military History Publications, Loftus, NSW, 1998
Serle, R. P. (ed.), *The Second Twenty-Fourth Australian Infantry Battalion of the 9th Australian Division: A History*, Jacaranda Press, Brisbane, 1963
Share, P. (ed.), *Mud and Blood: Albury's Own Second Twenty-Third Australian Infantry Battalion, Ninth Australian Division*, Heritage Book Publications, Frankston, Vic., 1978
Sheffield, G. D. (ed.), *Leadership and Command: The Anglo-American Military Experience Since 1861*, Brasseys, London, 1997
Speed, F. W., *Esprit de Corps: The History of the Victorian Scottish Regiment and the 5th Infantry Battalion*, Allen & Unwin, Sydney, 1988
Spencer, W. B., *In the Footsteps of Ghosts: With the 2/9th Battalion in the African Desert and the Jungles of the Pacific*, Allen & Unwin, Sydney, 1999
Stanley, P., *Tarakan: An Australian Tragedy*, Allen & Unwin, Sydney, 1997
Sublet, F. H., *Kokoda to the Sea: A History of the 1942 Campaign in Papua*, Slouch Hat Publications, Rosebud, Vic., 2000
Taylor, G., *Infantry Colonel*, Self Publishing Association, Upton-upon-Severn, 1990
Thomas, R., *42nd Infantry Battalion, the Capricornia Regiment: History, Badges, Medals & Insignias*, Ron Thomas, Geelong, Vic., 1996
Thomson, A., *Anzac Memories: Living with the legend*, Oxford University Press, Melbourne, 1994
Trigellis-Smith, S., *Britain to Borneo: A History of 2/32 Australian Infantry Battalion*, 2/32 Australian Infantry Battalion Association, Sydney, 1993
—— *All the King's Enemies: A History of the 2/5th Australian Infantry Battalion*, 2/5th Battalion Association, Melbourne, 1988
Tsuji, M., *Singapore: The Japanese Version*, Ure Smith, Sydney, 1960
Turell, A. N., *Never Unprepared: A History of the 26th Australian Infantry Battalion (AIF) 1939–1946*, 26th Battalion Reunion Association, Wynnum, Qld, 1992

Unit History Editorial Committee (ed.), *White over Green: The 2/4th Battalion*, Angus & Robertson, Sydney 1963
Uren, M., *A Thousand Men at War: The Story of the 2/16th Battalion, AIF*, William Heinemann, Melbourne, 1959
Van Creveld, M. L., *Command in War*, Harvard University Press, Cambridge, MA, 1985
—— *Fighting Power: German and US Army Performance, 1939–1945*, Arms and Armour Press, London, 1983
Wahlert, G. *The Other Enemy? Australian Soldiers and the Military Police*, Oxford University Press, Melbourne, 1999
—— (ed.), *Australian Army Amphibious Operations in the South-West Pacific: 1942–1945. Edited Papers of the Australian Army History Conference held at the Australian War Memorial, 15 November 1994*, Australian Army Doctrine Centre, Georges Heights, NSW, 1995
Wall, D. & Seale, C., *Singapore and Beyond: The Story of the Men of the 2/20 Battalion told by the Survivors*, 2/20 Battalion Association, East Hills, NSW, 1985
Warren, A., *Singapore 1942: Britain's Greatest Defeat*, Hambledon & London, London, 2002
Watson, P., *War on the Mind: The Military Uses and Abuses of Psychology*, Basic Books, New York, 1978
Watt, J., *History of the 61st Australian Infantry Battalion (AIF): Queensland Cameron Highlanders 1938–1945*, Australian Military History Publications, Loftus, NSW, 2001
Wick, S., *Purple over Green: The History of the 2/2 Australian Infantry Battalion 1939–1945*, 2/2 Australian Infantry Battalion Association, Sydney, 1977
Wigmore, L. G., Harding, B. A., Williams, J. & Staunton, A., *They Dared Mightily* (2nd edn), Australian War Memorial, Canberra 1986
Wilcox, C., *For Hearths and Homes: Citizen Soldiering in Australia 1854–1945*, Allen & Unwin, Sydney, 1998
Yeates, J. D. and Loh, W. G. (eds), *Red Platypus: A Record of the Achievements of the 24th Australian Infantry Brigade Ninth Australian Division, 1940–45*, 24th Australian Infantry Brigade, Perth, 1946

Index

21A course, 35, 36, 335
  standards achieved, 37
2nd AIF
  See Australian Imperial Force, 2nd

Abau, 260
Abbott, John Noel
  awarded DSO, 267
  exhaustion of, 267
  replaced as CO of 26th Battalion, 238
adjutants, 120
  role of, 64
administrative platoon, 83
aerial photography, 214
age
  8th Division COs, 133
  average of COs in 1918, 33
  average of COs in AIF in February, 111
  COs in 1945, 238
  COs in British Army in 1941, 174
  COs in Papua, 179
  declining trend of COs, 312
  for retirement, pre war, 33
  influence on authority, 241, 248
  maximum for lieutenant colonels in 2nd AIF, 54
  militia age limits lowered, 173
  militia and 2nd AIF COs in 1942, 167
  replacement militia COs, 174
aggression, 28, 70, 96, 100, 127, 128, 152, 156, 185, 259–60, 272, 273
Ainslie, Robert Inglis, 221, 373
  address prior to Tarakan landings, 11
  cautious approach on Tarakan, 258
  personality of, 219
  resists pressure from Wootten, 219
  takes command of 2/48th Battalion, 232
  utilises skills of company commanders, 212
air support, 36, 93, 164, 190, 200, 250, 253, 260
Aitape–Wewak campaign, 236, 243, 258, 259
  nature of, 262
Allen, Arthur Samuel 'Tubby', 50, 51, 70, 72, 101
Ama Keng, 158
amateurism, 47, 235, 273
Ambon, 11, 162
AMF
  See Australian Military Forces
Amies, Jack Lowell, 174
AMR&O
  See Australian Military Regulations and Orders
Anderson, Arthur Jeffery 'Fix it', 246
Anderson, Charles Groves Wright, 136, 271, 387
  abandons A Echelon, 160
  as 2iC of 2/19th Battalion, 135
  assessment of by Bennett, 137
  awarded Victoria Cross, 156
  becomes member of parliament, 274
  command style, 156
  evacuated to hospital, 146
  knowledge of close country tactics, 128–54
  leaves wounded behind at Parit Sulong, 160
  moral courage of, 160, 161
  personality of, 135, 156, 160

INDEX 409

relationship with Galleghan, 140
uses carriers for shock action, 150
Anderson, Raymond Keith
dies of wounds at Tobruk, 349
anti-aircraft platoon, 83
removed from battalion, 203
anticipation by COs, 210, 214
anti-tank platoon, 113, 114, 203
alternate roles of, 203
renamed tank-attack, 203
anti-tank weapons, 126, 147
2-pounders, 114, 203, 251, 345
shortcomings, 115
use in Malaya, 150
6-pounders, 114, 250
alternative uses of, 88
Boys rifles, 83, 88, 96
lack of effective examples in battalion, 88
non-standard ordnance, 113, 352
PITA (PIAT), 251
use of captured weapons, 88, 113
Anzac legend, 27, 235
influence of, 170
Aola, 186
appointments
duration of, 296, 313
influence of First World War experience, 33
political influence on, 134
pre-war duration, 32
pre-war procedures, 32
procedures for, 77
ARA
See Australian Regular Army
archival collections
relating to COs, 14, 278, 313
armour, 69
at Balikpapan, 1, 2, 250
at El Alamein, 123, 127, 128
at Labuan, 251
at Sattelberg, 219
'Frog' flame-throwers, 250, 251
German
at El Alamein, 126
at Tobruk, 88
at Vevi, 107
impervious to 2-pounder, 115
in jungle warfare, 251
in pre-war training, 35, 36, 40, 69

Italian
at Bardia, 88
at Tobruk, 95
Japanese
at Milne Bay, 185
in Malaya, 147, 149
lack of in pre-war AMF, 36
Matilda II, 250
armoured cars
use in Malaya, 151, 360
Arneil, Stanley Foch, 53, 140, 142
artillery
at Balikpapan, 250, 253
at El Alamein, 125–7, 355
at Khirbe, 99
at Tobruk, 91, 93, 96
German
at El Alamein, 125
in attack on Mount Shiburangu, 256
in jungle warfare, 203, 254
in Malaya, 144, 150, 152, 160
in New Guinea, 221, 260
in Papua, 190
in pre-war training, 36
in southern Bougainville, 265
lack of for Wadi Muatered attack, 97
shortages of ammunition in 1945, 236
sniping guns, 253
Artillery Hill, 254, 263
Ashkanasy, Maurice, 360
Assheton, Charles Frederick, 145
appointed to command 2/20th Battalion, 138
as 2iC of 2/18th Battalion, 136
command style, 157
killed on Singapore, 157
loses control of 2/20th Battalion, 157
personality of, 136
Auchinleck, Claude John Eyre, 111
Australian Comforts Fund, 66
Australian Imperial Force, 2nd, 48
officers discouraged from joining, 341
organisation, 80
selection of COs, 56
selection of officers, 68
Australian Instructional Corps, 56

Australian Military Forces
  anti-tank regiments
    2/4th, 149, 360
  armies
    First, 169, 238, 259
  artillery
    2/7th Field Regiment: ammunition allocation on Tarakan, 254
    2/10th Field Regiment, 144
    65th Field Battery, 160
  as a citizens' army, 274, 277, 289
  brigades
    4th: capability of in 1945, 235, 380; dismissal of COs over 35, 238; in New Guinea, 198
    7th, 228, 248: 35 maximum age for COs in, 238; on Bougainville, 258, 263
    8th, 52: capability of in 1945, 235; in New Guinea, 198, 259, 376; preparing COs for operations, 202
    10th, 170
    11th, 170: standard of, 178
    14th: little chance for training in Papua, 171; poor standard of, 169
    15th: ammunition expenditure on Buin Road, 253; capability of in 1945, 235, 380; in New Guinea, 198; in the Finisterres, 210, 215; in Wau–Salamaua operations, 213, 222; on Bougainville, 248, 258
    16th, 293: in Papua, 183, 186; in the Aitape–Wewak campaign, 260, 262; selection of COs, 50, 53; training of COs, 72
    17th: in Wau–Salamaua campaign, 209, 369; selection of COs, 53; training in, 69
    18th, 52: at Tobruk, 78, 95; in Great Britain, 71; in Papua, 193; in the Finisterres, 207, 210; in the United Kingdom, 283; standard of battalions, 71
    19th: at Tobruk, 90; command relationships in, 99; formation of, 54
    20th, 246: at El Alamein, 119; in New Guinea, 216–18; selection of COs for, 54
    21st, 246: in Papua, 183, 193; on the Kokoda Track, 180; selection of COs, 54
    22nd, 131: in the fighting for Singapore, 130, 145; training in Malaya, 144
    23rd, 131, 174: poor standards in, 169
    24th, 131: at Tobruk, 110; in British North Borneo, 249–50; joins 9th Division, 132, 356; on Labuan, 251
    25th, 246: at Gona, 189; in Papua, 183, 193; preparing COs for operations, 202
    26th: at Sattelberg, 207; on Tarakan, 258
    27th: command relationships in, 140, 143, 162; formation of, 132; in Malaya, 142, 143; in the fighting for Singapore, 162
    29th: capability of in 1945, 380; improving standard of, 178; on Bougainville, 248, 263; poor training in, 170
    30th: little opportunity for training in Papua, 171; on the Kokoda Track, 180; poor standard of, 169
  capability in 1945, 383
  casualties, 4
  command crisis in, 248, 382
  commands
    Northern, 165
    Southern, 167
    Western Command, 167
  corps
    I, 236, 283
    II, 239
  Directorate of Military Training, 199, 201
  divisions
    2nd, 337
    3rd, 174, 337: in New Guinea, 211
    4th, 40: poor-quality officers in, 170

6th: communications in, 89; drain of experienced officers, 246; formation of, 48; heavy casualties in 1941, 110; in Greece and Crete, 78; in Libya, 78; in Papua, 178; organisation of, 80; regeneration of command structure in, 77; returns to Australia, 111; role of PMF officers in, 57; selection of COs, 54; standard of, 76, 343; training of, 71

7th, 339: at Balikpapan, 250–1; drain of experienced officers, 246; formation, 54, 57, 76; in Lebanon, 78; in New Guinea, 198, 215; in Papua, 178; innovation in, 111; returns to Australia, 111; selection of COs, 55; training of, 71

8th, 164, 339, 356: Bennett appointed as GOC, 142; command relationships in, 146, 162; competence of COs, 154, 164; in the fighting for Singapore, 130; Japanese assessment of, 152; leadership in, 139; nature of COs, 134, 146; organisation of, 131; proficiency, 146; selection of COs, 134, 146; use of carriers in, 151; withdrawal in Singapore, 363

9th, 244: at Tobruk, 78; command relationships in, 98; equipment shortages at Tobruk, 89; formation of, 78, 132, 356; heavy casualties in 1941, 110; in New Guinea, 198, 206, 209, 215, 219; selection of COs for, 129; standard of COs in late 1942, 129; suspicions of, 231

enlistments in, 4
ethos, 41, 45, 235
forces
 Northern Territory, 178
headquarters
 Advanced Land Headquarters, 201
 Land Headquarters, 229, 230

independent companies
 2/2nd, 238
 2/3rd, 226
inexperience of, 270
infantry battalions
 3rd, 51: at Gona, 188–90; in Papua, 173; on the Kokoda Track, 365
 4th, 52: Kwama River crossing, 221–2
 5th, 174
 6th: reaction to Egan, 177
 7th, 168, 174, 175: experience of command team in 1945, 244; in Bougainville, 256–7
 8th, 58, 105, 174
 9th, 248: captures Artillery Hill, 254; on Bougainville, 257, 265; rested, 385
 12th, 295
 12/40th, 295
 12/50th, 133
 14th, 174
 15th, 174: on Bougainville, 249
 16th, 40
 17th: attempt to have it transferred to 2nd AIF, 134
 18th, 42, 44, 51, 136, 138
 20th, 53
 22nd, 228
 22/29th, 33
 24th, 37, 246: cohesion developed by Smith, 228; in Wau–Salamaua operations, 210
 24/39th, 34, 40
 25th: casualties in 1945, 262
 26th, 238, 239, 267
 27th, 44, 175
 28th, 168
 29/22nd, 38
 29/46th: training of, 177
 30th: operations on the Ramu, 259–60
 31st, 335: poor training of, 169
 31/51st, 238: casualties in 1945, 262; experience of command team in 1945, 244
 33rd, 34
 35th: on Bougainville, 249

Australian Military Forces (*cont.*)
  36th, 33, 51, 61: in Papua, 173; mass absence without leave, 175
  37th, 42, 44, 168, 175: poor training of, 171
  37/39th, 37, 40
  39th, 33, 45: and the League of National Security, 46, 337; at Gona, 183, 190; on the Kokoda Track, 190, 193–4; reinforced with 2nd AIF officers, 173
  40th, 295
  41st, 336
  43rd, 175: standard of, 171
  46th, 334
  47th: on Bougainville, 249
  49th: assessed the worst battalion in Australia, 169; in Papua, 173
  51st, 334
  53rd: in Papua, 173, 184
  55th: in Papua, 173; merged with 53rd Battalion, 184
  55/53rd, 51
  56th, 51, 133, 336
  57/60th, 38, 168: fails to capture Yaula, 213; in the Finisterres, 373; on Bougainville, 248; reaction to Webster, 239
  58th, 55, 170: morale in, 171
  58/59th, 271: age profile in 1945, 238; Bobdubi Ridge operations, 214, 224; casualties in 1945, 262; experience of command team, 244; in Bougainville, 244; in Wau–Salamaua operations, 226–7; poor standard of, 224
  59th, 38
  61st, 33, 228: mutiny in, 8, 266; on Bougainville, 266
  2/1st, 60: at Bardia, 89, 100; at Eora Creek, 186; expereinced command team in 1945, 244; in the Aitape–Wewak campaign, 258, 260, 263; casualties in, 262; indiscipline in, 61; training of, 72
  2/2nd, 11, 72, 75, 274, 293: at Bardia, 89, 94; at Eora Creek, 186; at Pinios Gorge, 88, 101, 105, 345
  2/3rd, 246, 339: at Bardia, 95; at Eora Creek, 186, 192; at Pinios Gorge, 91; at Tobruk, 267; morale in Aitape–Wewak campaign, 262
  2/4th, 71, 72, 75: in the Aitape–Wewak campaign, 253, 259
  2/5th, 72: discipline in, 60; on Mount Tambu, 213, 374
  2/6th, 29, 60, 63, 75, 103: age profile in 1945, 238; at Bardia, 98, 211, 226, 244, 349, 378; at Lababia Ridge, 212; command relationships in, 64; experienced command team in 1945, 243; experienced personnel in 1945, 242; in Libya, 88; in the Aitape–Wewak campaign, 37; reaction to Wood, 233; standard of in 1940, 343
  2/7th, 53, 63: experienced personnel in 1945, 247; in Wau–Salamaua operations, 225; on Crete, 101, 103; reaction to Walker, 102
  2/8th, 63: at Bardia, 351; at Mount Shiburangu, 255–6; at Vevi, 109; in Libya, 106, 351; nature of records relating to Greek campaign, 351
  2/9th: at Buna, 187, 192; at Sanananda, 183; attacks Salient at Tobruk, 92, 95; captures Crater Hill, 213
  2/10th, 65, 230, 274, 284: at Balikpapan, 1, 242, 244–6, 255; at KB Mission, 182, 185–6; at Tobruk, 182; command relationships in, 244–5; experience of command team in 1945, 247; reaction to Daly, 239, 241
  2/11th, 238: in Aitape–Wewak campaign, 259; reaction to Green, 240
  2/12th, 53, 63, 74, 91, 92, 339: at Tobruk, 347; attacks Prothero, 212; attacks Salient at Tobruk, 92; in the Finisterres, 207

2/13th: at El Alamein, 120, 125, 127, 128, 216; at Jievenang, 218; in advance on Lae, 216; standard of prior to El Alamein, 117
2/14th, 55, 61, 109, 238, 331, 339: at Gona, 183; at Manggar, 246, 251–3, 255; experienced command team in 1945, 243; on the Kokoda Track, 179, 190
2/15th, 10, 63, 293: at El Alamein, 115, 217; at Kumawa, 209; at Nanda, 219, 373; at Tobruk, 109; brawls in Darwin, 61; character of, 217; first 2nd AIF battalion with PMF CO, 56; in Huon Peninsula operations, 217
2/16th, 64, 98, 109, 110: at Gona, 189, 191
2/17th, 115, 129, 238: at El Alamein, 217; at Jievenang, 209, 217, 218; experienced command team in 1945, 244
2/18th: at Nithsdale Estate, 152, 155, 162; discipline in, 137; in the fighting for Singapore, 150, 151, 158
2/19th, 133: at Bakri, 151, 154, 156; cohesion of, 135; command relations in, 137, 357; destroyed in the fighting for Singapore, 146, 150; discipline in, 138; in withdrawal from Bakri, 150, 154–6, 159–61; loses Anderson, 146; training of, 135, 357
2/20th, 137, 146, 304: command relationships in, 137–8; disintergrates, 157; in Malaya, 144; in the fighting for Singapore, 157; indiscipline in, 138, 144
2/21st, 133, 280, 284: religious adherence in, 291
2/22nd, 133
2/23rd: at El Alamein, 116, 118, 123, 353; at Tobruk, 92, 347; attack on Salient at Tobruk, 80–96; on Tarakan, 11

2/24th, 271: at El Alamein, 113, 118, 125, 127, 355; at Tobruk, 91, 96; casualties in 1945, 262; experienced personnel in 1945, 242; on Tarakan, 11; reaction to Warfe, 231
2/25th, 246
2/26th, 137: at Gemas, 151; at Namazie Estate, 151; fights for Kranji, 145; loss of Boyes, 161
2/27th, 65, 175, 230: at Gona, 182, 188, 193, 195; in Lebanon, 93, 94; loses confidence in Cooper, 195; on the Kokoda Track, 185, 195
2/28th, 206, 214: at Ruin Ridge, 118, 221; crosses the Busu, 209, 221; in advance on Lae, 207, 219; on Labuan, 249–50
2/29th: at Bakri, 145, 157, 160; in the fighting for Singapore, 157
2/30th, 134: as prisoners of war, 160; assessments of, 142; at Gemas, 138–55; at Namazie, 157, 160; cohesion of, 142; command relationships in, 139; in the fighting for Singapore, 162; indiscipline in, 140; training of, 357
2/31st, 174, 246, 283: in Lebanon, 93, 99; on the Kokoda Track, 194
2/32nd, 283
2/33rd, 246, 283
2/40th, 133, 280, 283: religious adherence in, 291
2/43rd, 175: at El Alamein, 127; at Tobruk, 109, 346; command relationships in, 109, 232; on Labuan, 250; poor reaction to Joshua, 233
2/48th, 4: at El Alamein, 116, 120, 122, 125–8, 344, 352, 355; at Sattelberg, 212, 219, 221; at Tel el Eisa, 117; at Tobruk, 94, 98, 109; battle drills, 122; experienced personnel in 1945, 242; fights for Trig 29, 122, 126; on Sattelberg, 232, 373; reaction to Ainslie, 232; training in Syria, 115

Australian Military Forces (*cont.*)
  Papuan Infantry Battalion, 221
  light horse regiments
    20th, 53
    21st, 51, 337
  logistics units
    14th Line of Communications Sub-Area, 175
    Australian Overseas Base, 51, 338
  machine gun battalions
    2/2nd, 10
  medical units
    2/4th Australian General Hospital, 305
  overstretch of, 131, 133–4, 143, 269, 307
  posting paths in, 301
  pre-war standards, 31, 37
  proficiency of, 71, 76, 146, 235, 237, 242, 270, 380
  provost units
    1st Australian Detention Barracks, 339
  record keeping practices, 278–9, 304, 329
  regional recruitment policy, 283
  schools
    Australian Training Centre (Jungle Warfare), 200
    Command and Staff School, 31, 70, 168, 202, 204, 365: establishment of, 38, 40; impact of, 71
    Land Headquarters Tactical School, 197, 202, 225, 238, 239, 255, 365, 371: attendees at, 202; British assessment of, 202; instructional philosophy, 201; syllabus of, 200–1
    Regimental Commanders School, 200, 202, 365
  training units
    8th Infantry Training Battalion, 175
  university regiments
    elite ethos of, 293, 388
    Melbourne University Rifles, 177, 293, 334, 337, 388
    Sydney University Regiment, 39, 51, 293, 334

  urban–rural mix of personnel, 284–5
Australian Military Regulations and Orders, 59–61, 66, 68, 358
Australian Regular Army formation of, 274
Australian Staff Corps, 56
  See also Permanent Military Forces
Australian War Memorial, 14
authority, 16, 24
  delegation of, 26, 28, 122, 272
automatic weapons
  Bren guns, 82, 83, 345
    shortcomings, 87
  Owen guns, 219
  use of captured weapons, 87–8, 114
  Vickers guns, 82, 87, 88, 114, 151, 180, 203, 223, 256, 345
Awala, 179
awards
  Distinguished Service Order, 100–1, 112, 212, 216, 225, 228, 266, 301
  for El Alamein, 112
  Military Cross, 175, 232, 244, 266, 267
  Victoria Cross, 156, 373
Ayer Bemban, 152
Ayer Hitam, 152

Bakri, 147–9, 151, 152, 156, 157, 159
Baldock, Albert, 175
Balfe, John Walter
  wounded at El Alamein, 355
Balikpapan, 1, 2, 236, 241, 242, 244–7, 250–3, 255, 352, 383
Bardia, 88–90, 94–5, 98–9, 267, 349
Barham, Lindley de Lisle, 336
Barker, Harold Arthur, 169
Barrett, John
  We were There, 10
Barter, Margaret Ann
  Far Above Battle, 11
Bartholomew Committee, 87, 345
Bartlett, Frederick Charles
  Psychology and the Soldier, 24, 26
Bartley, Thomas William
  longest serving CO, 295

bases of social power
    See leadership – power bases
batmen, 82
battalion associations, 7
battalion character, 217, 233
battalion commanders
    appointment procedures, 313
        See also Australian Military
            Forces – record keeping
            practices
    attributes of 1945 generation, 238
    biographies of, 6
    cohorts of, 312
    common origins of, 290
    competence of, 304
    confidants, 64, 66
    criticism of, 7
    dismissals of, 72, 195, 222–5, 229,
        233, 249, 258–9, 266, 297,
        304, 365, 369, 374
    educational standards, 55, 287
    experience of in 1945, 243
    ignored by authors, 12
    inexperience of, 96, 270
    influence on war dairies, 331
    information networks, 66, 85–6,
        122
    limited perspective, 188
    longevity of, 201
    military experience pre war, 295
    neglect in historiography, 13–14
    occupations, pre war, 45, 56,
        287
    overseas born, 280
    places of birth, 280
    position in battle, 92, 123, 155–9,
        182, 211, 257, 355
    post-command appointments, 301
    postings prior to appointment,
        300
    post-war careers, 274
    pre-war standards, 31, 34
    ranks attained pre war, 295
    relationship with subordinate
        officers, 67, 271
    religious adherence, 292
    researching, 14
    responsibilities, 59, 64
    role in post-war army, 274
    role of, 5, 13, 54, 63–4, 68, 85, 118,
        122, 193, 222, 250, 255, 305,
        385
    rural origins, 291
    shortage of, 246
    standards expected of COs by
        troops, 62, 340
    states of birth, 280, 291
    states of residence, 280–1
    studies of, 6
    tactical ability, 70, 72–154, 272
    training of, 70–1, 75, 117, 147,
        168, 172, 197, 202, 365,
        370
    treatment by battalion histories, 8,
        171, 177
    treatment in Australian military
        historiography, 12
    treatment in Australian war
        literature, 10
    urban–rural mix of, 284–5
battalion headquarters
    advisers attached to, 255
    main, 120
    rear, 120
    tactical, 120, 255
battle drills, 2, 94, 113, 242–3
    at El Alamein, 116, 122
    in jungle warfare, 206
    in Papua, 183
    should not replace battle procedure,
        206
battle procedure, 94–5
    at El Alamein, 113
    in jungle warfare, 206
    in Papua, 183, 185, 190, 195
    poor at Tobruk, 94
    thorough in 2/2nd Battalion at
        Bardia, 94
Baxter-Cox, Alfred Richard, 109
    concerns about his weight, 105
    relationship with 2iC, 64
BCOF
    See British Commonwealth
        Occupation Force
Bean, Charles Edwin Woodrow,
    8
Beaumont, Joan, 163, 280
    Gull Force, 11
Beenleigh, 199
Bennett, Henry Gordon, 134, 135,
    140, 142, 143, 339
    appointment to 8th Division,
        356
    erratic decisions, 146

416 INDEX

Bennett, Henry Gordon (*cont.*)
  feud with Taylor, 143–5
  membership of Old Guard, 337
  overlooked for 2nd AIF commands, 142
  personality of, 143
Best, Francis Mayfield
  introduces Scottish dress to 27th Battalion, 44
  recruiting technique, 45
Bierwirth, Rudolph, 75, 339
Binks, Hector McClean
  dismissal of, 259
Bishop, John Ackland
  appointment to command 2/27th Battalion, 230
  'gloom' at returning to staff appoitment, 230
  memoirs, 379
Bismarck Sea, 198
Blackburn, Arthur Seaforth, 362
Blair, Dale
  Dinkum Diggers, 277
Blamey, Thomas Albert, 42, 48–55, 80, 100, 102, 103, 106, 108–10, 126, 129, 131, 132, 144, 163, 174, 184, 193, 201, 230, 231, 356, 379
  angered by Jackson, 230
  influence on appointment of COs, 50, 51, 55, 129, 230–1
  Koitaki address, 192
  membership of White Army, 337
  raising the 2nd AIF, 49
Bobdubi Ridge, 206, 213, 214, 222, 223
Borneo
  British North, 236, 246, 249
  Dutch, 236
  See also Balikpapan
Bosman, 259
Bougainville, 236, 238, 248, 256–9, 263, 267, 384
  ammunition shortages on, 254
Bourne, Charles Cecil Francis, 207, 212
Boyes, Arthur Harold 'Sapper', 137, 143–5, 339
  command cabal with Galleghan, 144, 145
  command style, 137
  removed from command, 145, 161, 359
Bradley, Bert Henry James, 136
bravado, 101, 226
Bray, Alan Claude 'Donkey', 245
Bren gun carriers, 83, 90, 101, 111, 113, 120, 126, 203
  employment in Malaya, 151, 360
  misuse of, 96, 150
Brennan, Pat, 277
Bridge, Carl, 27
Bridgeford, William, 384
Brisbane, 199
British Army
  1930s battalion command course, 37
  armies
    Eighth: revision of doctrine in, 113, 120
  Australian officers on exchange with, 38
  battalion command in Normandy, 385
  battle procedure in, 273
  command in, 261
  declining age of COs, 242
  divisions
    1st Armoured, 127
    7th Armoured, 261
  doctrine, 86
  leadership in, 273
  regiments
    Black Watch: 2nd Battalion, 74
    Gordon Highlanders: 1st Battalion, 273
    Green Howards: 6th Battalion, 260, 262
    Royal Tank Regiment: 46th Battalion, 123
    Sherwood Rangers: 1st Battalion, 105, 107
  replaces overage battalion commanders, 174
  schools
    impact of, 73, 110, 163
  sensitivity to casualties in Normandy, 261
  Territorial Army
    class restriction on officer corps, 337
    pre-war efficiency, 41, 337

INDEX   417

training organisation in Malaya, 147
Broadbent, John Raymond, 244, 293
Brocksopp, Arthur Edward, 212
Brocksopp, William Arthur 'Bill', 245
Brougham, Kenneth Langloh 'Ken', 243
Brown Spur, 251, 253
Brune, Peter, 182, 192, 194, 329
  nature of work, 12
  significance of scholarship, 11
  treatment of COs, 12
Brunei, 267
Buin Road, 244, 253
Bulhem River, 216
Bulim, 151
Bulimba, Operation, 217
Buna, 181, 184, 187
Bundey, Eric William, 336
Burns, James McGregor
  on leadership, 26
Burrows, Frederick Alexander 'Bull', 33, 51
  wounded at Tobruk, 349
Busu River, 206, 209
Butler, David, 239

Caldwell, William Blythe 'Bull', 174
Callinan, Bernard James
  epitomises 1945 COs, 238
  personality of, 238
calmness, 104, 195, 273
camaraderie, 101, 104, 135, 193, 236, 263, 275
Cameron, Allan Gordon, 194
  acting command of 39th Battalion, 194
  reputation of, 190
  witholds 3rd Battalion at Gona, 188–90
Cameron, Claude Ewen, 42, 44, 51
  as commander 8th Brigade, 221
  not appointed to 2nd AIF, 52
Campbell, Ian, 139, 142, 151, 161
  'A model for battalion command', 6
Campbell, Ian Ross, 293
  captured on Crete, 349
Canadian Army
  declining age of COs, 242
  leadership in, 273
  officer-men relations, 273

Canning's Saddle, 207
Cannon, William George, 109, 110
  appointed to 2/14th Battalion, 55
  changes battalion war diaries, 331
  evacuated due to exhaustion, 109
  removed from command, 352
Canungra, 200
Cape Endaiadere, 186
CARO
  See Central Army Records Office
Caro, Albert Edward, 191
  dismissal of, 195, 369
  reaction to casualties at Gona, 192, 195
Carr, Howard Hammond, 133, 334
carrier platoon, 83
casualties
  COs, 278, 312
  at El Alamein, 123, 305
  battle, 305
  comparison between theatres, 198
  in 1941, 100
  in Malaya and Singapore, 305
  in New Guinea, 212, 305
  in Papua, 182, 305
  non-battle, 307
  proportion suffered by infantry battalions, 4
  COs in Papua, 184
  impact on control, 183
  in Australian battalions in 1945, 262
  in Papua, 195
  psychological, 305
    at Tobruk, 305
    COs, 304, 363
    in Papua and New Guinea, 305
    incidence of, 268
    on Tarakan, 268, 305
  sensitivity to, 220, 222, 253–4, 257, 376, 384
caution, 258, 262, 384
  dangers of, 259
censuses
  1933 Commonwealth, 279
  AMF, 280
Central Army Records Office, 278
  archival practices, 13
Challen, Hugh Bradbury, 183

418 INDEX

chaplains
  conflict with COs, 66
  role of, 66
character
  See personality
Chilton, Frederick Oliver 'Fred', 5, 11, 39, 60, 66, 67, 75, 88, 239, 240, 271, 276, 288, 293, 340
  as commander 18th Brigade, 207, 246
    praises Daly, 245
    seeks Daly for 2/10th Battalion, 230
  at Bardia, 89, 94
  at Pinios Gorge, 91, 345
  attends Command and Staff School, 39
  award of DSO, 11, 13
  calmness in battle, 101
  discussion of by Barter, 11
  origins of, 290
  personality of, 11–101
  post-war career, 274
  reaction to casualties, 104
  reflections on battalion command, 274
  relationship with sergeants, 67
  takes command of 2/2nd Battalion, 67–75
  thoughts on discipline, 60, 61
  thoughts on jungle warfare, 207
Choa Chu Kang Road, 157
Citizen Military Forces, 274
  role of wartime COs in, 274
Clarebrough, James Augustus 'Jack'
  membership of the White Army, 337
  recruiting strategy, 39–45
class
  See social class
Clothier, Leslie Allen, 118
CMF
  See Citizen Military Forces
Coates, Henry John, 217, 219, 231, 233
  Bravery Above Blunder, 12, 212
  significance of scholarhship, 12
coercion
  See power – coercive
Cohen, Paul Alfred
  See Cullen, Paul Alfred

cohesion
  See unit cohesion
Colvin, George Edward 'Gorgeous George', 120, 128, 216, 246
  awarded DSO, 216
  coolness under attack at El Alamein, 126
  initiative during advance on Lae, 216
  personality of, 216
  wounded at El Alamein, 355
combat exhaustion
  See exhaustion
combined arms warfare, 236, 250–6
  in New Guinea, 221, 376
  in Papua, 187
  in the jungle, 206, 220
  training for, 74, 200, 201, 255
    in Malaya, 144
    pre-war, 36, 69
    prior to El Alamein, 117
Combined Operations Training Centre, 73, 111
command
  application of civilian skills to, 288
  assessing effectiveness of, 11–20, 217
  Australian style of, 29, 159, 202, 271–2
    not unique, 272–3
  characteristics required, 331
  definitions of, 16
  functions of, 18, 25, 84, 270
  in British Army, 273
  in Canadian Army, 18–28, 273, 388
  in captivity, 163
  in German Army, 28, 333
  in New Zealand Army, 28, 272, 273
  loneliness of, 239
  modern Australian Army definition, 16, 28
  modern British Army model, 16–17
  nature of, 16, 17
  physicality of, 179, 195–7, 215, 224, 226, 238, 271, 278, 312, 350, 365
  skills, 45, 54, 75
  strain of, 188, 196, 201, 225, 237, 238, 260, 268, 271, 278
command post exercises, 40

command relationships, 70, 95, 100, 109, 117, 131, 135, 137, 166, 169, 215, 232, 236, 244–6
  development of, 201, 211, 255, 371
  influence of personality on, 223, 236, 248–50
command systems, 18, 85, 91, 113, 129, 146, 196
  inadequacy of, 78, 131, 164, 234, 270
commanding officers
  See battalion commanders
commando squadrons
  role of, 5
communications, 85, 234
  difficulties in Lebanon, 93
  field telephones, 212, 346
  improvements in at El Alamein, 113
  in jungle warfare, 182, 204, 215
  in Papua, 180
  influence on control, 89, 120, 122, 155, 214, 346
  poor standards of, 88, 91, 93, 270
  runners, 82, 88, 91, 92, 120, 181, 182, 215
  use of enemy equipment, 91
company commanders
  declining age of, 238
  role of, 213, 242, 244–5, 250
competence
  definition of, 20
confidence, 195
Connell, John William Dennis
  loses control at Yaula, 213
Conroy, Thomas Mayo
  at Mount Tambu, 213
  dismissal of, 374
  loses confidence of company commanders, 213, 374
control, 17, 71, 85
  at El Alamein, 120, 125
  at Tobruk, 93, 95, 182, 346, 347
  in battle, 94, 113, 156, 159
  in jungle warfare, 183, 206
  loss of in Malaya and Singapore, 158, 159, 161
Cook, Francis William, 245
Cook, Thomas Page, 72, 75
  awards, 110
  captured during exercise, 342

Cooper, Geoffrey Day Thomas 'Geoff', 182, 188, 189, 193, 195, 196, 330, 367
  actions on Kokoda Track questioned, 185
  at Gona, 182, 188, 193
  calming influence of, 193
  decisiveness questioned, 195, 369
  dismissal of, 195
  post-war career, 274
  thoughts on jungle warfare, 182
  thoughts on orders, 188
Cootamundra, 44
COs
  See battalion commanders
Cotton, Thomas Richard Worgan, 246
  thoughts on post-war service, 274
courage, 25, 100, 112, 188, 193, 273
Cox, Geoffrey Souter
  personality of, 259
CPX
  See command post exercises
Crawford, John Wilson, 51, 92, 170
Crellin, William Wauchope 'Bill', 339
  laissez-faire command style, 109
Crete, 78
Crisot, 261, 262
Crosky, Percy William
  indecisive at Kwama River, 221–2
Cullen, Paul Alfred, 74, 101, 244, 262, 287
  actions at Eora Creek, 186, 193
  assessment of battalion justice, 60
  changes name, 292
  emotions at Eora Creek, 188
  and Nambut Ridge operations, 260
  post-war career, 274
  sensitivity to casualties, 258, 260
  thoughts on jungle warfare, 181
  thoughts on standard of 2/1st Battalion, 235
culture, battalion
  See battalion character
Cummings, Clement James, 194
  calm demeanour of, 193

Cummings, Clement James (*cont.*)
  at Buna, 186–7, 193
Curtin, John
  declares the 'battle for Australia', 184
Cusworth, Kenneth Stanley 'Ken'
  reaction to in 29/46th Battalion, 177

Daly, Thomas Joseph 'Tom', 65, 109, 242, 273, 275, 290
  ability of, 1–3, 241
  appointed Chief of the General Staff, 274
  appointed to command 2/10th Battalion, 230
  as adjutant 2/10th Battalion, 65, 244
  relationship with Verrier, 65
  as BM 18th Brigade, 88, 109, 245, 347
  critical of Salient attacks at Tobruk, 98
  assessment of 2/10th Battalion in 1945, 247
  at Balikpapan, 1, 244, 246, 255
  known as the 'Boy Bastard', 239
  personality of, 3, 241
  praised by Chilton, 245
  reflections on battalion command, 274
  relationship with command team, 244
  relationship with soldiers, 62, 241, 275
  thoughts on discipline, 54–60
Damour, 93, 94
Dampier Strait, 198
Darwin Mobile Force, 31, 61
Darwin Purge
  See militia – replacement of COs
decision-making, 17, 270
Defence Act 1903, 32, 33, 35, 48, 59
delegation
  See authority – delegation of
demographic analysis, utility of, 277
Derham, Francis Plumley 'Frank'
  member of the White Army, 337

Derna, 106
Derrick, Thomas Currie 'Diver', 373
desperation
  effect on command decisions, 96, 112, 127, 150, 157–9, 186
Deverell, Cyril John, 40
Devine, William
  The Story of a Battalion, 1
devolved command
  See mission command
Dexter, David St Alban, 9, 222, 224
Dexter, Walter Roadknight 'Bill', 267
  appointed to command 61st Battalion, 229
  assessment of 2/7th Battalion, 224
  awarded DSO, 212
  breakdown of, 8, 266
  commands defence of Lababia Ridge, 212
  personality of, 211, 266, 373
directive control
  See mission command
discipline
  enforcement of, 25, 61, 271, 272, 339
  of officers, 68, 177
  sanctions available to battalion commanders, 60
  soldiers' attitudes to, 61
  standards, 62, 177
Disher, Harold Clive, 105
Dobbs, James Gordon 'Jimmy', 182, 330
  dismissal of, 195
  haunted by experience of Papua, 196
  loses control at KB Mission, 182, 185–6
  personal example at Buna, 193
doctrine
  concept of fire support, 251
  development in the Australian Military Forces, 111
  modification of British doctrine for Australian conditions, 204
  pre-war, 36
Dougherty, Ivan Noel 'Doc', 34, 75, 87, 99, 101, 103, 170, 175, 190, 195, 246, 247, 351
  as commander 21st Brigade, 185, 230

adverse report on Caro, 369
critical of fire support at Gona, 253
criticises Cameron, 189
praised by Rhoden, 246
as commander 23rd Brigade, 169, 172
at Tobruk, 90, 99
concern for men on Crete, 101
loses confidence in Cooper, 195
personality of, 102
post-war career, 274
'ruthless' removal of COs, 172
DSO
See awards – Distinguished Service Order
Dunbar, Colin Campbell, 182
dismissal of, 194, 195
Dunkley, Howard Leslie Ewen 'Harry'
awarded MC, 244
enlists as a private, 244
frustration at role on Bougainville, 256-7
Duntroon
See Royal Military College Duntroon
Dwyer, Michael John, 256, 257

Eather, Kenneth William 'Ken', 51, 52, 70
as commander 25th Brigade, 246
concerns about loss of experienced personnel, 247
at Bardia, 89, 100
comments on standard of 2/1st Battalion, 76
ignores burning canteen, 61
performance in training, 72
relationship with adjutant, 64
echelons, 84-5, 120, 241
Edgar, Bill, 329
Edgar, Cedric Rupert Vaughan 'Boss'
dismisses all COs over 35, 238
educational standards
See battalion commanders – educational standards
Egan, Eugene William 'Bill', 176
command style, 178
dismissed from command of 35th Battalion, 249
reaction to in 6th Battalion, 177

El Alamein, 110, 128, 272
nature of fighting at, 120, 125
England, Vivian Theophilus 'The Black Panther', 6, 51, 72
loses control at Bardia, 95
reputation as harsh disciplinarian, 339
Eora Creek, 186
equipment
shortages of, 71, 167, 236
Erskine Creek, 223
Evans, Bernard, 116
addresses 2/23rd battalion before El Alamein, 118, 354
as commander 24th Brigade, 219
appoints Joshua to 2/43rd Battalion, 232
inexperience of, 98
patrolling policy at Alamein, 117
personality of, 96, 98
personally leads attack at El Alamein, 123
poor battle procedure at Tobruk, 96
poor control at Tobruk, 347
questions orders at Tobruk, 98
thoughts on battle drills, 116
Evans, Michael, 10
examinations, 34-5
subjects of, 36
example, 26, 28, 85, 177, 272
at El Alamein, 125
in Malaya and Singapore, 156, 157
in New Guinea, 198
in Papua, 186, 193
on the battlefield, 11, 123, 125, 186, 193, 226
exhaustion, 109, 195, 197, 224-5, 262, 268, 271
among American troops in north-west Europe, 266
experience, 28
influence of, 126, 178, 216, 221
role of, 221, 241-8, 265, 270-2

Falconner, Alexander Hugh 'Alex', 7
dismissal of, 227
Farquhar, Lindsay Keith, 175
Farrell, Nevis William Parkington
dismissal of, 259
fear, 104

Fergusson, Maurice Alfred 'Ant Eater', 129
　as commander 8th Brigade, 384
Field, John, 63, 69, 74, 95, 98
　appointed to 2/12th Battalion, 53
　as commander 7th Brigade, 173, 228, 230, 238, 248, 265
　　criticism of Monaghan, 249
　　reports adversely on Dexter, 266
　concerns about training, 69
　eager to attack at Tobruk, 96
　enforces discipline, 339
　struggles for information at Tobruk, 92, 347
　wins USI essay prize, 37
Field Service Regulations, 84–6, 94
field telephones, 215
Finisterre Range, 198, 207, 246
Finschhafen, 198, 206, 212, 352
First World War experience of COs, 293
　relevance to battalion command, 34
First World War veterans
　age, 33
　alternative roles, 110, 175
　appointed to militia commands, 167
　blocking promotion of younger officers, 33–172
　generation gap with younger soldiers, 170
　lack of drive, 170, 175
　number in command pre-war, 32, 33
　outdated tactical knowledge, 72, 168, 170, 172, 186
　role in 8th Division, 132
　unsuitability for battalion command, 109, 168
flame-throwers, 250, 256
followership, 21, 26
formal groups, 22, 59, 62, 105
Franklin, Leslie, 223
French, Bevan John, 243
French, David, 3, 125
French, John
　'Bases of Social Power', 23
Freyberg, Bernard Cyril, 101
Frith, Clifford, 175, 365
Frost feature, 251, 253
Fuller, John Frederick Charles on age and command, 32–3

Gabriel, Richard
　definition of military competence, 20
Galleghan, Frederick Gallagher 'Black Jack', 142, 144, 153
　2/30th Battalion's loyalty to, 142
　abandons wireless equipment, 154
　command cabal with Boyes, 144, 145
　criticism of Wigmore's *Japanese Thrust*, 10, 21
　dismisses anti-tank guns, 147
　draws revolver on own troops, 157
　evacuated to hospital, 145
　fails 2nd AIF medical examination, 134
　flawed plan at Gemas, 152, 155
　high standards set by, 139, 357
　hypocrisy of, 134–40, 142
　limited use of carriers, 150, 360
　mishandles artillery, 150, 160
　overlooked for 2nd AIF command, 53, 134
　personality of, 139
　relationship with Anderson, 140
　relationship with Maxwell, 140, 359
　relationship with officers, 139
　reputation of, 139
　thoughts on command, 142, 158
　under strain in Malaya, 160
garrison battalions, 168, 301
Geard, Charles John
　dismissal of, 229
Gemas, 131, 147, 149, 152, 153, 155, 160
Gemencheh, Sungei (River)
　See Gemas
German Army
　attacks at Vevi, 107
　infantry sections, 83
　mortars, 87
　tactics in the desert, 125
Gippsland, 42
Goble, Norman Leopold
　dismissed from command of 47th Battalion, 249
Godfrey, Arthur Harry Langham, 64, 72, 88, 96–9, 348, 378
　commands 24th Brigade, 110
　concerns about his weight, 105

INDEX 423

performance during attack over
  Wadi Muatered, 98, 348
  personality, 64, 65
  reaction to casualties, 104
  thoughts on discipline, 60
Gona, 182, 183, 185, 190, 195
  ammunition expenditure at, 253
  assessment of COs at, 12
  operations conducted in haste, 251
Gosford, 42
Grace, Colin Henry, 246, 293, 375
  awarded DSO, 217
  cautious command style, 217, 219, 373
  orders withdrawl of 2/15th
    Battalion at El Alamein, 217
  personality of, 216
Grant, John MacDonald, 33
Greece, 78, 87–8, 91, 101, 163
Green, Charles Hercules 'Charlie', 259
  command style, 240
  concious of age, 240
  loss of lamented by 2/2nd Battalion, 240
  nicknamed 'Chuckles', 239
  not initially welcomed by 2/11th
    Battalion, 240
  personality of, 239, 240
  youngest Australian CO of the war, 239
Green Spur, 251
Greta, 61
Grey, Jeffrey, 31
Griffiths-Marsh, Roland, 4
group dynamics, 22, 142, 231
Guadalcanal, 179
Guinn, Henry George
  appointed Commandant of LHQ
    Tactical School, 225
  awarded DSO, 225
  command style, 225
  dismissal of, 225
  exhaustion of, 224–225, 378
Gullett, Henry Baynton Somer 'Jo', 343

Haifa, 73
Hale, Frederick William Harvey, 171
  personality of, 170
Hamilton, Adrian Kelso, 74, 342
Hamilton, Ian Standish Monteith, 14

Hammer, Heathcote Howard 'Tack',
  115–17, 120, 122–4, 127, 272,
  330, 373
  as commander 15th Brigade, 223
  assessment of Guinn, 224
  assessment of Warfe, 226
  clashes with COs on Bougainville, 248
  doubts about Starr, 224
  praise for Smith, 228
  praises battalions in 1945, 242, 380
  thoughts on command in jungle warfare, 257
  boozy interview with David Dexter, 377
  command philosophy, 122
  commended by Morshead, 126
  concerns about 2/48th's role on 30
    October 1942, 128
  orders carrier-mounted attack on
    Trig 29, 126
  personality of, 231
  reputation among troops, 116
  solo patrol at El Alamein, 124
  training regime, 116
  wounded at El Alamein, 124, 355
Hansa Bay, 259
Harper, Glyn John, 27
Harris, Hubert Cochrane, 334
Hastings, Max, 260
Hawthorne Studies, 332
Hay, David Osborne, 29, 242, 243, 343
  Nothing Over Us, 235
Hayes, Geoffrey, 27
headquarters company, 84, 344
Hearder, Rosalind, 163
Heckendorf, Erwin Ernest 'Heck', 141
Helwan, 72
Hendry, James Gordon
  awarded MC, 267
  suicide of, 268
Henning, Peter, 163, 280
  Doomed Battalion, 277
Heraklion, 101
Herring, Edmund Francis 'Ned', 172,
  175, 231
  member of the White Army, 337
hierarchical leadership
  See leadership – institutional

Hill, Deane Pender, 212
Hill 87, 1, 2
Hingston, David, 272
Hinson, James, 124
Hiscock, Walter Richard Grenville, 75
Hongorai River, 248
Honner, Ralph 'Jump', 194, 271, 329
  actions at Gona, 188, 190
  calm demeanour of, 194
  influence of, 193
  post-war career, 274
  requests permission to remain at Isurava, 190
  thoughts on confidence, 195
Horner, David Murray, 6
  Crisis of Command, 172
Howard, Michael Eliot
  thoughts on leadership, 21, 257
Howden, Walter Stace, 256
  at Mount Shiburangu, 255, 257
  sensitivity to casualties, 257
Hughes, William Morris 'Billy', 134, 140
Huon Peninsula, 198, 210, 217, 219
Hutchison, Ian 'Hutch'
  determined personality, 192

Imita Ridge, 179
indecision, 99, 195, 222
independent companies, role of, 5
Indian Army
  at Slim River, 147
  Australian officers on exchange with, 38
  brigades
    45th, 156, 160
  corps
    III, 130
  divisions
    11th, 162
infantry battalions
  operating procedures, 120
  organisation, 86, 115, 343
  strength of, 3, 80
    weakness as a result of, 125
  tropical organisation, 206
infantry divisions
  role of, 4
  strength of, 4
  tropical organisation, 203
informal groups, 22, 67
Ingleburn, 58

innovation
  See tactical innovation
institutional leadership
  See leadership – institutional
insubordination, 98, 161, 363
integrity, 24
intelligence
  in jungle warfare, 214
intelligence officers, 84, 86, 90, 120
Ioribaiwa, 183
Isaachsen, Oscar Cedric
  post-war career, 274
Isaksson, Olof Hedley, 212
Isurava, 190, 193, 194, 272

Jackson, David Arion Collingwood, 240
Jackson, Donald Robert 'Don', 76, 268
  as 2iC of 2/28th Battalion, 249
  as adjutant 2/1st Battalion, 64
  as BM 24th Brigade, 127
  earns Blamey's ire by seeking transfer, 230
  training at RMC Duntroon, 36
Japanese Army
  assessments of, 164
  defensive tactics, 206, 250, 254
  dispositions at Eora Creek, 186
  dispositions at Manggar, 251
  driving charge tactic, 130, 156
  fire support, 180
  poor British knowledge of, 147
  tactics in New Guinea, 215
  tactics in Papua, 183
  use of captured vehicles in Malaya, 151
  use of radio direction finding equipment, 362
Java, 236
Jeanes, Mervyn Roderick, 9, 250
  criticism of Long's *Final Campaigns,* 9
Jeater, William David, 137–9, 144, 145, 304
  loses grip on command, 138
  psychological illness, 359
  removed from command, 138, 144
jeeps, 120, 123, 203, 264
Jemaluang, 137, 151
Jivevenang, 209

Johnston, Mark, 27, 60, 266, 355
  Alamein, 11, 118
  At the Front Line, 10
Johore, 144, 146, 152, 164
Johore Bahru, 144
Joshua, Robert 'Bob'
  awarded MC, 232
  becomes member of parliament, 274
  clashes with 2/43rd Battalion, 232–3
  dismissal of, 8, 233
Joynt, William Donovan, 337
judgement, 270
jungle warfare, 25
  attributes for, 226
  Australian confidence in, 197, 206, 235, 369, 380
  Australian reluctance to operate at night, 210, 373
  communications in
    See communications – in jungle warfare
  comparisons with Middle East fighting, 179, 182, 208, 305
  difficulties of control in, 183, 211
  dispersion in, 210, 211
  employment of artillery
    See artillery – in jungle warfare
  intelligence in
    See intelligence – in jungle warfare
  logistics in
    See logistics – in jungle warfare
  nature of, 180, 183, 197, 254–62
  organisation for
    See infantry battalions – tropical organisation and infantry divisions – tropical organisation
  patrols, role of, 214, 228, 251
  physicality of, 179, 365
  scale of, 208, 253, 257, 262–352
  tactics, 152–4, 182–3, 190, 198, 207, 209, 220
  terrain, impact of, 207, 209
  training for
    See training – for jungle warfare
  tropical disease, 307
  unnerves COs, 209
Jungle Warfare (AMF Christmas book), 67

Kabrit, 73
Kappe, Charles Henry, 144
Keegan, John
  The Mask of Command, 25
Kehoe, John Francis, 118
Kelly, Joseph Lawrence Andrew 'Joe', 244
Key, Arthur Samuel
  holds mass orderly room, 339
  personality of, 194
Khirbe, 99
King, Roy
  as commander 3rd Brigade, 172
  holds mass orderly room, 60
Koitaki Factor, 192–3
Kokoda Track, 110, 179, 180, 182, 184–6, 193, 194, 196, 219, 246, 253, 305, 365
Ku-ring-gai, 42, 43
Kwama River, 221

Lababia Ridge, 212, 224, 373
Labuan, 249–50
Lae, 206, 207, 215, 216, 219, 231, 246
Laffin, John Alfred Charles, 27
Lamb, Donald James, 91
  draws pistol on New Zealand anti-tank gunners, 103
Lambert, Eric Frank
  military service, 10
  Twenty Thousand Thieves, 10, 62, 80
Lao, Tzu
  Tao Teh Ching, 276
Lavarack, John Dudley, 36, 38, 39, 57
  as GOC 7th Division, 54
  as GOC First Army, 169
  views on the standard of the militia, 169
leadership, 27, 68
  and command, 21
  and popularity, 116
  contingency theory, 22, 104, 272, 332
  definitions, 21
  evolutionary, 104, 271
  in Papua, 192–5
  in the 8th Division, 139
  in the Mediterranean theatre, 105
  institutional, 24

leadership *(cont.)*
  modern Australian Army definition, 24
  power bases, 23, 26, 171
  styles, 24, 26, 272, 333
  trait theory, 21
League of National Security, 46, 337
Leane, Raymond Lionel 'Ray'
  personality of, 8
Lebanon, 78, 94, 98, 109
left out of battle group, 84, 120, 344
liaison officers, 85, 86, 90, 92, 120, 159, 344
Libyan campaign, 78
Lind, Edmund Frank, 133
Lindsay, Martin, 273
Little George Hill, 257, 263
Lloyd, John Edward
  as commander 16th Brigade, 186
  ignores protests of COs at Eora Creek, 186
Lloyd, John Kevin, 40
Lloyd, Paul Acton, 184
logistics
  in jungle warfare, 203
Long, Gavin Merrick, 8, 173, 207, 216, 231, 238, 239, 241, 244, 246, 247, 250, 260, 262, 265, 267
  interviews conducted by, 15
Louch, Thomas Steane 'Tom', 72, 110, 342
  as commander 29th Brigade, 170
  loses control at Bardia, 90
  views on official history, 9
Loughran, Lewis John
  dismissed from 8th Battalion, 58
Loughrey, Jack, 122
  personality of, 119
  rebuilding the 2/28th Battalion, 119
loyalty, 190

MacArthur, Douglas, 58, 184
MacDonald, Alexander Bath 'Bandy', 109
  appointed Deputy Director of Military Training, 110
  as second in command 2/27th battalion, 65
  personality of, 9, 65
  struggles in Lebanon, 109

machine gun battalions, 87, 250
  role of, 5, 21
machine gun platoon, 203, 251
  formation of, 114
  retention of, 114
machine guns
  See automatic weapons
McAllester, James Clulow, 331
McArthur, Neil Alexander
  membership of the White Army, 337
McCarter, Lewis 'Lew'
  addresses 2/28th Battalion before attack on Ruin Ridge, 118
  personality of, 119
McCarthy, Dudley
  assessment of Hutchison, 192
  South-West Pacific Area – First Year, 168
McCormack, Patrick John, 44
McDonald, Herbert Hector
  Monaghan seeks to dismiss, 249
McElwain, Roger, 241, 261
Mackay, Iven Gifford, 44, 167, 169, 171, 223
McKinna, John Gilbert
  personality of, 228
Madang, 198, 221, 376
Magno, Charles Keith Massey
  killed at El Alamein, 355
Maitland, James Angus, 171
Makotowa, 263
Malahang Anchorage, 207
Malayan campaign, 139, 142–64, 270
  summary of, 131
Malin, 260
Mandai Road, 162
Manggar Airfield, 246, 251–3, 255
  ammunition expenditure at, 253
Marangis, 259
Markham Valley, 198
Marlan, Robert Francis 'Spike', 61, 63, 339
  appointed to 2/15th Battalion, 56
  captured during 'Benghazi Handicap', 349
Marson, Richard Harold 'Drover Dick', 246
Marston, Robert Raymond 'Happy Bob', 373
  personality of, 239

Martin, James Eric Gifford 'Sparrow', 96, 98
  inexperience of, 98
mask of command, 25, 104, 142, 192, 194
mateship, 22, 236, 263
Mat Mat, 212
Matthews, Geoffrey Ronald 'Geoff'
  as platoon commander 2/10th battalion, 65
  attitude to combat exhaustion, 263
  clashes with Webster on Bougainville, 248
  disputes seniority in brigade, 248
  frustration at ammunition allocation on Bougainville, 254
  loses control in Bougainville, 264
  personality of, 248
  rebuilds morale of 9th Battalion, 265, 385
  reflections on Dexter's dismissal, 266
  stands back at Little George Hill, 257
Maughan, David Wilfrid Barton, 347
Mawaraka, 263, 264
Maxwell, Duncan Stuart 'Little'
  affection for 2/19th Battalion, 161
  appointed to command 2/19th Battalion, 133
  as commander 27th Brigade, 140, 143–5, 161
  Bennett's assessment of, 135
  command of 56th battalion, 44
  command style, 136, 143–4
  conflict between medical career and military command, 44, 133, 135, 161
  disobeys orders, 161, 363
  lack of command experience, 133, 135, 143–4, 161
  personality of, 135, 143
  pleads for surrender, 161
  relationship with 2iC, 136
  role in the fall of Singapore, 161
Mayberry, William Maurice, 244, 387
  clashes with Hammer, 248
Mediterranean campaigns
  summary of, 80
Melbourne, 33, 45, 46, 49, 50, 71, 163, 174, 228, 284

Menari, 185
Menzies, Robert Gordon
  decrees all 2nd AIF commands go to militia officers, 57, 65
Merdjayoun, 101
meritocracy, 29, 77, 164, 273
Merritt, William Gerard Thomas, 194
METS
  See Middle East Tactical School
Middle East Staff School, 73, 75, 230
Middle East Tactical School, 75, 110, 112, 129, 200, 272, 365
Miles, Edward Samuel, 228
military competence
  See competence
militia
  experience of in 1945, 244, 380
  improving standards of, 178, 227, 235
  inferior status to 2nd AIF, 297
  poor standards of, 165, 172–3, 364
  pre-war recruiting, 30, 31
  pre-war standards, 37, 336
  relations with the community, 44
  replacement of COs, 165, 174–5, 178, 196, 198, 223, 227–8, 295
  training of, 176
Milne Bay, 173, 179, 182, 185–6, 193, 231
mission command, 26, 122, 183, 213, 246, 256–7
Mission Ridge, 195
Mitchell, John Wesley, 63, 67, 72, 105–10
  appointed to 2/8th Battalion, 53
  at Tobruk, 90
  charged with embezzlement, 338
  performance in Libya, 106–7, 351
  personality of, 105
  poor performance at Vevi, 109, 351
  relationship with officers, 67
Miteiriya Ridge, 127
Miyeoumiye, 93
Monaghan, Raymond Frederic 'Mad Mick'
  dismissed, 38
  erratic command of 29th Brigade, 248–9, 383

Montgomery, Bernard Law, 113, 115–17
  on the role of the battalion commander, 117
Montgomery, Keith Hamilton, 174
morale
  contribution of chaplains, 66
  in 1945, 262–6
  influence of COs, 118, 171, 184, 194, 266, 271
  influence of oratory on, 119
  role of dress in, 44
  role of sport in building, 44, 62
Morphett, Hurtle Cummins, 212
Morris, Basil Moorhouse, 169
Morshead, Leslie James, 2, 96, 98, 117, 126, 129, 207
  as commander 18th Brigade, 52, 65, 69, 341
  on the role of COs at El Alamein, 117
mortar platoon, 83
  strength increased, 113, 203
mortars, 219, 253, 256
  2-inch, 82, 87
    capacity to generate smoke screen, 345
  3-inch, 83, 87, 113, 180, 203, 250, 345
    shortcomings, 87
  4.2-inch, 250
  81 mm, 87
Mosigetta Road, 263
Moss, Richard George, 168, 364
Moten, Murray John, 195
  as commander 17th Brigade, 213, 214, 222
  assessment of Guinn, 225
  personality of, 65, 222
  preparation for battle of Damour, 94
  relationship with 2iC, 65
motorised transport, 69, 80, 83, 147
  employment in Malaya, 151
  reduced scales in tropical organisations, 203
Mount Shiburangu, 255–6
Mount Tambu, 213
Muar, 130, 156
Muatered (Wadi), 98–9, 349
Mubo, 211

MUR
  See Australian Military Forces – university regiments
mutinies
  in 9th Battalion, 265
  in 61st Battalion, 8, 266
  in 2/20th Battalion, 138
  in 2/30th Battalion, 140

Nadzab, 198
Namazie, 151, 152, 157, 160
Nambut Ridge, 260
National Archives of Australia, 14, 278
National Archives of the United Kingdom, 14
Nauro, 185
naval gunfire, 1, 250, 253
Nelson, Hank, 163
New Britain, 215, 236
New Guinea campaign
  nature of COs in, 34
  outline of, 199
New South Wales, 51, 52, 54, 58, 281, 337
New Zealand Army
  at El Alamein, 272
  at Pinios Gorge, 103
  authority of younger COs in, 241
  battle procedure in, 272
  casualties among COs, 305
  declining age of COs, 242
  leadership in, 273
  occupations of battalion commanders pre war, 287
  sensitivity to casualties in Italy, 261
Nicholson, M. J., 22
Nithsdale Estate, 152
Norman, Colin Hugh Boyd, 371
  command style, 249
  decision to cross Busu, 221
  dismissed from command of 2/28th Battalion, 249
  favours tight defensive perimeters, 209
  military education of, 370
  personality of, 249
  poor relationship with Porter, 249–50
  refuses to advance without ammunition resupply, 219

seeks to influence Official History, 250
thoughts on jungle warfare, 206, 214
uses infiltration tactics, 207
Normandy campaign, 261, 262, 385
North, Francis Roper, 335
Northcott, John, 356
Northern Territory, 19, 169, 171, 172, 174
'Notes for Commanding Officers', 66, 85
Numa Numa Trail, 257

O'Connor, John Christian Watson 'Burn-em-up'
personality of, 228
Oakes, Roland Frank 'Roly', 135, 136, 161, 359
appointed to command 2/26th Battalion, 145
views on fall of Singapore, 9, 164
obedience, 98, 128, 162, 163, 188, 192
expectations of, 98, 140, 163, 166, 185, 190, 194, 217
Oboe operations
See Balikpapan, Borneo, Tarakan
observation posts, 92, 209, 228, 253
Official History of Australia in the War of 1914–1918
treatment of COs, 8
Official History of Australia in the War of 1939–1945
criticisms of, 10
editorial policy, 9
treatment of COs, 10
Ogle, Robert William George 'Bob', 109, 355
training regime, 115
Old Guard, 337
'old man', the, 5, 268
Old Vickers, 214, 223
Oliver, William
feud with 2iC, 169
oral history
nature of, 15
orders, 170, 188, 190, 195, 196, 201, 250, 271, 273
orders group, 86
Owen, William Taylor, 173

Pabu, 209
padres
See chaplains
Palazzo, Albert, 40
Papua, 184
campaign summary, 179
inexperienced officers in, 184–5
logistics in, 179
nature of battalion commanders in, 179
poor standard of original garrison, 169
Parit Sulong, 160, 161
massacre at, 160
Parramatta feature, 1, 2, 245, 246
Parry-Okeden, William Nugent, 370
limitations imposed on Ramu operations, 259–60, 384
personality of, 259, 260
reaction to orders at Sanananda, 192
takes command of 2/9th Battalion, 184
thoughts on discipline, 60
thoughts on fear, 193
Parsons, Percival Augustus, 51, 72, 73, 75
patrols, 183, 214, 228, 251, 253, 260, 262
patronage
and selection of battalion officers, 60–8
influence on appointment of COs, 50, 54, 77, 108, 133, 175, 230, 273, 293, 379
influence on pre-war appointments, 45
Paul, Albert Thomas
dismissal of, 365
Permanent Military Forces
command of 2nd AIF battalions, 50
command of militia units, 43
criticisms of, 57, 142, 274, 339
poor performance in command, 58
pre-war strength, 30
staff role of, 58, 297
training for command appointments, 31, 38
wartime commands, 58, 133, 137, 143, 228, 230, 241, 244, 277, 292, 297, 313, 339

personal example
  See example
personality
  influence on command style, 119, 216–18
  traits of battalion commanders, 31–42, 62
Peterson, Leslie James 'Les', 45, 241
Picken, Keith Sinclair, 247, 266
Pinios Gorge, 88, 91, 101, 103–5, 272
pioneer battalions, role of, 5
pioneer platoon, 83, 86
PMF
  See Permanent Military Forces
Pond, Samuel Austin Frank
  loses 2/29th Battalion on Singapore, 157
Pooler, Jeffery, 170
Pope, Alexander, 175
Porter, Selwyn Havelock Watson Craig 'Uncle Bill', 93, 94, 99, 169
  as commander 24th Brigade and Joshua, 233
    seeks to dismiss Norman, 249
  indecision at Khirbe, 99
  personality of, 249
  seeks to influence Official History, 38
  shot in buttocks, 101, 349
Potts, Arnold William 'Pottsie', 58, 98
  as 2iC 2/16th Battalion, 64
  as commander 21st Brigade, 195
  assessments of, 110
  feels fear, 104
  personality of, 98, 329
  resents relinquishing acting command, 109
power
  coercive, 23–6, 28, 59–61, 135, 137, 142, 159, 170, 272
  expert, 23–5, 65, 104, 163, 170, 177, 233, 271
  legitimate, 23, 24, 59, 135, 137, 170
  referent, 23, 24, 27, 28, 59, 62, 65, 104, 135, 137, 157, 226, 233, 271
  reward, 23, 24, 62, 104
predictability, 25, 54, 60, 67, 156, 177, 193

Price, Ernest William Alfred 'Ernie', 243
pride, 98, 101, 127, 145, 193
PRO
  See National Archives of the United Kingdom
professionalism, 47, 235, 273, 277, 283
promotion
  from the ranks, 295
  pre-war criteria, 35
Promotion and Selection Committee, 32
Promotions and Selection Board, 166
Prothero, 207, 373
Proud, Alfred Joseph, 168
psychological casualties
  See casualties – psychological
  See also exhaustion
Public Record Office
  See National Archives of the United Kingdom
Purge, the
  See militia – replacement of COs
Purser, Muir, 61
  removed from command of 36th Battalion, 175

Queensland, 200

radio, 70, 74, 89, 122, 147
  18 set, 94, 120
  101 set, 120, 361
  108 set, 93, 120, 154, 361
  assists control, 94, 212
  employment in Malaya, 154, 361
  employment in New Guinea, 212
  SCR 536 'walkie-talkie', 204, 215
Rai Coast, 259
Ramsay, Alan Hollick, 380
Ramsay, George Ernest 'Gentleman George', 141, 162
  as 2iC 2/30th Battalion, 142
  personality of, 141, 162
  relationship with Galleghan, 141
Ramu River, 259
Ramu Valley, 198
Randwick Barracks, 39
RAP
  See Regimental Aid Post
Ras el Medauuar, 91

Raven, Bertram
  'Bases of Social Power', 23
reconnaissance
  role of, 213, 217
Reconnaissance Group, 85
Reddick, Walter 'Wally', 243
Reformatory Road, 150
Refshauge, William Dudley, 223
  assessment of 24th Battalion, 227
regeneration
  of command structure, 70, 76–7, 111, 129, 131–2, 134, 135, 172, 248
Regimental Aid Post, 84
regimental medical officers, 65, 86
  conflict with COs, 66
  role of, 66
regimental police, 67
regimental sergeants major, 67
regimental traditions, 300
  pre-war, 32
regional identities, 175, 283, 381, 387
resource management, 17, 37, 250
respect, 23, 135, 137, 138, 142, 156, 157, 233, 271, 272, 275
responsibility, 17
Retimo, 244
R Group
  See Reconnaissance Group
Rhoden, Phillip Edington 'Phil', 246, 247, 288
  at LHQ Tactical School, 201, 202
  cautious approach at Manggar, 246, 251–3, 255, 262
  post-war career, 274
  reflections on battalion command, 276
  reflections on Gona, 251
  relationship with regimental sergeant major, 67
  sense of achievement in Papua, 196
  thoughts on jungle warfare, 179, 180
  thoughts on qualities of a CO, 62, 67–271
rifle companies, 83
risk management, 112, 126, 147, 149, 150, 222, 234, 250, 255, 259, 270
  practised by Japanese, 130
RMO
  See regimental medical officers

Roach, Leonard Nairn 'Len', 11, 133
  dismissal of, 163
  membership of the White Army, 337
  questions defence of Ambon, 162
Robertson, Andrew Esmond, 164
  acrimony with officers of 2/19th Battalion, 146
  uses carrriers for shock action, 150
Robertson, Horace Clement Hugh 'Red Robbie', 99, 106, 351
Robertson, John Charles, 145
  dismisses anti-tank guns, 147–9, 360
  killed at Bakri, 147, 157
Robson, Ewan Murray, 246
Robson, Leslie Lloyd
  'Origin and character of the First AIF', 277
Rocke, Mark
  definition of trust, 24
Rourke, Henry Gordon, 143
Rowan, John George
  awarded Military Cross, 175
  reaction on taking command of 37th Battalion, 175
Rowell, Francis Albert, 244
Rowell, Sydney Fairbairn, 110
Royal Military College Duntroon, 35, 36, 52, 56, 290, 294
RSM
  See regimental sergeants major
ruthlessness, 192

Sadler, Rupert Markham, 168, 174–6
Salamaua, 206, 213, 228, 257
Salvation Army, 66
Sanananda, 179, 184, 192, 195
Sanderson, Roger Warren, 245
Sandover, Raymond Ladais, 387
  captured on Crete, 349
Sattelberg, 212, 219, 231, 373
Saucer, the, 122, 125, 127
Savige, Stanley George 'Stan', 53, 76, 91, 99, 106, 222
  as GOC 3rd Division, 174, 249
  requests Dexter for 61st Battalion, 228
  attitude towards dismissed COs, 224
Scarlet Beach, 217

Scotch College, 45
Scott, William John Rendel
  Beaumont's assessment of, 11
  membership of the White Army, 337
seconds in command
  older in 1945, 239
  role of, 64–5, 85, 241, 266
secret armies, 46, 337
sections, 83, 343
  fire and movement, 83
Segamat, 130
self-education, 37
sensitivity to casualties, 261
Sepik River, 259
Seymour, 63
Sfakia, 102
Shaggy Ridge, 198, 373
Sheffield, Gary, 16
  on command systems, 18
signals
  See communications
Simpson, Colin Hall
  membership of the White Army, 337
Simpson, Noel William 'the Red Fox', 115, 216–18
  appointed to command 2/43rd Battalion, 233, 380
  at El Alamein, 217
  command style, 218
  one of 2nd AIF's best COs, 217
  personality of, 217
  reaction of 2/17th Battalion to, 218
  story of nickname, 218, 376
  training regime, 115
sincerity, 24, 119
Singapore, 130, 131, 145, 150, 156, 157, 162, 270
  Australian withdrawal in, 363
Sio, 217
situational leadership
  See leadership – contingency theory
Slim River, 147
Smith, George Frederick 'Larry the Bat', 378
  as 2/6th Battalion battle 2iC at Bardia, 98, 349
  at LHQ Tactical School, 201
  awarded DSO, 228
  command style, 228

leaves command due to injury, 228
'legendary kindness' of, 228
personality of, 227
story of nickname, 228
Smith's Weekly, 170
social class
  a 'command class', 56, 280, 285, 289, 291, 293
  influence on pre-war appointments, 45–6, 289
South Australia, 55, 284, 387
Spence, Iain, 27
Spowers, Allan 'Jiggy'
  attacks tanks with Bren gun carriers, 96
Squires, Ernest Ker, 31
Squires Report, 31, 33
Stanley, Peter, 235, 258, 262, 355
  Alamein, 11, 118
  Tarakan, 11
Starr, Patrick Daniel Sarsfield 'Danny'
  as a company commander, 222
  defence of 58/59th Battalion's performance, 223
  dismissal from 2/5th Battalion, 214, 223
  dismissal from 58/59th Battalion, 224, 226
  loses contact with the Japanese, 214
  personality of, 223
Stevens, Jack Edwin Stawell, 50, 170
  as commander 21st Brigade, 55, 69, 70, 98
  dismisses Cannon, 352
  as GOC 6th Division, 246
  on the dangers of tentative action, 258
  personality of, 98
Stevenson, John Rowlstone
  wounded in Lebanon, 349
Stewart, Albert William John
  relationship with men, 171–2
  removed from command, 168
  re-raising the 37th Battalion, 42
Stewart, Mick, 243
Sturdee, Vernon Ashton Hobart, 102, 132–6, 142, 143, 163, 169, 356
  criteria for selection of battalion commanders, 132

Sublet, Frank Henry, 37, 191, 246, 247
　influence of experience on writing, 12
　Kokoda to the Sea, 12
　origins of, 290
　ruthless attitude, 191–2
　thoughts on jungle warfare, 181
succession planning
　See regeneration – of command structure
SUR
　See Australian Military Forces – university regiments
surprise, 86
Sydney, 39, 40, 43, 70, 71, 132, 175, 339
Syria, 115
Syrian campaign
　See Lebanon

tactical exercises without troops, 35, 36, 38, 40, 41, 65, 69, 70, 74, 116
tactical headquarters
　See battalion headquarters – tactical
tactical innovation, 126, 154, 190
tactics
　defensive, 152
　offensive, 153
Tank Hill, 237
tank-attack platoon
　See anti-tank platoon
tanks
　See armour
Tarakan, 11, 232, 236, 258, 262
　ammunition shortages on, 254
　cautious operations on, 258
Tasmania, 53, 133, 283, 284
Taylor, George, 385
　Infantry Colonel, 273
Taylor, Harold Burfield, 133, 143, 146
　assessment of 2/19th Battalion, 135
　assessment of Jeater, 137, 139
　assessment of Maxwell's command, 136
　feud with Bennett, 143–5
　orders Varley to withdraw from Nithsdale, 162
　reflections on Maxwell's appointment, 133
　role in defence of Singapore, 363
　training program, 144
Tel el Eisa, 116, 126
Tengah, 157
tentative action
　See caution
Territorial Army (UK)
　See British Army
TEWT
　See tactical exercises without troops
Thomas, Kevan Brittan, 212
Thomson, Alistair
　Anzac Memories, 15
Thyer, James Hervey 'Jim', 163, 164
Timor, 238, 295
Tobruk, 78, 87, 88, 90, 91, 93, 95–6, 98, 126, 209, 267, 346
training, 28, 270
　for jungle warfare, 199
　influence on proficiency of AMF, 202
　of battalions, 69
　in 1945, 242
　in militia, 170–1, 176
　prior to El Alamein, 117
　of officers, 68, 117
　role of, 22
trust, 24–5, 28, 60, 62, 67, 100, 105, 133, 135, 137, 156, 157, 177, 183, 210, 231, 233, 235, 243, 246, 263, 271, 272
　lack of at Tobruk, 98
Tucker, Frederick Alfred George 'Fag', 237
　cautious approach on Tarakan, 258
　distribution of Japanese swords at Tarakan, 11
Turner, Robert William Newton, 118, 127
　addresses 2/13th Battalion before El Alamein, 118
　killed at El Alamein, 127, 216, 355
　proposes foolhardy daylight attack at El Alamein, 127

unit cohesion, 71, 142, 193, 263, 281
　demonstrations of, 125, 228
　development of, 62, 272

United States Army, 198
    communications practices, 215
    divisions
        82nd Airborne, 272
    exhaustion in north-west Europe, 258–66
    in New Britain, 215
    purge of over-age officers, 272
    style of warfare, 251
'unneccessary campaigns', the, 236

van Creveld, Martin
    Command in War, 15
    command model, 17
    on assessing command effectiveness, 20
    on command systems, 18
Varley, Arthur Leslie, 143, 150, 152, 155, 158, 162
    abandons company at Nithsdale, 162
    flawed plan at Nithsdale, 152
    loses control of 2/18th Battalion on Singapore, 158
    personality of, 136
    relationship with 2iC, 136
    uses carriers for shock action, 150
Vasey, George Alan 'Bloody George', 105, 109, 194
    as GOC 7th Division, 185
Vasey Highway, 251
VDC
    See Volunteer Defence Corps
Verrier, Arthur Drummond
    capabilities, 62–5, 109, 340
    relationship with adjutant, 65
    reputation among troops, 65, 109, 340
    struggles at Tobruk, 109
Vevi, 105, 107–9
Vickers guns
    See automatic weapons
Victoria, 49, 54, 55, 67, 132, 133, 281, 337
Vincent, Thomas Guy 'Tom', 136, 146, 164
Vitiaz Strait, 198
Volunteer Defence Corps, 175, 301

Wain, William John 'Hells Bells'
    concerns about attack on Meteiriya Ridge, 128
    concussed at El Alamein, 355
    dismissed from 2/43rd Battalion, 232
    popularity of, 232
Waite's Knoll, 251, 253
Wakin, Malham, 25
Walker, Allan Seymour
    Middle East and Far East, 304
Walker, Theodore Gordon 'Myrtle', 34, 63, 271
    appearance, 102
    appointed to 2/7th Battalion, 53
    Blamey's assessment of, 102
    calm demeanour, 104
    captured on Crete, 349
    good control at Bardia, 91
    leadership during Greece and Crete campaigns, 103–4
    leadership on Crete, 101
    no First World War experience, 34
Wallis, Robert Jonathan, 174
Walters, Vernon Morris 'Mick', 213, 374
war diaries, 14, 331
    as a forum for criticism, 128
Ward, Kenneth Harry
    killed in Papua, 184
    not appointed to 2nd AIF, 52
Warfe, George Radford 'Warfie', 271
    24th Battalion's reaction to, 231
    actions on Tarakan, 11
    aggressive command style, 223–7, 378
    cautious approach on Tarakan, 258
    clashes with Hammer, 248
    personality of, 226, 378
    post-war career, 274
    reaction in 58/59th Battalion to, 227
Watkins, Leslie William 'Les', 118
Watson, Peter
    War on the Mind, 20
Wau, 214, 224
Webster, Peter Glynn Clifton, 8
    clashes with Matthews on Bougainville, 248
    personality of, 239
Weir, Charles Gladstone, 113, 118, 123, 124, 127, 272
    accompanies patrol at El Alamein, 124

addresses 2/24th Battalion before El Alamein, 113, 118
severely wounded at El Alamein, 124
wounded at El Alamein, 355
Western Australia, 40, 54, 55, 280, 381
Wewak
  ammunition expenditure at, 253
White, Cyril Brudenell Bingham, 356
White Army
  See League of National Security
Whitehead, David Aide 'Torpy', 337
Wigmore, Lionel Gage, 9, 142, 164
Williamson, Harry James, 125
Willoughby, 43
Wilmoth, John Alfred
  appointed to command 7th Battalion, 174
  attitude of troops to, 177
  command style, 177
  membership of the White Army, 337
  post-war career, 274
Wilson, Henry Hughes, 25
Windeyer, William John Victor, 51, 98, 109
  as commander 20th Brigade, 216, 217, 219, 246, 293
  criticism of Dexter's *New Guinea Offensives*, 9
  medically unfit, 52
  personality of, 231
  reputation as CO of Sydney University Regiment, 52

wireless
  See radio
Wise, Reginald John, 255, 256
Wood, Frederick George 'Fighting Freddie', 229, 235, 243
  allows Dexter to conduct Lababia battle, 212
  reputation for being excitable, 103
  response to in 2/6th Battalion, 233
Wootten, George Frederick, 51, 70, 72, 75, 98
  as commander 9th Division, 110
  as commander 18th Brigade, 98, 110, 186
  as GOC 9th Division
    dismisses Norman, 249
  concerns about weight, 50, 52, 72
  criticism of Long's *To Benghazi*, 9
  membership of the Old Guard, 337
  personality of, 98
  pre-Second World War career, 52
  reaction to new recruits, 59
Wrigley, Hugh, 75, 227
  as second in command 2/6th battalion, 64
  collapses in Greece, 101
  personality, 65
  wounded, 349

Yong Peng, 161
Youl, Geoffrey Arthur Douglas
  appointment to 2/40th Battalion, 133
  prior military service, 356